Methods in Molecular Biology

Series Editor
John M. Walker
School of Life and Medical Sciences
University of Hertfordshire
Hatfield, Hertfordshire, UK

For further volumes:
http://www.springer.com/series/7651

For over 35 years, biological scientists have come to rely on the research protocols and methodologies in the critically acclaimed *Methods in Molecular Biology* series. The series was the first to introduce the step-by-step protocols approach that has become the standard in all biomedical protocol publishing. Each protocol is provided in readily-reproducible step-by-step fashion, opening with an introductory overview, a list of the materials and reagents needed to complete the experiment, and followed by a detailed procedure that is supported with a helpful notes section offering tips and tricks of the trade as well as troubleshooting advice. These hallmark features were introduced by series editor Dr. John Walker and constitute the key ingredient in each and every volume of the *Methods in Molecular Biology* series. Tested and trusted, comprehensive and reliable, all protocols from the series are indexed in PubMed.

T-Cell Receptor Signaling

Methods and Protocols

Edited by

Chaohong Liu

Department of Pathogen Biology, Huazhong University of Science and Technology, Wuhan, China

Humana Press

Editor
Chaohong Liu
Department of Pathogen Biology
Huazhong University of Science and Technology
Wuhan, China

ISSN 1064-3745 ISSN 1940-6029 (electronic)
Methods in Molecular Biology
ISBN 978-1-0716-0265-2 ISBN 978-1-0716-0266-9 (eBook)
https://doi.org/10.1007/978-1-0716-0266-9

This Humana imprint is published by the registered company Springer Science+Business Media, LLC, part of Springer Nature.
The registered company address is: 233 Spring Street, New York, NY 10013, U.S.A.

Preface

T cell is an important component of adaptive immune system together with B cells. Besides the difference that T cells are originated from thymus and B cells are developed from bone marrow, the function and diversity of T cells is much more complicated than B cells. TCR signaling is important for the fulfillment of the positive and negative selection of T cells in the thymus as well as the T cell activation. T cells have $CD4^+$ and $CD8^+$ T cells according to the specificity of MHCI and MHCII. $CD4^+$ T cells consist of Th1, Th2, Th17, follicular helper T cells, and regulatory T cells according to the master transcriptional factors as well as the cytokine productions. Nowadays, some new subsets of $CD4^+$ T cells have been identified, such as Th9. The differentiation of different T cell subsets is highly correlated with the TCR signaling. Lots of new advanced technologies have been applied to identify new T cell subsets and functions of T cells such as mass spectrometry, single cell technique, and CRISPR/Cas9. Additionally, the differentiation and expansion of different T cell subsets are set up with different techniques. This volume of *Methods in Molecular Biology* focuses on various aspects of T cells. Chapters 1 and 2 provide protocols to explore the T cell diversity using mass cytometry. Chapters 3 and 4 present protocols to analyze the T cells from single cell level. Chapter 5 provides CRISPR/Cas9 technique to study the T cell activation. Chapters 6–11 numerate all kinds of techniques to set up the differentiation of different T cell subsets such as Tregs, CD4+ T cells, T_{fh} cells, and stem cell memory-like T cells *in vitro and in vivo*. Chapters 12–15 establish the procedures of artificial antigen presentosomes for T cell activation, imaging chimeric antigen receptor (CAR)-triggered T cell activation, the impact of phytochemicals on T cell activation and TB vaccine-activated T cells. Chapters 16 and 17 provide techniques to study the T cell development as well as positive and negative selection in thymus. Chapter 18 establishes two-photon microscopy to study T cell receptor signaling dynamics. Chapters 19 and 20 offer protocols to study the ubiquitination and central carbon metabolism in T cells. Chapter 21 summarizes the technique to study peripheral T cell homeostasis in mice. The last chapter gives a technique to isolate MAIT cells. I have to admit that this volume might not provide complete methods to study T cell biology. There are many new methods coming out to study the T cells and it requires many volumes to cover this topic, and I have tried to offer a glimpse of the current advanced protocols.

I hope the scientific community working in the T cell field can benefit from the protocols published in this volume. I thank all the contributors for their time and contributions to this volume. Additionally, I would like to thank John Walker, Senior Editor, and the staff at Springer Nature for all their support to publish these protocols. Last but not least, I would like to thank Xizi Sun and Yue Wen for their support as well.

Wuhan, China *Chaohong Liu*

Contents

Contributors

CECILE ALANIO • *Department of Systems Pharmacology and Translational Therapeutics, Institute for Immunology, Perelman School of Medicine, University of Pennsylvania, Philadelphia, USA; Parker Institute of Cancer Immunotherapy, University of Pennsylvania, Philadelphia, PA, USA*

MAKOTO ANDO • *Department of Microbiology and Immunology, Keio University School of Medicine, Shinjuku-ku, Tokyo, Japan*

BERTRAM BENGSCH • *Department of Medicine II, Gastroenterology, Hepatology, Endocrinology, and Infectious Diseases, University Medical Center Freiburg and Signaling Research Centres BIOSS and CIBSS, University of Freiburg, Freiburg im Breisgau, Germany*

CHIEN-CHUNG CHANG • *Institute of Molecular and Cellular Biology, National Tsing Hua University, Hsinchu, Taiwan*

KONG CHEN • *Division of Pulmonary, Allergy, and Critical Care Medicine, Department of Medicine, University of Pittsburgh Medical Center, Pittsburgh, PA, USA*

NICOLAS CHEN • *Département de Biochimie et médecine moléculaire, Université de Montréal, Montréal, QC, Canada*

XIAO LEI CHEN • *State Key Laboratory of Cellular Stress Biology, Innovation Center for Cell Signaling Network, School of Life Sciences, Xiamen University, Xiamen, China*

XIN CHEN • *State Key Laboratory of Quality Research in Chinese Medicine, Institute of Chinese Medical Sciences, University of Macau, Macau SAR, China*

XUEYING CHEN • *Department of Rheumatology and Immunology, State Key Laboratory of Biotherapy and Collaborative Innovation Center for Biotherapy, West China Hospital, Sichuan University, Chengdu, China*

XUYONG CHEN • *Center for Childhood Cancer & Blood Diseases, Hematology/Oncology & BMT, The Research Institute at Nationwide Children's Hospital, Ohio State University, Columbus, OH, USA*

CHON-KIT CHOU • *State Key Laboratory of Quality Research in Chinese Medicine, Institute of Chinese Medical Sciences, University of Macau, Macau SAR, China*

MARIA CIOFANI • *Department of Immunology, Duke University Medical Center, Durham, NC, USA*

HUA DING • *Cell Pathology Division, Department of Pathology and Laboratory Medicine, The Children's Hospital of Philadelphia, Philadelphia, PA, USA*

MENGQI DONG • *Immunology-Oncology Unit, Maisonneuve-Rosemont Hospital Research Center, Montreal, QC, Canada; Département de Microbiologie, Infectiologie et Immunologie, Université de Montréal, Montreal, QC, Canada*

ZHIQIANG DU • *Innovation Center, Shanghai Benemae Pharmaceutical Corporation, Shanghai, China*

JOSEPH M. DYBAS • *The University of Pennsylvania, Philadelphia, PA, USA; Division of Protective Immunity, Department of Pathology and Laboratory Medicine, The Children's Hospital of Philadelphia, Philadelphia, PA, USA*

IVAN LILYANOV DZHAGALOV • *Institute of Microbiology and Immunology, National Yang-Ming University, Taipei, Taiwan*

Carla A. Erb • *Division of Pulmonary, Allergy, and Critical Care Medicine, Department of Medicine, University of Pittsburgh Medical Center, Pittsburgh, PA, USA*
Jaclyn R. Espinosa • *Department of Immunology, Duke University Medical Center, Durham, NC, USA*
Annabel Ferguson • *Division of Pulmonary, Allergy, and Critical Care Medicine, Department of Medicine, University of Pittsburgh, Pittsburgh, PA, USA*
Natania S. Field • *The University of Pennsylvania, Philadelphia, PA, USA; Division of Protective Immunity, Department of Pathology and Laboratory Medicine, The Children's Hospital of Philadelphia, Philadelphia, PA, USA*
Marilaine Fournier • *Immunology-Oncology Unit, Maisonneuve-Rosemont Hospital Research Center, Montreal, QC, Canada*
Guo Fu • *State Key Laboratory of Cellular Stress Biology, Innovation Center for Cell Signaling Network, School of Life Sciences, Xiamen University, Xiamen, China; Cancer Research Center of Xiamen University, Xiamen, China*
Simon-David Gauthier • *Département de Microbiologie, Infectiologie et Immunologie, Université de Montréal, Montréal, QC, Canada*
Martin Guimond • *Division Immunologie-Oncologie, Centre de Recherche de l'Hôpital Maisonneuve-Rosemont, Montréal, QC, Canada; Département de Microbiologie, Infectiologie et Immunologie, Université de Montréal, Montréal, QC, Canada*
Sarah E. Henrickson • *Department of Systems Pharmacology and Translational Therapeutics, Institute for Immunology, Perelman School of Medicine, University of Pennsylvania, Philadelphia, USA; Division of Allergy Immunology, Department of Pediatrics, Children's Hospital of Philadelphia, Philadelphia, PA, USA*
David W. Hoskin • *Department of Microbiology and Immunology, Dalhousie University, Halifax, NS, Canada; Department of Pathology, Dalhousie University, Halifax, NS, Canada; Department of Surgery, Dalhousie University, Halifax, NS, Canada*
Chia-Lin Hsu • *Institute of Microbiology and Immunology, National Yang-Ming University, Taipei, Taiwan*
Hongbo Hu • *Department of Rheumatology and Immunology, State Key Laboratory of Biotherapy and Collaborative Innovation Center for Biotherapy, West China Hospital, Sichuan University, Chengdu, China*
Lei Huang • *State Key Laboratory of Cellular Stress Biology, Innovation Center for Cell Signaling Network, School of Life Sciences, Xiamen University, Xiamen, China*
Mari Ikeda • *Department of Microbiology and Immunology, Keio University School of Medicine, Shinjuku-ku, Tokyo, Japan*
Mangalakumari Jeyanathan • *Department of Pathology and Molecular Medicine, McMaster Immunology Research Centre, McMaster University, Hamilton, ON, Canada; Michael G. DeGroote Institute for Infectious Disease Research, McMaster University, Hamilton, ON, Canada*
Xian Jia • *State Key Laboratory of Cellular Stress Biology, Innovation Center for Cell Signaling Network, School of Life Sciences, Xiamen University, Xiamen, China*
Shan Jiang • *State Key Laboratory of Cellular Stress Biology, Innovation Center for Cell Signaling Network, School of Life Sciences, Xiamen University, Xiamen, China*
Taisuke Kondo • *Department of Microbiology and Immunology, Keio University School of Medicine, Shinjuku-ku, Tokyo, Japan*
Dominique Leboeuf • *Skolkovo Institute of Science and Technology, Moscow, Russia*
Kendra A. Libby • *Department of Cell Biology, Yale School of Medicine, New Haven, CT, USA*
Yu Liu • *Department of Immunology, School of Basic Medicine, Tongji Medical College, Huazhong University of Science and Technology, Wuhan, China*

XUEHUI LONG • *Department of Immunology, Nanjing Medical University, Nanjing, Jiangsu, China*

JIADI LUO • *Division of Pulmonary, Allergy, and Critical Care Medicine, Department of Medicine, University of Pittsburgh Medical Center, Pittsburgh, PA, USA*

SASIKANTH MANNE • *Department of Systems Pharmacology and Translational Therapeutics, Institute for Immunology, Perelman School of Medicine, University of Pennsylvania, Philadelphia, USA*

HEATHER J. MELICHAR • *Immunology-Oncology Unit, Maisonneuve-Rosemont Hospital Research Center, Montreal, QC, Canada; Département de Médecine, Université de Montréal, Montreal, QC, Canada*

MOUTUAATA M. MOUTUOU • *Département de Microbiologie, Infectiologie et Immunologie, Université de Montréal, Montréal, QC, Canada*

RYUNOSUKE MURO • *Department of Immunology, Graduate School of Medicine and Faculty of Medicine, The University of Tokyo, Tokyo, Japan*

TAKESHI NITTA • *Department of Immunology, Graduate School of Medicine and Faculty of Medicine, The University of Tokyo, Tokyo, Japan*

KAITLIN C. O'BOYLE • *Department of Systems Pharmacology and Translational Therapeutics, Institute for Immunology, Perelman School of Medicine, University of Pennsylvania, Philadelphia, USA*

TAKUYA OHTANI • *Department of Systems Pharmacology and Translational Therapeutics, Institute for Immunology, Perelman School of Medicine, University of Pennsylvania, Philadelphia, USA*

CLAIRE E. O'LEARY • *Department of Medicine, University of California-San Francisco, San Francisco, CA, USA*

PAULA M. OLIVER • *The University of Pennsylvania, Philadelphia, PA, USA; Division of Protective Immunity, Department of Pathology and Laboratory Medicine, The Children's Hospital of Philadelphia, Philadelphia, PA, USA*

YI-GENG PANG • *Institute of Molecular and Cellular Biology, National Tsing Hua University, Hsinchu, Taiwan*

MELANIE R. POWER COOMBS • *Department of Biology, Acadia University, Wolfville, NS, Canada*

WANJING SHANG • *School of Life Science and Technology, ShanghaiTech University, Shanghai, China; Institute of Biochemistry and Cell Biology, Shanghai Institutes for Biological Sciences, Chinese Academy of Sciences, Shanghai, China; University of Chinese Academy of Sciences, Beijing, China*

JOHN WILLIAM SHERMAN • *Center for Childhood Cancer & Blood Diseases, Hematology/Oncology & BMT, The Research Institute at Nationwide Children's Hospital, Ohio State University, Columbus, OH, USA*

JAMIE B. SPANGLER • *Department of Biomedical Engineering, Johns Hopkins University, Baltimore, MD, USA; Department of Chemical and Biomolecular Engineering, Johns Hopkins University, Baltimore, MD, USA*

XIAOLEI SU • *Department of Cell Biology, Yale School of Medicine, New Haven, CT, USA*

HIROSHI TAKAYANAGI • *Department of Immunology, Graduate School of Medicine and Faculty of Medicine, The University of Tokyo, Tokyo, Japan*

JAKUB TOMALA • *Department of Biomedical Engineering, Johns Hopkins University, Baltimore, MD, USA; Department of Chemical and Biomolecular Engineering, Johns Hopkins University, Baltimore, MD, USA; Institute of Microbiology of the Czech Academy of Sciences, Prague, Czech Republic*

FEI WANG • *School of Life Science and Technology, ShanghaiTech University, Shanghai, China*
HAOPENG WANG • *School of Life Science and Technology, ShanghaiTech University, Shanghai, China*
LIANGYU WANG • *School of Life Science and Technology, ShanghaiTech University, Shanghai, China*
RUONING WANG • *Center for Childhood Cancer & Blood Diseases, Hematology/Oncology & BMT, The Research Institute at Nationwide Children's Hospital, Ohio State University, Columbus, OH, USA*
WEI WANG • *Department of Immunology, School of Basic Medicine, Tongji Medical College, Huazhong University of Science and Technology, Wuhan, China*
XIAOMING WANG • *Department of Immunology, Nanjing Medical University, Nanjing, Jiangsu, China*
JIANNAN WENG • *State Key Laboratory of Cellular Stress Biology, Innovation Center for Cell Signaling Network, School of Life Sciences, Xiamen University, Xiamen, China*
XIUFANG WENG • *Department of Immunology, School of Basic Medicine, Tongji Medical College, Huazhong University of Science and Technology, Wuhan, China*
JOSHUA D. WHEATON • *Department of Immunology, Duke University Medical Center, Durham, NC, USA*
E. JOHN WHERRY • *Department of Systems Pharmacology and Translational Therapeutics, Institute for Immunology, Perelman School of Medicine, University of Pennsylvania, Philadelphia, USA; Parker Institute of Cancer Immunotherapy, University of Pennsylvania, Philadelphia, PA, USA*
XIONGWEN WU • *Department of Immunology, School of Basic Medicine, Tongji Medical College, Huazhong University of Science and Technology, Wuhan, China*
CHANGCHUN XIAO • *State Key Laboratory of Cellular Stress Biology, Innovation Center for Cell Signaling Network, School of Life Sciences, Xiamen University, Xiamen, China; Cancer Research Center of Xiamen University, Xiamen, China*
ZHOU XING • *Department of Pathology and Molecular Medicine, McMaster Immunology Research Centre, McMaster University, Hamilton, ON, Canada; Michael G. DeGroote Institute for Infectious Disease Research, McMaster University, Hamilton, ON, Canada*
WENYONG YANG • *Department of Rheumatology and Immunology, State Key Laboratory of Biotherapy and Collaborative Innovation Center for Biotherapy, West China Hospital, Sichuan University, Chengdu, China*
AKIHIKO YOSHIMURA • *Department of Microbiology and Immunology, Keio University School of Medicine, Shinjuku-ku, Tokyo, Japan*
LEI ZHANG • *State Key Laboratory of Cellular Stress Biology, Innovation Center for Cell Signaling Network, School of Life Sciences, Xiamen University, Xiamen, China*
YANG ZHANG • *Department of Immunology, Nanjing Medical University, Nanjing, Jiangsu, China*
HAIPING ZHENG • *State Key Laboratory of Cellular Stress Biology, Innovation Center for Cell Signaling Network, School of Life Sciences, Xiamen University, Xiamen, China*
TYNG-AN ZHOU • *Institute of Microbiology and Immunology, National Yang-Ming University, Taipei, Taiwan*
XIAOJUAN ZHOU • *State Key Laboratory of Cellular Stress Biology, Innovation Center for Cell Signaling Network, School of Life Sciences, Xiamen University, Xiamen, China*
QI ZHU • *School of Life Science and Technology, ShanghaiTech University, Shanghai, China*

Chapter 1

Exploration of T-Cell Diversity Using Mass Cytometry

Kaitlin C. O'Boyle, Takuya Ohtani, Sasikanth Manne, Bertram Bengsch, Sarah E. Henrickson, E. John Wherry, and Cecile Alanio

Abstract

T-cell diversity is multifactorial and includes variability in antigen specificity, differentiation, function, and cell-trafficking potential. Spectral overlap limits the ability of traditional flow cytometry to fully capture the diversity of T-cell subsets and function. The development of mass cytometry permits deep immunoprofiling of T-cell subsets, activation state, and function simultaneously from even small volumes of blood. This chapter describes our methods for mass cytometry and high-throughput data analysis of T cells in patient cohorts. We provide a pipeline that includes practical considerations when customizing a panel for mass cytometry. We also provide protocols for the conjugation and titration of metal-labeled antibodies (including two T-cell panels) and a staining procedure. Finally, with the aim to support translational science, we provide R scripts that contain a detailed workflow for initial evaluation of high-dimensional data generated from cohorts of patients.

Key words Mass cytometry, CyTOF, T cells, Systems biology, High-throughput analysis, R, Clustering, Data processing

1 Introduction

The quantification of the remarkable variety of T-cell subsets and the range of activation states benefit from deep immunoprofiling strategies. High-dimensional flow cytometry has led to rapid progress in our understanding of the range of T-cell functional (and dysfunctional) states. However, mass cytometry, with the potential to simultaneously measure at least twice to three times as many cell surface and intracellular markers, allows us to further deepen our understanding of T-cell function.

Mass cytometry is a variation of flow cytometry in which antibodies are labeled with heavy metal ions rather than fluorochromes [1]. Overcoming the issue of spectral overlap, this advance allows for the routine use of more than 40 simultaneous probes. Mass cytometry is now widely used to characterize different cell types and tissues in health and disease [2, 3]. As it generates large complex

Chaohong Liu (ed.), *T-Cell Receptor Signaling: Methods and Protocols*, Methods in Molecular Biology, vol. 2111,
https://doi.org/10.1007/978-1-0716-0266-9_1,

datasets, deep immunoprofiling via mass cytometry must be combined with systems immunology through data analytic strategies and a number of sophisticated methods have been proposed to support researchers in this regard [4]. However, there is a need for an integrated approach for the analysis of large datasets.

This chapter describes our methods for mass cytometry and high-throughput data analysis of T cells in large cohorts of patients. We discuss practical considerations in panel design. We provide protocols for the conjugation and titration of metal-labeled antibodies, two sample T-cell panels, and a staining procedure. Finally, we provide R scripts that contain a detailed workflow for initial assessment of high-dimensional data generated from large datasets.

2 Materials

2.1 Antibody Conjugation

1. Maxpar Antibody Labeling Kit (Fluidigm).
2. Purified antibody (glycerol free and carrier free).
3. Filters: 30 kDa or 50 kDa and 3 kDa Amicon Ultra-0.5 centrifugal filter unit (Millipore Sigma; *see* **Notes 1** and **2**).
4. 0.5 M TCEP solution (Bond-Breaker, Thermo Fisher).
5. NanoDrop (Thermo Fisher).
6. Antibody stabilizer (PBS base, Boca Scientific) or PBS with 0.05% sodium azide.

2.2 Staining

1. Complete medium: RPMI 1640 with L-glutamine, 10% heat-inactivated fetal bovine serum (FBS), 100 U/mL penicillin-streptomycin, 2 mM L-glutamine.
2. Stain buffer: PBS, 1% FBS.
3. 1× permeabilization buffer: Dilute 10× permeabilization buffer (eBioscience, Thermo Fisher) to 1× with deionized H_2O.
4. Live/dead stain: Dissolve maleimido-mono-amide-DOTA (Macrocyclics) in L-Buffer (Fluidigm) to 1 mM and then add isotopically purified La139 (Trace Sciences) to 0.5 mM. Dilute mmDOTA-La139 1:400 in PBS.
5. Surface antibody cocktail: Surface antibodies diluted in stain buffer (*see* **Note 3**).
6. Intracellular antibody cocktail: Intracellular antibodies diluted in 1× permeabilization buffer (*see* **Note 3**).
7. Permeabilization solution: 3 parts Foxp3/transcription factor fixation/permeabilization concentrate and 1 part diluent (eBioscience, Thermo Fisher).
8. Fixative: PBS, 1.6% paraformaldehyde (Alfa Aesar, Fisher Scientific), 125 nM iridium (Fluidigm).

2.3 Acquisition and Normalization

1. Acquisition solution: Milli-Q H_2O (Millipore Sigma), 10% EQ Four Element Calibration Beads (Fluidigm).
2. CyTOF Helios (Fluidigm).
3. CyTOF Software v6.7 (Fluidigm).

2.4 Analysis

1. FlowJo v10 (TreeStar).
2. RStudio v1.1.383.
3. R v3.5.1.
4. R scripts and other resources (https://github.com/wherrylab/Cytof_analysis_calanio).

3 Methods

3.1 Panel Design

We provide two panels for T-cell analysis in humans (Tables 1 and 2) with the aim to provide an in-depth analysis of the T-cell compartment. While our T-cell panel in Table 1 is particularly suited for deep analysis of T-cell differentiation and exhaustion in CD4 and CD8 T cells, our general immunophenotyping panel in Table 2 allows a broader investigation of other cell types as well as subsets of T helper cells.

Designing a panel for mass cytometry is a very similar process as designing a panel for flow cytometry conceptually, but it is mechanistically different. In flow cytometry, abundant markers should be paired with "dim" fluorochromes, while less abundantly expressed markers should be paired with "bright" fluorochromes. In mass cytometry, the terms "bright" and "dim" can be translated into the analogous characteristic for metal ions—sensitivity. Less abundant markers should be placed on more sensitive channels, and more abundant markers should be placed on low sensitivity channels. Within the atomic mass window of AM 89-209, channels in the middle of the mass range are most sensitive, while upper and lower channels are less intense.

There are a few crucial components that every mass cytometry panel must have. Particles must be metal labeled in some fashion to be counted as an event by the mass cytometer. For this, cells are stained with DNA intercalators containing iridium (Ir191 and Ir193). To further assess viability, maleimido-mono-amide-DOTA (in our case loaded with La139) or cisplatin (best detected on the 195 channel) may be used. La139 is a less pure isotope which we have chosen to utilize for live/dead discrimination and/or as a "dump" channel (containing antibodies that identify cells we want to exclude from the analysis, such as B cells and monocytes in a purely T-cell-focused panel). Although there is almost no spillover between channels due to the mode of detection as in fluorescent cytometry, there can be contamination of the signal

Table 1
T-cell panel

Isotope channel	Antibody/ reagent	Clone	Source	Order #	Commercial or in-house conjugation	Stain	Category
113 In	CD4	RPA-T4	BioLegend	300502	In-house	Surface	Lineage
115 In	CD57	TB01	Thermo Fisher	MA5-16948	In-house	Surface	Differentiation
139 La	CD14	63D3	BioLegend	367102	In-house	Surface	Dump
139 La	CD19	HIB19	BioLegend	302202	In-house	Surface	Dump
140 etc	Beads	n/a	Fluidigm	201078	n/a	n/a	QC
141 Pr	CD3	UCHT1	BioLegend	300443	In-house	Surface	Lineage
142 Nd	CD26	BA5b	BioLegend	302702	In-house	Surface	Differentiation/activation
143 Nd	CD95	DX2	BioLegend	305602	In-house	Surface	Differentiation
144 Nd	CTLA4	BNI3	BioLegend	369602	In-house	Intracellular	Exhaustion
145 Nd	CD49d	9F10	BioLegend	304302	In-house	Surface	Differentiation
146 Nd	CD8a	RPA-T8	BioLegend	301002	In-house	Surface	Lineage
147 Sm	CD45RA	H100	BioLegend	304102	In-house	Surface	Differentiation
148 Nd	CD103	Ber-ACT8	BioLegend	350202	In-house	Surface	Differentiation
149 Sm	TCF-1 (TCF7)	7F11A10	BioLegend	655202	In-house	Intracellular	Differentiation
150 Nd	CD127/IL-Ra	HIL-7R-M21	BD Biosciences	552853	In-house	Surface	Differentiation
151 Eu	CD39	A1	BioLegend	328202	In-house	Surface	Exhaustion
152 Sm	Granzyme B	CLB-GB11	Thermo Fisher	MA1-10338	In-house	Intracellular	Activation
153 Eu	Tim-3	F38-2E2	BioLegend	345002	In-house	Surface	Exhaustion
154 Sm	Granzyme K	GM6C3	Thermo Fisher	MA1-17755	In-house	Intracellular	Exhaustion
155 Gd	CD27	O323	BioLegend	302802	In-house	Surface	Differentiation

156 Gd	Helios	22F6	BioLegend	137202	In-house	Intracellular	Exhaustion
158 Gd	PD-1	EH12.2H7	BioLegend	329902	In-house	Surface	Activation/exhaustion
159 Tb	CCR7	G043H7	BioLegend	353202	In-house	Surface	Differentiation
160 Gd	Tbet	4B10	BioLegend	644802	In-house	Intracellular	Differentiation
161 Dy	CD28	CD28.2	BioLegend	302902	In-house	Surface	Differentiation
162 Dy	CD69	FN50	BioLegend	310902	In-house	Surface	Differentiation/activation
163 Dy							
164 Dy	Ki67	Ki67	BioLegend	350502	In-house	Intracellular	Activation/exhaustion
165 Ho	Eomes	WD1928	Invitrogen	14-4877-82	In-house	Intracellular	Differentiation/exhaustion
166 Er	CD85j	GHI/75	BioLegend	333702	In-house	Surface	Differentiation
167 Er	CD38	HIT2	BioLegend	303502	In-house	Surface	Differentiation/activation
168 Er	TOX	Rea473	Miltenyi Biotec	n/a	In-house	Intracellular	Exhaustion
169 Tm	TIGIT	MBSA43	Invitrogen	16-9500-82	In-house	Surface	Exhaustion
170 Er	CXCR5	J52D4	BioLegend	356902	In-house	Surface	Differentiation
171 Yb							
172 Yb	2B4	C1.7	BioLegend	329502	In-house	Surface	Exhaustion
173 Yb							
174 Yb	HLA-DR	L243	Biolegend	307602	In-house	Surface	Differentiation/activation
175 Lu	LAG3	11C3C65	BioLegend	369302	In-house	Surface	Exhaustion
176 Yb	CXCR3	5E8	BioLegend	310702	In-house	Surface	Differentiation/activation
191/193	Iridium	n/a	Fluidigm	201192B	n/a	Fixative	DNA
209 Bi	CD16	3G8	Fluidigm	3209002B	Commercial	Surface	Lineage
89Y	CD45	HI30	Fluidigm	3089003B	Commercial	Surface	Lineage

Table 2
General immunophenotyping panel

Isotope channel	Antibody/reagent	Clone	Source	Order #	Commercial or in-house conjugation	Stain	Category
113 In	CD45RO	UCHL1	BD	555491	In-house	Surface	Differentiation
115 In	CD123	6H6	BioLegend	306002	In-house	Surface	Lineage
139 La	L/D MM-DOTA	n/a	Macrocyclics	B-272	In-house	Live/dead	Viability
140 etc	Beads	n/a	Fluidigm	201078	n/a	n/a	QC
141 Pr	CD3	UCHT1	BioLegend	300443	In-house	Intracellular	Lineage
142 Nd	CD26	BA 5b	BioLegend	302702	In-house	Surface	Differentiation/activation
143 Nd	CD4	RPA-T4	BioLegend	300502	In-house	Intracellular	Lineage
144 Nd	CD11b/Mac-1	ICRF44	Fluidigm	3144001B	Commercial	Surface	Lineage
145 Nd	CD19	HIB19	BioLegend	302202	In-house	Surface	Lineage
146 Nd	CD8a	RPA-T8	BioLegend	301002	In-house	Intracellular	Lineage
147 Sm	CD14	M5E2	BioLegend	301810	In-house	Surface	Lineage
148 Nd	CD56	HCD56	BioLegend	318345	In-house	Surface	Lineage
149 Sm	CD11c	3.9	Tonbo	70-0116-U100	In-house	Surface	Lineage
150 Nd	FceRI	AER-37 (CRA-1)	Fluidigm	3150027B	Commercial	Surface	Lineage
151 Eu	CD39	A1	BioLegend	328202	In-house	Surface	Exhaustion
152 Sm	Granzyme B	CLB-GB11	Thermo Fisher	MA1-10338	In-house	Intracellular	Activation
153 Eu	CD45RA	HI100	Fluidigm	3153001B	Commercial	Surface	Differentiation
154 Sm	CD335/NKp46	9E2	BioLegend	331904	In-house	Surface	Lineage
155 Gd	CD27	L128	Fluidigm	3155001B	Commercial	Surface	Differentiation
156 Gd	Helios	22F6	BioLegend	137202	In-house	Intracellular	Activation/exhaustion
158 Gd	PD-1/CD279	EH12.2H7	BioLegend	329902	In-house	Surface	Activation/exhaustion

159 Tb	CCR7/CD197	G043H7	Fluidigm	3159003A	Commercial	Surface	Differentiation
160 Gd	Tbet	4B10	Fluidigm	3160010B	Commercial	Intracellular	Lineage/activation
161 Dy	CD152/CTLA-4	14-D3	Fluidigm	3161004B	Commercial	Intracellular	Exhaustion
162 Dy	Foxp3	PCH101	Fluidigm	3162011A	Commercial	Intracellular	Lineage
163 Dy	CD294/CRTH2	BM16	Fluidigm	3163003B	Commercial	Surface	Lineage
164 Dy	CD161	HP-3G10	Fluidigm	3164009B	Commercial	Surface	Lineage
165 Ho	Eomes	WD1928	Invitrogen	14-4877-82	In-house	Intracellular	Activation/exhaustion
166 Er	TCF-1 (TCF7)	7F11A10	BioLegend	655202	In-house	Intracellular	Differentiation
167 Er	CD38	HIT2	DVS	3167001B	Commercial	Surface	Differentiation/activation
168 Er	CD138	DL-101	Fluidigm	3168009B	Commercial	Surface	Lineage
169 Tm	TIGIT	MBSA43	Invitrogen	16-9500-82	In-house	Surface	Activation/exhaustion
170 Er	CXCR5	RF8B2	BD	552032	In-house	Surface	Lineage
171 Yb	IL-33Ra/ST2	B4E6	MD Biosci	101002	In-house	Surface	Lineage
172 Yb	Ki67	B56	Fluidigm	3172024B	Commercial	Intracellular	Activation/exhaustion
173 Yb	HLA-DR	L243	Fluidigm	3173005B	Commercial	Surface	Activation
174 Yb	TCRgd	B1	Biolegend	331202	In-house	Surface	Lineage
175 Lu	IgD	IA6-2	BioLegend	348235	In-house	Surface	Lineage
176 Yb	CD127/IL-Ra	A019D5	Fluidigm	3176004B	Commercial	Surface	Differentiation
191/193	Iridium	n/a	Fluidigm	201192B	n/a	Fixative	DNA
209 Bi	CD16	3G8	Fluidigm	3209002B	Commercial	Surface	Lineage
89Y	CD45	HI30	Fluidigm	3089003B	Commercial	Surface	Lineage

due to metal impurities, sample impurities, and oxidation products. One such example is gadolinium 157 (Gd157), which should be avoided due to lack of purity [5].

Lineage markers—CD45, CD3, CD4, and CD8—are integrated in order to identify T cells. The remaining channels are used to develop an in-depth analysis of the T-cell compartment. Our T-cell panel incorporates core markers for T-cell differentiation, T-cell activation, and T-cell exhaustion (Table 1). The remaining channels in this panel can be utilized to customize the panel (e.g., integration of metal-labeled tetramers).

3.2 Antibody Conjugation

The vast selection of antibody-metal conjugates that are commercially available is constantly expanding. Compared to in-house conjugates, less batch effect is expected between lots of commercial antibodies, and therefore commercially conjugated antibodies are preferred. If there is no commercial option, in-house conjugation adds flexibility to customize your panel.

The protocol below is based closely on Fluidigm's Maxpar Antibody Labeling Kit v11 protocol, but includes a few key modifications and suggestions (*see* **Notes 4** and **5**).

1. Equilibrate the polymer to room temperature and spin down before opening.
2. Resuspend the polymer in 95 μL L-Buffer and mix well to dissolve the polymer completely.
3. Add 5 μL lanthanide solution to the polymer tube (final concentration: 2.5 mM) and mix well (*see* **Note 6**).
4. Incubate at 37 °C for 30–40 min to load the polymer with the metal.
5. During the incubation, add 300 μL R-Buffer and 100 μg antibody to a 30 kDa filter (*see* **Note** 7). Label the filter and tube with the antibody.
6. Centrifuge the 30 kDa filter at 12,000 × *g* for 10 min. Discard flow-through.
7. During the centrifugation, prepare 4 mM TCEP by mixing 8 μL 0.5 M TCEP with 992 μL R-Buffer.
8. After the centrifugation, add 100 μL 4 mM TCEP to the 30 kDa filter and mix well.
9. Incubate at 37 °C for 30 min to partially reduce the antibody. *Do not exceed* 30 min.
10. During the incubation, add 200 μL L-Buffer to a 3 kDa filter. Label the filter and tube with the metal.
11. After the 30–40 min incubation, retrieve the metal-loaded polymer (**step 4**), spin down, and transfer to the 3 kDa filter containing L-Buffer.

12. Centrifuge the 3 kDa filter at 12,000 × *g* for 25 min. Discard flow-through.
13. Add 400 μL C-Buffer to the 3 kDa filter and centrifuge at 12,000 × *g* for 30 min to purify the metal-loaded polymer.
14. After the 30 min incubation, retrieve the partially reduced antibody (**step 9**), and immediately add 300 μL C-Buffer to the 30 kDa filter to wash.
15. Centrifuge the 30 kDa filter at 12,000 × *g* for 10 min. Discard flow-through.
16. Add 400 μL C-Buffer to the 30 kDa filter and centrifuge at 12,000 × *g* for 10 min to purify the partially reduced antibody. Discard flow-through.
17. Transfer the metal-loaded polymer (~20 μL) from the 3 kDa filter to the 30 kDa filter containing the partially reduced antibody.
18. Rinse the 3 kDa filter with 60 μL C-Buffer, transfer to the 30 kDa filter, and mix well. Discard 3 kDa filter.
19. Incubate at 37 °C for 90–120 min to conjugate the antibody with the metal-loaded polymer.
20. After the incubation, add 200 μL W-Buffer to the 30 kDa filter, and centrifuge at 12,000 × *g* for 5 min to wash the metal-conjugated antibody. Discard flow-through.
21. Repeat wash four times, for a total of five washes. On the final wash, centrifuge for 10 min instead of 5 min.
22. Invert the 30 kDa filter over a collection tube, and centrifuge the inverted filter/collection tube assembly at 1000 × *g* for 1 min to recover the metal-conjugated antibody.
23. Add 45 μL W-Buffer to the 30 kDa filter and invert over the collection tube. Centrifuge the inverted filter/collection tube assembly at 1000 × *g* for 1 min.
24. Repeat with an additional 45 μL W-Buffer. The final recovered volume should be ~100 μL. Discard 30 kDa filter.
25. Quantify the metal-conjugated antibody by measuring the absorbance at 280 nm using a NanoDrop or equivalent spectrophotometer (*see* **Note 8**).
26. Adjust the antibody concentration to approximately 0.5 mg/mL with antibody stabilizer, and transfer to a screw-top tube to avoid evaporation (*see* **Note 9**). Store at 4 °C.

3.3 Antibody Titration

Validation of all newly purchased or conjugated antibodies is necessary before staining actual samples. Titrations to determine the optimal staining concentration should be performed on all newly conjugated antibodies, but is usually only performed on first-time

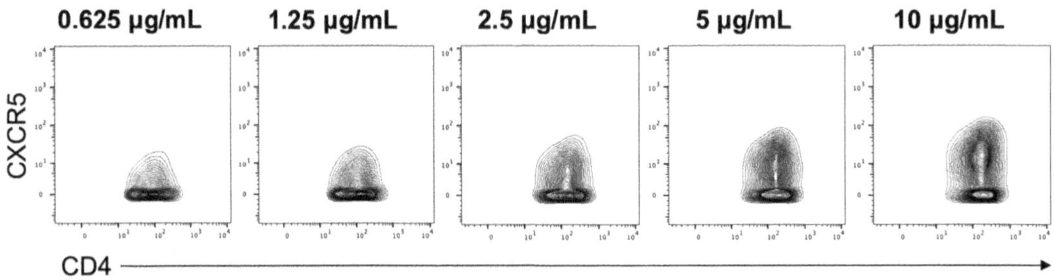

Fig. 1 Titration. An in-house conjugated CXCR5 antibody is stained at five concentrations to determine the optimal staining concentration

purchases of commercial antibodies, assuming variation between lots is not significant.

While titrating each antibody separately is ideal (only the antibody in question varies in concentration), titrating multiple antibodies simultaneously works fine. If the entire panel is not needed to gate down to the population(s) of interest, use an abbreviated panel that only includes the necessary lineage markers to allow you to identify the cell subset that your new antibody labels. To titrate an antibody, prepare five twofold serial dilutions of 0.625, 1.25, 2.5, 5, and 10 μg/mL (*see* **Note 10** and Fig. 1). After staining with a range of antibody concentrations and acquiring the data, analyze the FCS files in FlowJo. Contour plots (with outliers included) are preferred over histograms to plot the data.

There are two approaches to analyzing the titration data: (a) determine the stain indices and choose the dilution with the highest stain index or (b) "going by eye," which is, to say, choosing the dilution that has the most visual separation between negative and positive populations. For markers with bimodal distributions, the optimal concentration is that at which the positive and negative populations are most clearly separated. If there is little difference between concentrations, choose the lowest concentration that allows for that marker to be gated on. This is the most cost-effective choice and will reduce the frequency of conjugations necessary for a given epitope.

However, if there is not a clear difference between the various concentrations, stain index may be a helpful tool. Stain index is calculated as the difference of the geometric mean of mass intensity (MMI) of the positive population and the MMI of the negative population, divided by two times the robust standard deviation of the negative population (all of which can be determined in FlowJo). Calculating the stain indices is the preferred method for determining the optimal concentration of markers that produce a smeared (non-bimodal) stain with no clear positive or negative population.

$$\text{stain index} = \frac{\text{pos MMI} - \text{neg MMI}}{2 \times \text{neg SD}}$$

3.4 Staining

The procedure described below has been validated on cryopreserved human peripheral blood mononuclear cells (PBMCs), but could be adapted for freshly isolated PBMCs or samples from animal models. PBMCs from healthy donors were obtained by an experienced phlebotomist after written informed consent.

1. Thaw PBMCs. Swirl vial of cryopreserved PBMCs in 37 °C water bath until partially thawed. Add 1 mL warm complete medium to vial, and lift cells into a 15 mL tube containing 10 mL warm complete medium (*see* **Note 11**).
2. Centrifuge at 315 × *g* for 5 min. Aspirate supernatant.
3. Count the cells. Resuspend cells in 1 mL complete medium and count with a hemocytometer or automatic cell counter.
4. Centrifuge at 315 × *g* for 5 min. Aspirate supernatant.
5. Plate the cells. Resuspend cells to 5–15 × 10^6 cells/mL in complete medium. Plate 200 μL cell suspension per well in a 96-well round-bottom plate so there are 1–3 × 10^6 cells per well.
6. Rest cells by incubating at 37 °C for at least 1 h. During this time, prepare live/dead stain, surface antibody cocktail, and intracellular antibody cocktail (*see* **Note 3**).
7. After resting the cells, centrifuge the plate at 515 × *g* for 5 min. Remove supernatant (*see* **Note 12**).
8. For live/dead discrimination, resuspend cells in 50 μL live/dead stain per well. Incubate at room temperature for 5–10 min (*see* **Note 13**).
9. Wash with 170 μL stain buffer per well. Centrifuge at 515 × *g* for 5 min. Remove supernatant.
10. Stain for surface proteins by resuspending cells in 50 μL surface antibody cocktail per well. Incubate at room temperature for 30 min.
11. Wash with 170 μL stain buffer per well. Centrifuge at 515 × *g* for 5 min. Remove supernatant.
12. Repeat wash without resuspending the pellet. If not performing intracellular stain, proceed to **step 18**.
13. To fix and permeabilize the cells, resuspend in 50 μL permeabilization solution per well (use 100 μL per well for samples with cell clumps). Incubate at room temperature for 30 min.
14. Wash with 170 μL 1× permeabilization buffer per well. Centrifuge at 650 × *g* for 5 min. Remove supernatant.
15. Stain for intracellular proteins by resuspending cells in 50 μL intracellular antibody cocktail per well. Incubate at room temperature for 60 min.

16. Wash with 170 μL 1× permeabilization buffer per well. Centrifuge at 650 × *g* for 5 min. Remove supernatant.
17. Repeat wash twice without resuspending the pellet.
18. Resuspend cells in 100 μL fixative per well. Store overnight, or at least 1 h if acquiring the same day, at 4 °C, protected from light until acquisition.

3.5 Acquisition and Normalization

After fixation, wash cells twice with PBS and twice with Milli-Q H_2O. Resuspend cells in acquisition solution to count. Acquire samples on a CyTOF Helios at a speed below 400 events/s to reduce the probability of doublets. Closely monitor the speed and pressure of the CyTOF Helios sample loader during acquisition to detect any signs of clog. Every 2–4 h, recalibrate the instrument and raise detector voltages to limit the drift of signal intensity over time. Use Fluidigm's CyTOF Software to perform bead-based normalization of the data.

3.6 Analysis

Our analysis of T-cell diversity using mass cytometry is performed in two complimentary steps.

3.6.1 Predefined Analysis

The first step consists of a predefined approach, where we manually gate a large number of curated immunophenotypes and explore them using linear models as in Patin et al. [6].

We import the normalized FCS files into FlowJo (*see* **Note 14**) and identify live T cells using a gating strategy similar to that displayed in Fig. 2a.

After gating on CD4 and CD8 T cells, we identify 5–6 major T-cell differentiation states (Fig. 2b, c):

1. Naïve T cells (T_n): $CD45RA^+$ $CD27^+$ $CCR7^+$
2. Central memory T cells (T_{CM}): $CD45RA^-$ $CD27^+$ $CCR7^+$
3. Effector memory T cell 1 (T_{EM1}): $CD45RA^-$ $CD27^+$ $CCR7^-$
4. Effector memory T cell 2 (T_{EM2}): $CD45RA^-$ $CD27^-$ $CCR7^-$
5. Late effector memory T cells (T_{EMRA}): $CD45RA^+$ $CD27^-$

Stem cell memory T cells can be further separated from naïve T cells as being $CD95^+$ $CD49d^+$.

We recommend creating a FlowJo Table of the 14 immunophenotypes indicated in Table 3, Column 1 to display proportions of bulk CD4 and CD8 T cells, as well as proportions of each of the T-cell differentiation compartments. The remaining channels in this panel contain qualitative markers (examples of these stains in bulk CD8 T cells are provided in Fig. 2d). We determine their level of expression in each of the 14 quantitative T-cell subsets by determining the geometric mean of mass intensity (MMI) in FlowJo to produce $32*14$ qualitative immunophenotypes

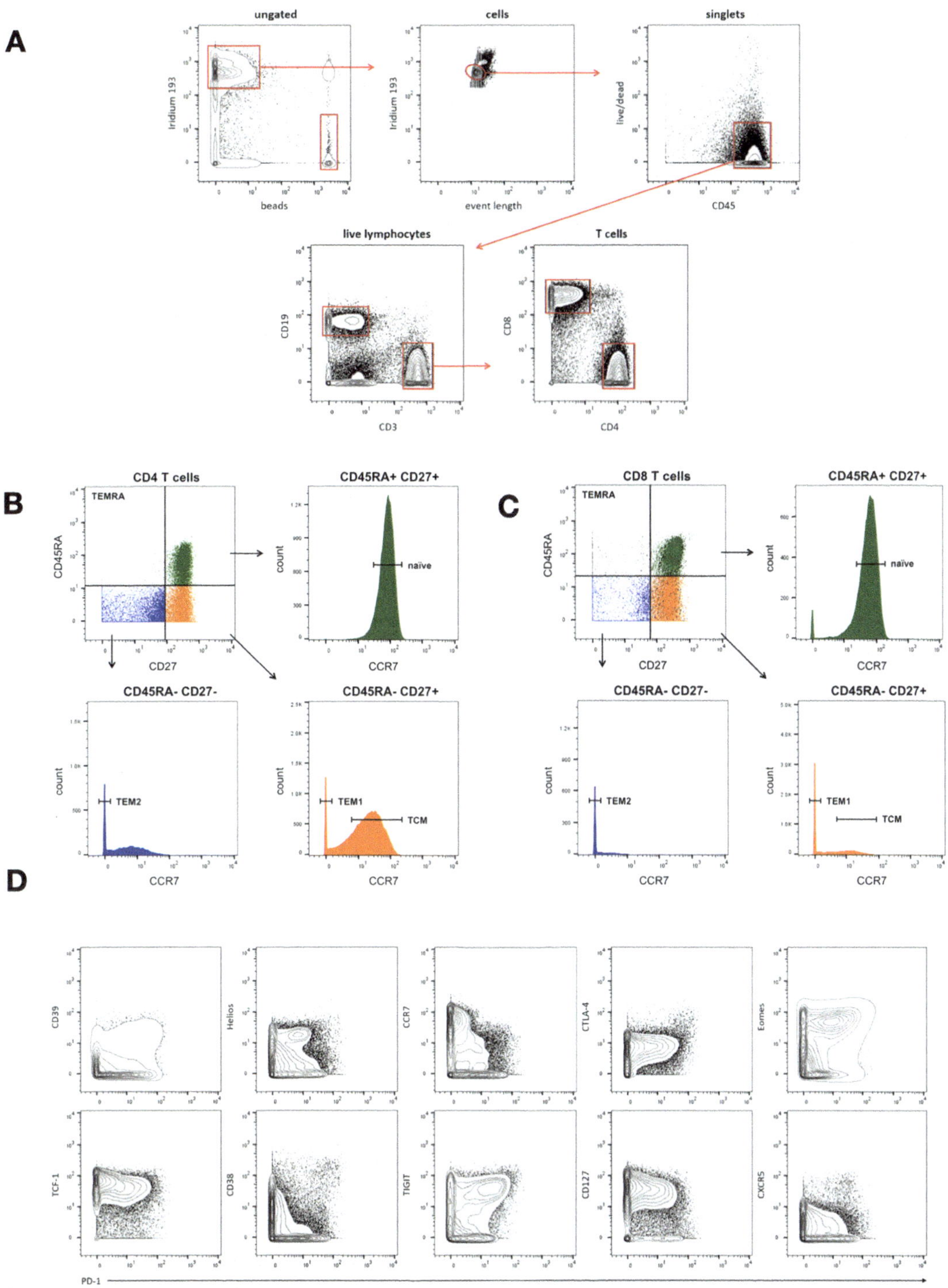

Fig. 2 Gating strategy. (**a**) Cryopreserved human PBMCs are stained for mass cytometry. Cells are identified as iridium^{+} beads^{-}; singlets are identified using event length, and live lymphocytes are CD45^{+} and unstained with the viability stain. T cells are identified as CD3^{+} CD19^{-}, and CD4 and CD8 T cells are further defined based on being exclusively positive for CD4 or CD8. (**b**) Predefined immunophenotypes are identified in CD4 T cells. (**c**) Predefined immunophenotypes are identified in CD8 T cells. (**d**) Expression of qualitative markers in bulk CD8 T cells is shown

Table 3
Immunophenotypes

Column 1 Percentage of	Column 2 In each Column 1 population, MMI of	
CD4 in CD3	CD16	LAG-3
Naïve in CD4	CD28	CD26
SCM in CD4	CD69	CD95
CM in CD4	Ki67	CTLA-4
EM1 in CD4	CD85j	CD49d
EM2 in CD4	CD38	CD103
EMRAin CD4	TOX	CD127
CD8 in CD3	CXCR5	CD45RA
Naïve in CD8	CD39	TCF-1
SCM in CD8	Tim3	Granzyme B
CM in CD8	CD27	Granzyme K
EM1 in CD8	Helios	CCR7
EM2 in CD8	PD-1	TIGIT
EMRA in CD8	Tbet Eomes CD57	2B4 CXCR3 HLA-DR

(Table 3, Column 2). We recommend encoding each immunophenotype as a statistic in a FlowJo Template for ease of processing.

We export the FlowJo Table as a .csv file which is then analyzed using our CyTOF_Analysis_Part1 R script. This first script is used to evaluate the association between factors such as treatment or disease and mass cytometry-generated immunophenotypes. It runs a linear regression for each of the collected immunophenotypes, with each factor being included. It can be customized to include important covariates such as age, viral serostatus, or BMI, as well as potential batch effects [6]. It calculates both the p-value of the association and the false discovery rate (FDR) when considering all of the tests as one multiple testing family. It plots the immunophenotypes that are significantly associated with the examined factors, using an adjustable threshold, while removing the effects of batch and confounding variables. An example of an effect size plot is provided in Fig. 3a.

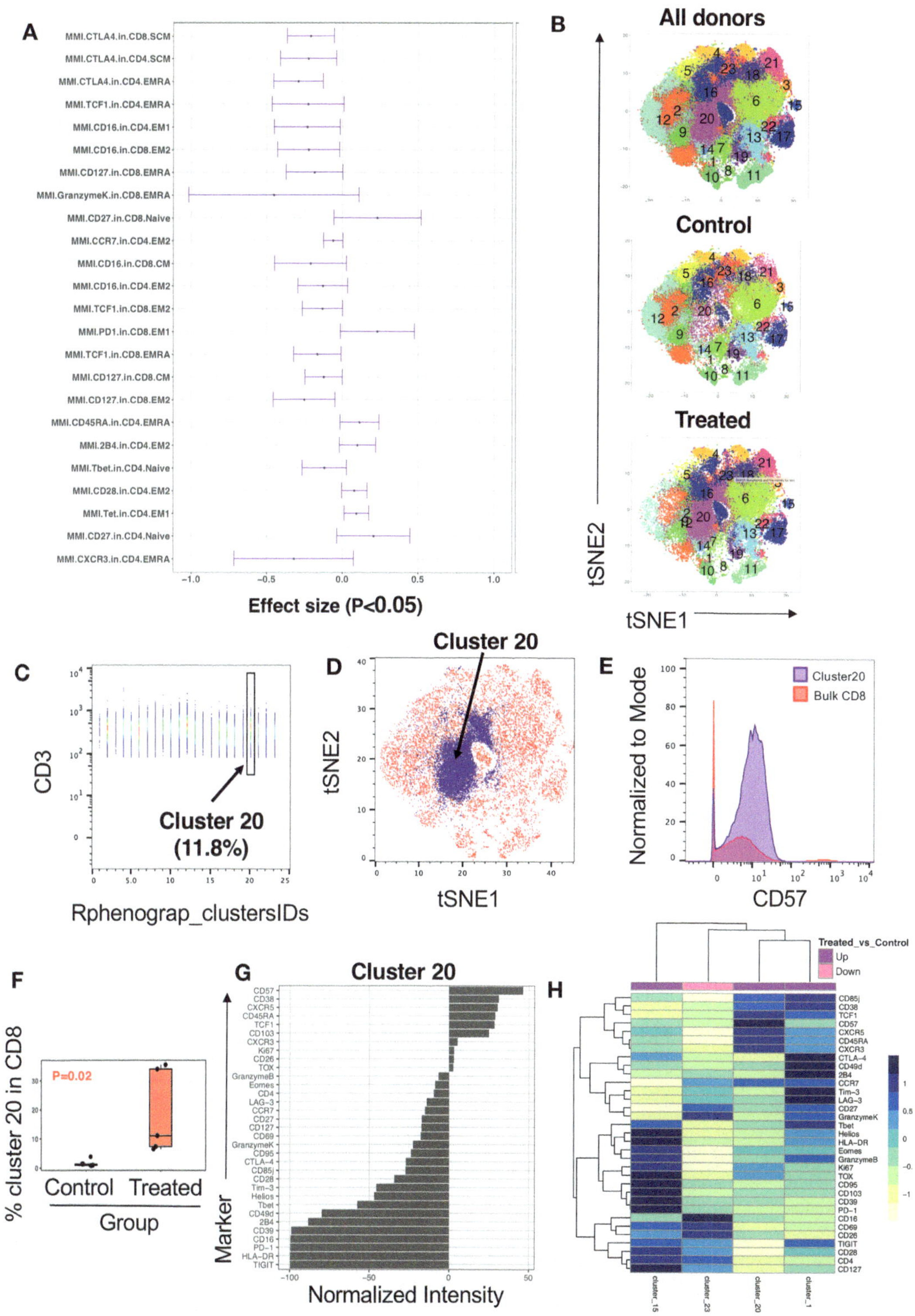

Fig. 3 Semi-biased analysis. (**a**) Effect sizes of significant associations (adj. $P < 0.05$) between group and immunophenotype in a sample cohort. Effect sizes were estimated in a linear mixed model, with immunophenotypes as response variables and group as the treatment variable. Dots represent the mean of the beta

The first part of the analysis offers a broad overview of the structural differences in cell composition across different groups of patients.

3.6.2 Semi-Biased Analysis

The second step of our analysis consists of a semi-biased approach, where we analyze the composition of the T-cell compartment independently of previously defined immunophenotypes.

From FlowJo, we export FCS files of the subpopulation of interest (e.g., CD8 T cells) using the following settings: FCS3, Include all events, and All uncompensated parameters.

In RStudio, we either use the cytofkit_GUI within the R package cytofkit [7] or run the function manually as proposed in the CyTOF_Analysis_Part2 R script. In both cases, the main steps are to (a) select the folder containing the exported FCS files and identify the files for analysis and (b) select the markers to be incorporated in the high-dimensional analysis. For the latter, we exclude the markers used to gate the subpopulation of interest (Iridium, Beads, CD45, Dump, CD3, CD8), as well as empty channels. The cytofkit function allows the selection of a number of features for analysis. In our routine practice, we transform data using cytofAsinh and merge using min, which samples the minimum number of cells among all the selected FCS files from each FCS file. This eliminates any skewing due to variations in cell number among the files. We select Rphenograph as the clustering method and tSNE for dimensionality reduction and cluster data visualization. We do not use cellular progression algorithms in our routine analysis.

This analysis creates a folder that contains 1/ newly generated FCS files that incorporate new dimensions such as tSNE or cluster IDs, 2/ an Rdata file, and 3/ a number of csv and pdf files.

Although a tSNE plot of the clusters is automatically generated by the cytofkit function, it can be further customized using the cytofkitShinyAPP function, as shown in Fig. 3b. In parallel, the tSNE plot can be recreated in FlowJo for more plotting options. For this, import the newly generated FCS files into a new workspace in the CyTOF_cytofkit folder. Create new FCS files by concatenating all files (all.fcs) as well as files for each group of donors (i.e., Control.fcs and Treated.fcs). Import these new files into a new

Fig. 3 (continued) estimate. Lines represent the 95% confidence intervals (CyTOF_Analysis_Part1 R script). (**b**) tSNE plot of clusters for CD8 T cells from all samples (control and treated), control samples, and treatment samples (cytofkitShinyAPP function, CyTOF_Analysis_Part2 R script). (**c**) Gating strategy for identification of clusters (FlowJo). (**d**) Overlay of cluster 20 (purple) on tSNE plot reconstituted in FlowJo using cytofkit-generated FCS files. (**e**) Histogram overlay of CD57 expression in cluster 20 (purple) compared to bulk CD8 T cells (red). (**f**) Boxplot of proportions of cluster 20 in subset of samples and the p-value of the association (CyTOF_Analysis_Part3 R script). (**g**) Normalized intensity of expression of qualitative markers for cluster 20 (CyTOF_Analysis_Part3 R script). (**h**) Heatmap of significant clusters. Top bar indicates the direction of the effect of the association with treatment condition (CyTOF_Analysis_Part3 R script)

FlowJo Workspace along with the newly generated FCS files from the cytofkit analysis. Edit columns to add an "Annotation" column. Annotate individuals and concatenated samples by group. Individual clusters can be identified and gated by plotting Rphenograph_-clusterIDs on the x-axis (Fig. 3c). In a FlowJo Layout, recreate the automatically generated tSNE plot using tsne_1_linear and tsne_2_linear dimensions for the all.fcs file. You can now overlay relevant clusters of interest (Fig. 3d) and investigate their pattern of expression (Fig. 3e).

You can proceed using the cytofkitShinyAPP function to visualize heatmaps of marker expression within clusters. However, for additional custom options, we suggest using our CyTOF_Analysis_Part3 R script.

The first part of the script allows you to generate a heatmap of cluster proportions across the different samples, as well as boxplots of proportions of each cluster in subgroups of samples (Fig. 3f). It also allows you to evaluate the statistical significance of the differences in proportions observed across groups of samples by using the indicated regression code. This code generates a table with p-values across groups for each cluster.

The second part of the script allows you to examine the composition of the clusters. This can be done by generating a bar plot of each cluster, with the normalized intensity of each marker along the *x*-axis (Fig. 3g). Alternatively, you can generate a heatmap with color-coded normalized differences to compare clusters (Fig. 3h). To facilitate interpretation of the heatmap, we use a column annotation indicating the direction of the estimated effect, represented as a color-coded bar above the heatmap.

The second part of the analysis allows you to visualize the data, evaluate and plot the significant differences, and investigate the composition of relevant clusters.

Our high-throughput analysis workflow offers a broad overview of mass cytometry-generated datasets, which aims to facilitate the generation of innovative scientific hypotheses.

4 Notes

1. The Fluidigm protocol calls for the use of a 50 kDa Amicon Ultra-500 μL V-bottom centrifugal filter; however, a 30 kDa filter works equally well.
2. When handling the filters, grasp by the top rim or by its solid sides avoiding contact with the holes on either side. Do not touch the inner filter portion of the tube (white portion) with pipette tip, because this may damage the integrity of the filter. The filters have a maximum capacity of 500 μL.

3. Use an Excel spreadsheet to calculate the volume of each antibody based on the optimal concentration determined by a titration. An example spreadsheet is provided (https://github.com/wherrylab/Cytof_analysis_calanio).

   Prepare 15% excess volume of antibody cocktails to account for pipetting error.

   Prepare the antibody cocktails on the day of the experiment to optimize stability.

4. To stay organized during the initial steps that deal with the metal and antibody separately, divide the workspace into two areas: the left side for the metal and the right side for the antibody. This way, L-Buffer will be used on the left side to load the metal onto the polymer and R-Buffer will be used on the right side to reduce the antibody. Also note that the 3 kDa filter is used for the metal (left) and the 30 kDa filter is primarily used for the antibody (right).

5. We suggest performing a maximum of eight conjugations at a time to avoid interfering with the careful timing of the protocol. Although it is not recommended, if performing more than ten conjugations, be sure to scale up when preparing 4 mM TCEP (*see* Subheading 3.2, **step 7**).

6. Although the protocol calls for 5 μL Lanthanide solution, performing the conjugation with as little as 3 μL (final concentration: 1.5 mM) has produced similar results.

7. Commercial antibodies are typically concentrated between 0.5 and 1.0 mg/mL. Often manufacturers will supply a slight excess of the product. If the supplied volume of antibody exceeds the reported volume, use the entire volume supplied instead of the exact volume. Starting with slightly more than 100 μg does not have any apparent effects on the conjugation reaction, but does increase the total possible yield of conjugated antibody.

   If the antibody concentration is less than 0.5 mg/mL, an alternative step must be performed after **step 4** of Subheading 3.2 to concentrate the antibody. Simply transfer the entire volume of antibody to the 30 kDa filter and centrifuge at 12,000 × *g* for 5 min. If the volume exceeds 500 μL (max. capacity of the filter), repeat this step until all of volume has been passed through the filter. Add 400 μL R-Buffer and proceed to **step 6**, Subheading 3.2.

   Although this protocol is written for the conjugation of 100 μg antibody, it has worked well with 50 μg antibody.

8. Calculate percent recovery by dividing the final amount of conjugated antibody by the initial antibody amount. A yield of above 100% is not unheard of if slightly more than 100 μg antibody was used. A typical yield is greater than 80%.

9. The number of samples that can be stained with an in-house conjugated antibody depends on the optimal staining concentration, which should be determined by a titration. For example, if the optimal staining dilution is found to be 1:200 and each sample will be stained in 50 μL, about 800 samples can be stained. However, the actual number of samples would be less than 800 if we account for the 15% excess volume of antibody cocktail prepared for each stain and any evaporation or volume loss that may occur.
10. To prepare five twofold serial dilutions, follow the protocol outlined in the staining section with the following specifications. In a 96-well round-bottom plate, prepare five wells of 1×10^6 PBMCs per well from a single healthy donor. The viability stain, permeabilization, and fixation should all be performed as outlined. For the surface and intracellular stains, prepare 100 μL of each antibody cocktail (this is enough for exactly two samples and excludes the usual 15% error by design). Use the optimal staining concentrations for the antibodies in the panel that are not being titrated, and for the antibodies that are being titrated, use a 1:50 dilution (final concentration: 10 μg/mL if stock antibody concentration is 0.5 mg/mL). When performing the surface and intracellular stains, resuspend the first well with 50 μL antibody cocktail, leaving 50 μL remaining. Dilute the antibody cocktail with 50 μL stain buffer (titrated antibodies are now at 1:100), and resuspend the second well. Repeat this process of diluting the antibody cocktail before staining each additional well to produce dilutions of 1:50, 1:100, 1:200, 1:400, and 1:800. It is important to note that this approach does dilute the antibodies in the panel that are not being titrated as well; however, this does not seem to interfere with determining the optimal dilution of the titrated antibodies.
11. If cells have been stored at −80 °C for longer than 1 year, include 0.1 mg/mL DNase I and 10 mM $MgCl_2$ in the medium during the thaw, and rest to help prevent cell clumps due to free DNA from cell lysis.
12. Flicking is the preferred method to remove the supernatant; however, doing so takes practice to avoid cross-contamination and should not be performed for the first time on irreplaceable samples. To flick off the supernatant, hold the plate in the dominant hand, and remove the lid with the other hand. In one quick motion, invert the plate over a biohazard sink to rapidly eject the supernatant.
13. As an alternative to mmDOTA-La139, cisplatin can be used for live/dead discrimination. To label with cisplatin, resuspend cells in 100 μL per well 5 μM cisplatin diluted in PBS. Incubate

at room temperature for 1 min. Wash with 100 μL stain buffer per well, and centrifuge at 515 × g for 5 min. Repeat wash with 170 μL stain buffer per well and proceed to surface stain.

14. If normalization did not occur during acquisition, we recommend using the normalizer_GUI within the R package premessa (https://github.com/ParkerICI/premessa). Import the FCS files, identify beads for each channel, and normalize. A "normed" folder containing the normalized FCS files will be created in the same location as the original FCS files.

Acknowledgements

We thank Divij Matthew, University of Pennsylvania, USA, for his help with the titration figure, as well as Jacob Bergstedt, Lund University, Sweden, for his help with the R scripts.

References

1. Brodie TM, Tosevski V (2018) Broad immune monitoring and profiling of T cell subsets with mass cytometry. Methods Mol Biol 1745:67–82. https://doi.org/10.1007/978-1-4939-7680-5_4
2. Bengsch B, Ohtani T, Khan O, Setty M, Manne S, O'Brien S, Gherardini PF, Herati RS, Huang AC, Chang KM, Newell EW, Bovenschen N, Pe'er D, Albelda SM, Wherry EJ (2018) Epigenomic-guided mass cytometry profiling reveals disease-specific features of exhausted CD8 T cells. Immunity 48 (5):1029–1045.e5. https://doi.org/10.1016/j.immuni.2018.04.026
3. Good Z, Borges L, Vivanco Gonzalez N, Sahaf B, Samusik N, Tibshirani R, Nolan GP, Bendall SC (2019) Proliferation tracing with single-cell mass cytometry optimizes generation of stem cell memory-like T cells. Nat Biotechnol 37(3):259–266. https://doi.org/10.1038/s41587-019-0033-2
4. Lakshmikanth T, Brodin P (2019) Systems-level immune monitoring by mass cytometry. Methods Mol Biol 1913:33–48. https://doi.org/10.1007/978-1-4939-8979-9_3
5. Leipold MD, Newell EW, Maecker HT (2015) Multiparameter phenotyping of human PBMCs using mass cytometry. Methods Mol Biol 1343:81–95. https://doi.org/10.1007/978-1-4939-2963-4_7
6. Patin E, Hasan M, Bergstedt J, Rouilly V, Libri V, Urrutia A, Alanio C, Scepanovic P, Hammer C, Jönsson F, Beitz B, Quach H, Lim YW, Hunkapiller J, Zepeda M, Green C, Piasecka B, Leloup C, Rogge L, Huetz F, Peguillet I, Lantz O, Fontes M, Di Santo JP, Thomas S, Fellay J, Duffy D, Quintana-Murci L, Albert ML (2018) Natural variation in the parameters of innate immune cells is preferentially driven by genetic factors. Nat Immunol 19 (3):302–314. https://doi.org/10.1038/s41590-018-0049-7
7. Chen H, Lau MC, Wong MT, Newell EW, Poidinger M, Chen J (2016) Cytofkit: a bioconductor package for an integrated mass cytometry data analysis pipeline. PLoS Comput Biol 12(9): e1005112

Chapter 2

A Carrier Strategy for Mass Cytometry Analysis of Small Numbers of Cells

Xian Jia, Xiaojuan Zhou, Haiping Zheng, Shan Jiang, Jiannan Weng, Lei Huang, Zhiqiang Du, Changchun Xiao, Lei Zhang, Xiao Lei Chen, and Guo Fu

Abstract

The recent launch of mass cytometry or cytometry by time of flight (CyTOF) has revolutionized flow cytometry. Similar to fluorescence flow cytometry, a key challenge for CyTOF is to analyze samples of limited amount or very rare cell populations under various experimental settings. Here we describe a carrier strategy that significantly reduces the required sample amount without losing analytical resolution. We were able to detect as few as 5×10^4 human peripheral blood mononuclear cells (PBMCs) using this method. This simple method thus enables the maximal usage of valuable clinical samples.

Key words Mass cytometry (CyTOF), PBMCs, Small number of cells, Carrier, EL4

1 Introduction

The complexity and heterogeneity of biological systems are two major challenges that often prevent scientists from answering many biological questions. For example, it is well known that the tumor microenvironment is a very complicated ecosystem, containing tumor cells, stromal cells, immune cells, soluble factors, extracellular matrix, and blood vessels [1]. Moreover, immune cells embedded in the tumor microenvironment as well as in other immune responses are notorious for their diverse lineages, functional status, and cytokine production [2]. All of these require a high-dimensional analysis platform. In response to this demand, mass cytometry or cytometry of time of flight (CyTOF) recently emerged as a powerful tool for high-dimensional analysis [3]. Utilizing heavy metal ion-conjugated antibodies, mass cytometry can theoretically measure up to 100 parameters of a single cell, far more than that measured by the traditional fluorescence flow cytometry. Moreover, the heavy metal tags used for antibody conjugation are

Chaohong Liu (ed.), *T-Cell Receptor Signaling: Methods and Protocols*, Methods in Molecular Biology, vol. 2111, https://doi.org/10.1007/978-1-0716-0266-9_2,

normally absent from biological specimens [4, 5], thus essentially avoiding the autofluorescence issue quite often encountered when using fluorescence flow cytometry. There is also no or minimal spillover between different metal tags due to their discrete atomic mass. Because of these advantages, CyTOF was quickly adopted for numerous applications including analyzing signaling network [4], vaccination-induced response [6], and deep profiling of tumor microenvironment [7, 8].

CyTOF is still at an early stage of development. Many efforts have been put to establish and recapitulate the assays widely used for fluorescence flow cytometry. For example, the classical CFSE-based cell proliferation assay was successfully adopted by CyTOF [9]. Detection of rare cell populations and analysis of samples with low cell numbers are two frequently encountered difficulties. One may need quite a few markers to define rare cell populations. In this scenario, the inherent high dimensionality and low spillover characters of CyTOF make it an ideal platform for this task. Low cell number is often encountered when dealing with clinical samples, which are generally precious and of limited amount. For example, usually only a few milliliters of blood can be drawn from a given patient and may subject to further PBMC isolation and cryopreservation. Upon recovery, the numbers of live PBMCs can drop significantly. In this case, CyTOF analysis is not only cost effective but may also be absolutely required, simply because there are just not enough samples to split for multiple analyses by fluorescence flow cytometry. On the other hand, the mechanical design of CyTOF is unfavorable for analyzing rare cell populations or low number of cells. When samples go through the spray chamber of CyTOF, only about 30–40% of nebulized cells can get into plasma for further analysis, while the remaining cells are lost [10]. Moreover, 10–50% of cells are lost during the multiple-step staining process, while the extent of cell loss depends on the starting cell numbers, the number of staining steps, and whether permeabilization is used in the process. Taken together, that means only 10–20% of starting cells are eventually analyzed. This is less problematic when the number of cells is plenty in the samples, but can be prohibitive when analyzing rare cell populations or samples with low cell numbers. For these reasons, we decide to establish a protocol compatible with low cell number analysis, to take advantage of the high dimensionality of CyTOF and in the meanwhile circumvent the problem of high cell loss rate during sample preparation and running.

In fluorescence flow cytometry, the use of "carrier cells" makes it possible to examine a few thousand or even a few hundred cells [11]. We adapted this approach and tested a few candidate cell lines as carrier cells. In this chapter, we described an experimental system using EL4 cells, a murine T-cell lymphoma cell line, as carrier cells to detect small numbers of human PBMCs from healthy donors

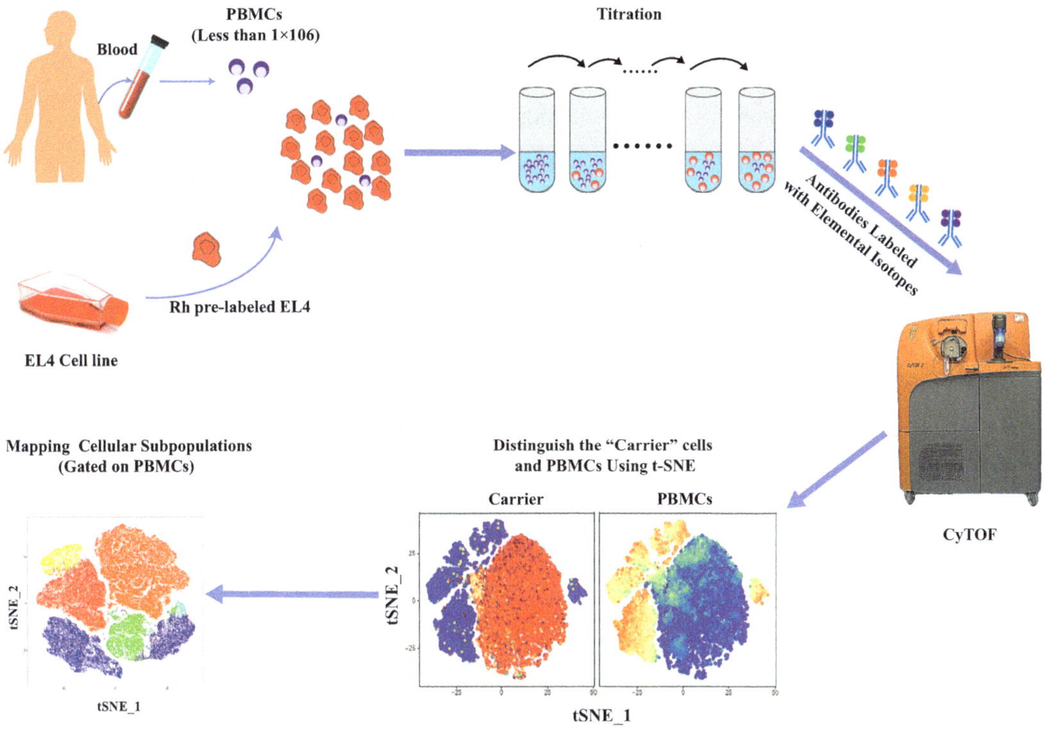

Fig. 1 Scheme of the carrier method for analyzing small numbers of cells by mass cytometry. Human PBMCs were titrated and mixed with Rh prelabeled EL4 carrier cells at different ratios, followed by staining with heavy metal ion-conjugated antibodies for human PBMCs. Samples were run on a CyTOF mass cytometer and analyzed by the t-SNE method. "Carrier" EL4 cells and human PBMCs can be clearly distinguished without overlapping. Subsequent gating on human PBMCs allowed further mapping of subpopulations of cells

(Fig. 1). We designed a 15-marker panel for mass cytometry analysis, comprised of human T, B, and myeloid lineage markers, as well as a DNA marker (Table 1). For sample running, the acquisition rate of CyTOF is much slower than fluorescence flow cytometry. For example, when the acquisition rate of CyTOF reaches 1000 cells/s, the number of doublets will increase significantly and severely compromise subsequent data analysis. A general trend is that the faster the sample acquisition, the lower the data quality. In our hands, acquisition at 300 cells/s is required to consistently obtain single cells.

For data analysis, we first tried the traditional method by manually gating cells of interest using FlowJo (v10.5.3; Tree Star Inc., Ashland, OR, USA) (Fig. 2). Human PBMCs can be clearly separated from carrier EL4 cells, and no interference was observed with subsequent cell subset identification. However, the increase in dimensionality makes the traditional manual gating method extremely labor intensive and time-consuming [12], with important information missed in some cases. Therefore, we analyzed the

Table 1
Antibody panel for the major cell subsets in PBMCs

No. #	Specificity	Clone	Tag	Dilution	Compartment	Purpose
1	CD45	HI30	89Y	1 μg/100 μL	Surface	Pan-leukocyte marker
2	CD19	HIB19	142Nd	1 μg/100 μL	Surface	B cells
3	CD123	6H6	143Nd	1 μg/100 μL	Surface	Plasmacytoid dendritic cells
4	CD38	HIT2	144Nd	1 μg/100 μL	Surface	Cell activation
5	CD8a	RPA-T8	146Nd	1 μg/100 μL	Surface	Cytotoxic T cells
6	CD11c	Bu15	147Sm	1 μg/100 μL	Surface	Dendritic cells
7	CD3	UCHT1	154Sm	1 μg/100 μL	Surface	T cells
8	CD20	2H7	171Yb	1 μg/100 μL	Surface	B cells
9	HLA-DR	L243	173Yb	1 μg/100 μL	Surface	MHC-II antigen presentation
10	CD14	M5E2	175Lu	1 μg/100 μL	Surface	Monocytes
11	CD4	RPA-T4	176Yb	1 μg/100 μL	Surface	T helper cells
12	CD16	3G8	209Bi	1 μg/100 μL	Surface	Natural killer cells, monocytes macrophages
13	Carrier cells nuclear acid		Rh103	2 μM	Intranuclear	Carrier cell labeling
14	DNA		Ir-191/193	125 nM/sample	Intranuclear	Cell identification
15	Cisplatin		195Pt	0.5 μM/sample	Intranuclear	Dead cell identification

same data set with two different algorithm-based methods, visualization by t-stochastic neighbor embedding (t-SNE) [13] (Fig. 3) and FlowSOM [14] (Fig. 2) using the Cytofkit package in R program. FlowSOM analyzes mass cytometry data using a self-organizing map. By comparing the t-SNE plots generated at each titration, we concluded that by using this carrier cell protocol we can readily identify major cell subsets in human PBMCs starting with as few as 5×10^4 cells without losing any resolution. Thus, this protocol can significantly reduce the starting sample amount required for CyTOF and facilitate additional assays in parallel, such as single-cell RNA-seq.

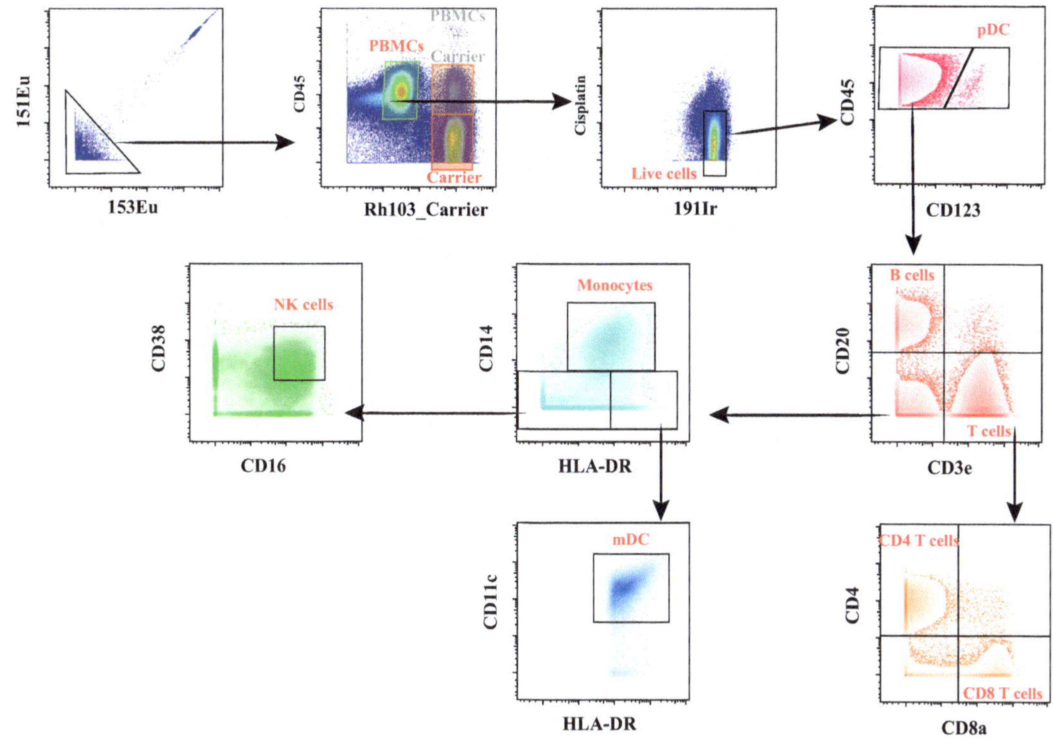

Fig. 2 Identification of major cell subsets in PBMCs by manual gating. Sequential manual gating to identify EQ. 1⁻ (excludes EQ normalization beads), CD45⁺carrier⁻ (excludes "carrier" cells and doublets), and cisplatin⁻ (excludes cisplatin-labeled dead cells) intact live cells. Within the live cells, plasmacytoid dendritic cells (pDC) were defined as CD45⁺CD123⁺. In the CD45⁺CD123⁻ population, T- and B-cell subsets were identified by CD3, CD4, CD8, and CD20. NK cells (CD14⁻HLA-DR⁻CD38⁺CD16⁺), monocytes (CD14⁺HLA-DR⁺), and myeloid DC (mDC, CD14⁻HLA-DR⁺CD11c⁺) were also identified [17]

2 Materials

All the reagents are stored at 4 °C unless otherwise indicated. Materials and consumables are kept in metal-free containers. New dust-free gloves are used throughout the procedure.

2.1 Preparation of Carrier Cells

1. Tissue culture-treated cell culture T25 flask with a vent cap.
2. Complete RPMI medium: RPMI 1640 supplemented with 10% heat-inactivated fetal bovine serum, 100 U/mL penicillin, 100 μg/mL streptomycin, and 10 mM HEPES.
3. Serum-free RPMI medium: RPMI 1640, 100 U/mL penicillin, 100 μg/mL streptomycin, and 10 mM HEPES.
4. EL4 cells (ATCC; TIB-39).

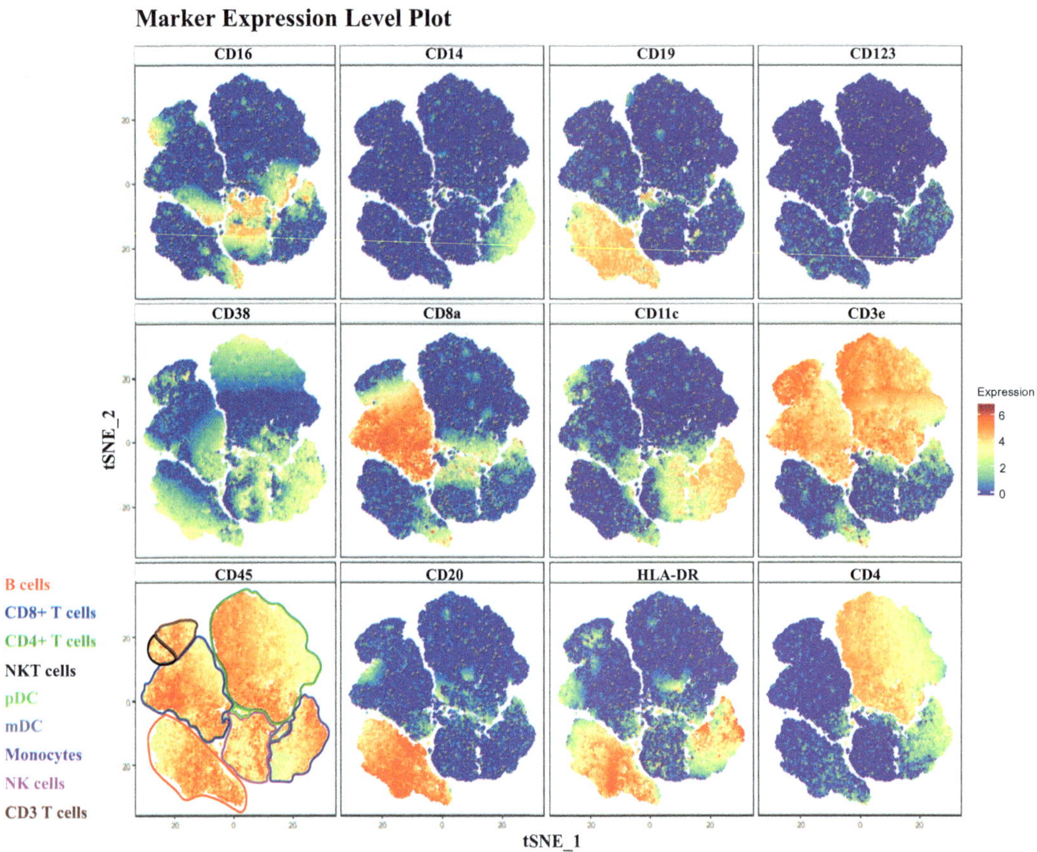

Fig. 3 Identification of cell subsets in PBMCs by t-SNE. t-SNE (t-stochastic neighbor embedding) is a dimensionality reduction algorithm, which is widely used in mass cytometry data analysis. Shown here is the expression pattern of each individual marker after excluding carrier cells and gating on PBMCs. Color code from dark blue to dark red indicates increased expression level of each marker. On the CD45 t-SNE plot, boundaries were drawn manually according to the automatically defined clusters and the marker expression patterns. Cell types were listed adjacent to this plot

5. Cell ID Rh-intercalator: 500 μM rhodium. Store aliquots at −20 °C and avoid repeated freeze/thaw cycles (*see* **Note 1**).
6. Maxpar Fix and Perm Buffer.
7. Maxpar Cell Staining Buffer.
8. Phosphate buffered saline (PBS) without heavy metal contaminants, pH 7.4.
9. 15 mL and 50 mL conical tubes.
10. Aerosol barrier pipette tips.

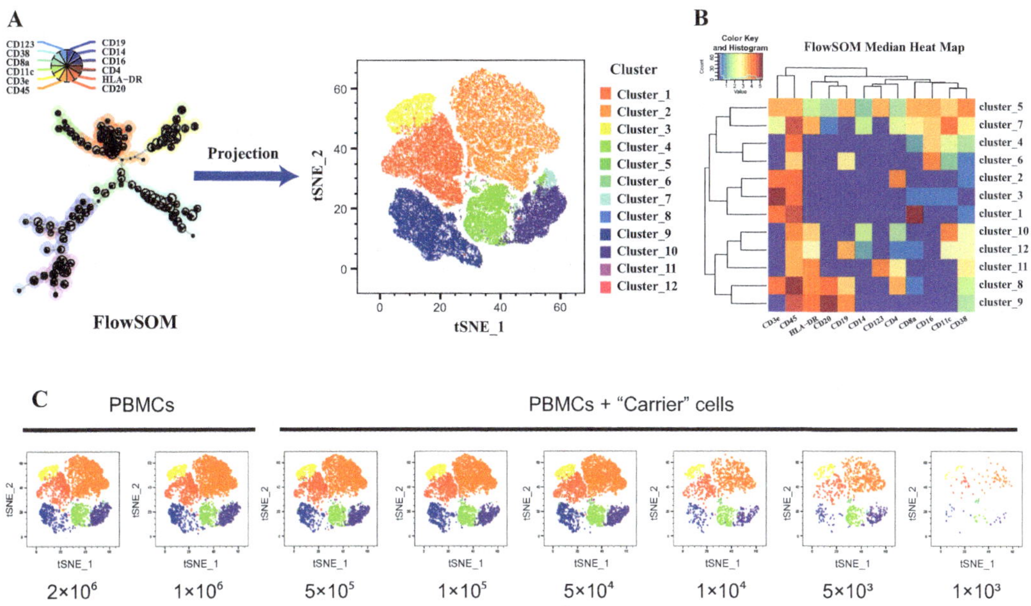

Fig. 2 FlowSOM combined with t-SNE for CyTOF data analysis. FlowSOM (flow cytometry data analysis using self-organizing maps), a clustering-based algorithm, can simultaneously cluster and visualize cytometry data in a two-dimensional grid of cell-type clusters. (**a**) Shown is the use of FlowSOM combined with t-SNE mapping to identify the major cell subsets in PBMCs in an unbiased avenues for visualization. (**b**) The identity of each cluster color coded in the t-SNE plot in (**a**) was further visualized using a FlowSOM heatmap, which can precisely exhibit the combinatory expression level of multiple markers. For example, Cluster_1, Cluster_2, and Cluster_9 can be readily identified as $CD8^+$ T cells, $CD4^+$ T cells, and B cells, respectively, by their expression patterns of CD8, CD4, CD19, and other markers. (**c**) The results of the "carrier" experiment were analyzed using the above method, and it was determined that as few as 5×10^4 PBMCs can be detected without losing resolution

2.2 Preparation of PBMCs

1. Medium with Benzonase for PBMCs: RPMI 1640 supplemented with 10% heat-inactivated fetal bovine serum, 100 U/mL penicillin, 100 μg/mL streptomycin, 10 mM HEPES, and 25 U/mL Benzonase. Warming up to 37 °C in a water bath prior to use (*see* **Note 2**).
2. Complete RPMI medium: RPMI 1640 supplemented with 10% heat-inactivated fetal bovine serum, 100 U/mL penicillin, 100 μg/mL streptomycin, and 10 mM HEPES.
3. Two 15 mL conical tubes for each donor PBMCs.
4. Healthy donor PBMCs.
5. Water bath set at 37 °C.
6. Trypan blue.
7. Hemocytometer.
8. Polypropylene round-bottom tubes with 35 μm cell strainer cap.

2.3 Mass Cytometry Cell Staining

1. 96-well U-bottom plates.
2. Fc Receptor Blocking Solution.
3. Cisplatin stock: 5 mM cisplatin. Store aliquots at −80 °C and avoid repeated freeze/thaw cycles (*see* **Note 3**).
4. Maxpar metal-conjugated antibodies: refer to Table 1.
5. Maxpar Fix and Perm Buffer.
6. Maxpar Cell Staining Buffer.
7. Cell ID Ir-intercalator: 125 μM iridium (191Ir/193Ir). Store aliquots at −20 °C and avoid repeated freeze/thaw cycles (*see* **Note 4**).
8. Purified deionized water, 18 MΩ-cm at 25 °C.
9. EQ™ Four Element Calibration Beads.
10. 16% formaldehyde (w/v), methanol free. Store aliquots at room temperature.
11. Mass cytometer (CyTOF2) (Fluidigm).
12. Vortex.
13. Polypropylene round-bottom tubes with 35 μm cell strainer cap.

3 Methods

3.1 Carrier Cell Labeling

1. Transfer EL4 cells from cell culture flasks into 15 mL conical tubes (*see* **Note 5**).
2. Centrifuge EL4 cells at 500 × g for 5 min at room temperature to pellet the cells.
3. Wash the pellet with 10 mL PBS and centrifuge again as above.
4. Prepare the cell intercalation solution by diluting rhodium (2 μM final concentration) with Maxpar Fix and Perm Buffer, and mix by vortexing.
5. Add 1 mL intercalation solution to the cells (not exceeding 10^7) and gently vortex. Incubate for 1 h at room temperature or leave overnight at 4 °C (*see* **Note 6**).
6. Wash cells by adding an equal volume of Maxpar Cell Staining Buffer.
7. Centrifuge as above and discard supernatant by aspiration.
8. Wash twice with prewarmed serum-free RPMI medium prior to use.

3.2 Thawing PBMCs

1. Warm medium with Benzonase for PBMCs in a 37 °C water bath.
2. Remove PBMC samples from liquid nitrogen, and keep them in the vacuum cup with liquid nitrogen or on dry ice (*see* **Note 7**).
3. Thaw 1–2 frozen vials at a time in the 37 °C water bath for 2–3 min, and remove the tube once the PBMC sample just thawed.
4. Add 1 mL of warmed PBMC Benzonase medium to the cryopreservation tube in a dropwise manner to retrieve all cells; repeat once to make sure all cells are collected. Then transfer cells to an appropriately labeled centrifuge tube and gently pipet up and down to mix cells.
5. Centrifuge cells at 300 × *g* for 5 min at room temperature.
6. Aspirate the supernatant and resuspend the cell pellet with 1 mL of warmed medium with Benzonase for PBMCs.
7. Filter cells through a 35 μm cell strainer if you observe any clumps.
8. Add 9 mL medium with Benzonase for PBMCs to the tube and gently pipet up and down to mix.
9. Centrifuge again at 300 × *g* for 5 min at room temperature.
10. Aspirate the supernatant and resuspend the cell pellet with 1 mL of prewarmed complete medium.
11. Count cells with a hemocytometer, and adjust the concentration to 5×10^6 cells/mL (or a maximum volume of 200 μL for samples with less than 1×10^6 cells) with prewarmed complete medium in 15 mL conical tubes (*see* **Note 8**).
12. Place the15 mL conical tubes in a 37 °C CO_2 incubator to rest cells for 15 min before staining (*see* **Note 9**).

3.3 Mix Carrier Cells and PBMCs

Cells will be transferred to a 96-well U-bottom plate from this step. Check the pellets, and quickly but gently flick the plate one time after centrifugation. The pellets may appear diffused and transparent after fixation and permeabilization.

1. Adjust the concentration of "carrier" cells to 1×10^7 cells/mL with prewarmed serum-free RPMI medium.
2. Wash the PBMCs with prewarmed serum-free RPMI medium, and adjust the concentration to 5×10^6 cells/mL or a maximum volume of 100 μL for samples with less than 1×10^6 cells.
3. Transfer 200 μL of PBMCs or all the cell suspensions (if less than 1×10^6 cells) together with 100 μL of "carrier cells" to a 96-well U-bottom plate (*see* **Note 10**).
4. Centrifuge again at 845 × *g* for 3 min at room temperature.

5. Wash the cells with 200 μL/well prewarmed serum-free RPMI medium and centrifuge as above.
6. Check the pellets and quickly but gently flick the plate at one motion to dump supernatant (*see* **Note 11**).

3.4 Cisplatin Labeling

1. Resuspend the cell pellet in 0.5 μM cisplatin (diluted in prewarmed serum-free medium), gently pipet up and down to mix, and incubate at room temperature for 2 min.
2. Centrifuge at 845 × *g* for 3 min at room temperature.
3. Check the pellets and quickly but gently flick the plate at one motion to dump supernatant.
4. Wash the cells with 200 μL/well of Maxpar Cell Staining Buffer and centrifuge as above.

3.5 Surface Staining of PBMCs

1. Add 50 μL/well of Fc Receptor Blocking Solution (0.5 mg/mL final concentration in Maxpar Cell Staining Buffer) to plate, and incubate for 10 min at room temperature.
2. Maxpar metal-conjugated antibodies are pooled into a cocktail in Maxpar Cell Staining Buffer with a 100 μL final reaction volume per well (50 μL of cell suspension +50 μL antibody cocktail).
3. Add antibody cocktail to each well without washing off Fc Receptor Blocking Solution. Gently pipetting up and down to mix. Incubate at 4 °C for 30 min. Gently mix samples every 15 min and avoid bubbles.
4. Wash the PBMCs with 150 μL of Maxpar Cell Staining Buffer; then centrifuge at 845 × *g* for 3 min at room temperature.
5. Check the pellets and quickly but gently flick the plate at one motion to dump supernatant.
6. Add 200 μL of fresh 1.6% formaldehyde solution to each well. Gently resuspend cells by pipetting up and down immediately (*see* **Note 12**).
7. Incubate at room temperature for 10 min.
8. Centrifuge at 845 × *g* for 3 min at room temperature.
9. Check the pellets; quickly but gently flick the plate at one motion to dump supernatant.

3.6 Intercalator Staining

1. Make 1:1000 dilution of Ir-intercalator solution to a final concentration of 125 nM with Maxpar Fix and Perm Buffer.
2. Add 200 μL of fresh Ir-intercalator solution to each well. Gently resuspend cells by pipetting up and down immediately, and avoid bubbles.
3. Incubate at room temperature for 1 h or leave overnight at 4 °C.

3.7 Loading Sample on CyTOF

1. Centrifuge at 845 × *g* for 3 min at room temperature to remove the supernatant.
2. Resuspend the cell pellets with Maxpar Cell Staining Buffer and gently mix samples to avoid bubbles.
3. Centrifuge at 845 × *g* for 3 min at room temperature.
4. Check the pellets and quickly but gently flick the plate to remove supernatant.
5. Wash twice with 200 μL/well deionized water, and centrifuge at 845 × *g* for 3 min at room temperature.
6. Resuspend the cell pellets in 100 μL/well deionized water (*see* **Note 13**).
7. Transfer cell suspension to round-bottom polypropylene tubes (*see* **Note 14**).
8. Resuspend with 1 mL of the 1:10 diluted bead solution prepared in deionized water (the concentration is about 6–8 × 10^5 cells/mL if the starting cell number is 1 × 10^6) (*see* **Note 15**).
9. Filter the cell suspension with 35 μm cell strainer cap.
10. Acquire data on a mass cytometer.

3.8 High-Dimensional Data Analysis

In this protocol, as an example for thorough evaluation, we analyzed the human PBMCs with labeled "carrier" cells by three methods: (1) traditional manual gating by FlowJo (Fig. 2), visualization by t-stochastic neighbor embedding [13] (Fig. 3), and FlowSOM [14] (Fig. 2) using the Cytofkit package in R program. The readers are recommended to choose their own analytical methods.

4 Notes

1. Rhodium, a cationic nucleic acid intercalator [15], is cell impermeable so that it can be used to either discriminate dead cells from live cells (cells were not fixed) or to identify nucleated cells like iridium. Here we utilize the former property to label the carrier cells by fixing and permeabilizing the EL4 cell line. Furthermore, the cells can be left at 4 °C in the intercalation solution for up to 48 h.
2. Benzonase treatment can reduce the viscosity and background by removing free DNA from lysed cells.
3. Storage at room temperature and multiple freeze/thaws will result in increased nonspecific binding, which can interfere with live/dead cell discrimination.

4. Both rhodium and iridium purchased from the supplier are often in highly concentrated metal intercalator solution and need to be diluted to a suitable concentration and avoid repeated freeze/thaw cycles.
5. EL4 is a murine T-cell lymphoma cell line. They are bigger than human PBMCs. The antibodies used for human PBMCs staining have no cross-activity with EL4 cells.
6. Cells can be left at 4 °C in the intercalation solution for up to 48 h. This step is optional because the Ir-intercalator solution at the final staining step can also distinguish the "carrier" cells from PBMCs since the staining intensity correlates with cell size and EL4 cells are much bigger than PBMCs. We recommend not skipping this step because the labeled "carrier" cells can be easily distinguished from EL4, CD45-, or dead cells when used together with cisplatin.
7. Thaw no more than 10 samples at a time.
8. It is common to recover $4–8 \times 10^6$ cells from one frozen vial with 10×10^6 cells/vial.
9. Resting cells are recommended for intracellular cytokine staining because it can increase sensitivity. For details, please *see* ref. 16.
10. The rate of cell acquisition using CyTOF is about 20–30%. So 200–300 events of PBMCs (the started number is 1000) can be acquired using the "carrier" method by adding 1×10^6 carrier cells in 1 mL volume (the concentration is about $6–8 \times 10^5$ cells/mL if the starting cell number is 1×10^6 to avoid the doublets). The increased "carrier" numbers will also increase the acquisition number of small numbers of PBMCs.
11. Flicking the plate to dump supernatant is an essential operation and needs some practice. When inverting the plate, be careful not to loosen or detach the cell pellet with any shaking movement.
12. Formaldehyde solution should be kept from air and light to remain stable and avoid contamination. Prepare fresh formaldehyde solution with PBS each time.
13. Washing with deionized water is critical for removing salt within the buffer. Residual salt can cause the current setting of CyTOF to drift when running the samples.
14. Leave the cell suspension without adding bead solution until ready to run.
15. These calibration beads contain natural abundance cerium (140/142Ce), europium (151/153Eu), holmium (165Ho), and lutetium (175/176Lu). They are used as an internal standard for normalization. Vigorously shake or vortex the

calibration bead bottle before use. After adding the bead solution, it is recommended to acquire data immediately because prolonged storage will result in adsorption of beads by the inner wall of polypropylene tubes, which may affect the signal collected.

Acknowledgments

This work was supported by the Natural Science Foundation of Fujian Province of China No.2018 J05065, to LZ, and National Natural Science Foundation of China grants 31770952, 31570911, and 2017ZX10202203-003- 001 to G.F.

References

1. Swartz MA, Iida N, Roberts EW et al (2012) Tumor microenvironment complexity: emerging roles in cancer therapy. Cancer Res 72(10):2473–2480
2. Satija R, Shalek AK (2014) Heterogeneity in immune responses: from populations to single cells. Trends Immunol 35(5):219–229
3. Bandura DR, Baranov VI, Ornatsky OI et al (2009) Mass cytometry: technique for real time single cell multitarget immunoassay based on inductively coupled plasma time-of-flight mass spectrometry. Anal Chem 81(16):6813–6822
4. Bendall SC, Simonds EF, Qiu P et al (2011) Single-cell mass cytometry of differential immune and drug responses across a human hematopoietic continuum. Science 332 (6030):687–696
5. Leipold MD, Newell EW, Maecker HT (2015) Multiparameter phenotyping of human PBMCs using mass cytometry. Methods Mol Biol 1343:81–95
6. Pejoski D, Tchitchek N, Rodriguez Pozo A et al (2016) Identification of vaccine-altered circulating B cell phenotypes using mass cytometry and a two-step clustering analysis. J Immunol 196(11):4814–4831
7. Chevrier S, Levine JH, Zanotelli VRT et al (2017) An immune Atlas of clear cell renal cell carcinoma. Cell 169(4):736–749.e718
8. Lavin Y, Kobayashi S, Leader A et al (2017) Innate immune landscape in early lung adenocarcinoma by paired single-cell analyses. Cell 169(4):750–765.e717
9. Good Z, Borges L, Vivanco Gonzalez N et al (2019) Proliferation tracing with single-cell mass cytometry optimizes generation of stem cell memory-like T cells. Nat Biotechnol 37 (3):259–266
10. Atkuri KR, Stevens JC, Neubert H (2015) Mass cytometry: a highly multiplexed single-cell technology for advancing drug development. Drug Metab Dispos 43(2):227–233
11. Leary JF (1994) Chapter 20 strategies for rare cell detection and isolation. Methods Cell Biol 42:331–358
12. Saeys Y, Van Gassen S, Lambrecht BN (2016) Computational flow cytometry: helping to make sense of high-dimensional immunology data. Nat Rev Immunol 16(7):449–462
13. Van Der Maaten L, Hinton G (2008) Visualizing data using t-SNE. J Mach Learn Res 9 (Nov):2579–2605
14. Van Gassen S, Callebaut B, Van Helden MJ et al (2015) FlowSOM: using self-organizing maps for visualization and interpretation of cytometry data. Cytometry A 87(7):636–645
15. Zeglis BM, Pierre VC, Barton JK (2007) Metallo-intercalators and metallo-insertors. Chem Commun (Camb) 28(44):4565–4579
16. Kutscher S, Dembek CJ, Deckert S et al (2013) Overnight resting of PBMC changes functional signatures of antigen specific T- cell responses: impact for immune monitoring within clinical trials. PLoS One 8(10):e76215
17. Yao Y, Liu R, Shin MS et al (2014) CyTOF supports efficient detection of immune cell subsets from small samples. J Immunol Methods 415:1–5

Chapter 3

Simultaneous Measurement of Surface Proteins and Gene Expression from Single Cells

Jiadi Luo, Carla A. Erb, and Kong Chen

Abstract

Single-cell transcriptomic analysis has become a new and powerful tool to study complex multicellular systems. Single-cell RNA sequencing provides an unbiased classification of heterogeneous cellular states at the transcriptional level, but it does not always correlate to cell-surface protein expression. A recently developed method called cellular indexing of transcriptomes and epitopes by sequencing (CITE-seq) simultaneously measures surface proteins and gene expression from single cells. Briefly, based on the existing single-cell sequencing technology, oligonucleotide-labeled antibodies and barcoded primer gel beads are used to bind to corresponding cell-surface proteins and mRNA, respectively. Further, libraries of labeled protein and RNA information are sequenced to integrate cellular protein and transcriptome reads together efficiently. CITE-seq is transforming comprehensive genomic studies into models of causal gene-protein investigation.

Key words Single-cell RNA sequencing, CITE-seq, Surface proteins, Gene expression, ADT library, 10× genomics

1 Introduction

Single-cell transcriptomics is opening a new and powerful platform for describing complex multicellular systems [1–4]. As the well-established methodology of single-cell transcriptomics, single-cell RNA-seq (scRNA-seq) simultaneously detects the mRNA concentration of numerous genes and measures the gene expression level of individual cells within a heterogeneous cell population [5]. The analysis of this sequencing data has made it possible to achieve unbiased cell-type classification and cellular developmental trajectories [6–11].

Single-cell RNA-seq, however, could not provide any phenotypic information at cell-surface protein levels. Traditionally, cell-surface protein levels were obtained by flow cytometry after staining with fluorescently labeled antibodies [12]. The flow cytometry data reflects the whole cell population state, but is unable to match

Chaohong Liu (ed.), *T-Cell Receptor Signaling: Methods and Protocols*, Methods in Molecular Biology, vol. 2111,
https://doi.org/10.1007/978-1-0716-0266-9_3,

the cell-surface proteins to the individual cells captured by single-cell RNA-seq. Here, we provide a detailed protocol on a recently developed method, cellular indexing of transcriptomes and epitopes by sequencing (CITE-seq), which simultaneously measures surface proteins and gene expression from single cells.

CITE-seq is developed on the foundation of single-cell RNA--seq technology, which is well established and widely used [13] (https://assets.ctfassets.net/an68im79xiti/UhAMGmlaEMmYMaA4A4Uwa/d65ff7b9bb5e88c2bb9e15e58f280e18/CG00052_SingleCell3_ReagentKitv2UserGuide_RevE.pdf). Essentially, CITE-seq is the process of single-cell RNA-seq in addition to the construction of antibody-derived tags library (ADT library) (Fig. 1). The key point during the construction of ADT library is the design of the special cell-surface antibodies. Similar to fluorescently labeled antibodies, the cell-surface antibodies for ADT library have been conjugated to readable oligonucleotides which contain a barcode for antibody identification and include a handle for PCR amplification (Fig. 2). In addition to the procedures of single-cell RNA-seq, extra steps must be added to the single-cell RNA-seq prep process to construct ADT library. Briefly, single cells are first stained with oligonucleotide-labeled antibodies which can bind to the target cell-surface proteins. Following washing steps to remove unbound antibodies, protein-labeled cells are then captured by barcoded primer gel beads and emulsified by oil to generate cell-gel bead-oil droplets, getting the antibody-derived tags and all the mRNA from the single cell annealed to oligo-dT from gel bead via their 3′ polyA tails. After the reverse transcription, mRNA generates cDNA indexed with unique barcode contained by gel bead, while the antibody-derived tags generate ADT-derived cDNA indexed with the same barcode as mRNA-derived cDNA (*see* **Note 1**). Amplify the mRNA-derived cDNA and ADT-derived cDNA mixture, and then separate the amplified mRNA-derived cDNA and ADT-derived cDNA by fragment size selection (ADT-derived cDNA <180 bp and mRNA-derived DNAs >300 bp). Then, the two separate cDNA types can be further converted into sequencing libraries independently. Analysis of the sequencing data of these two libraries can integrate the gene expression level and cell-surface protein information into each individual cell. CITE-seq is transforming the comprehensive genomic studies into models of causal gene-protein mechanism investigation [14].

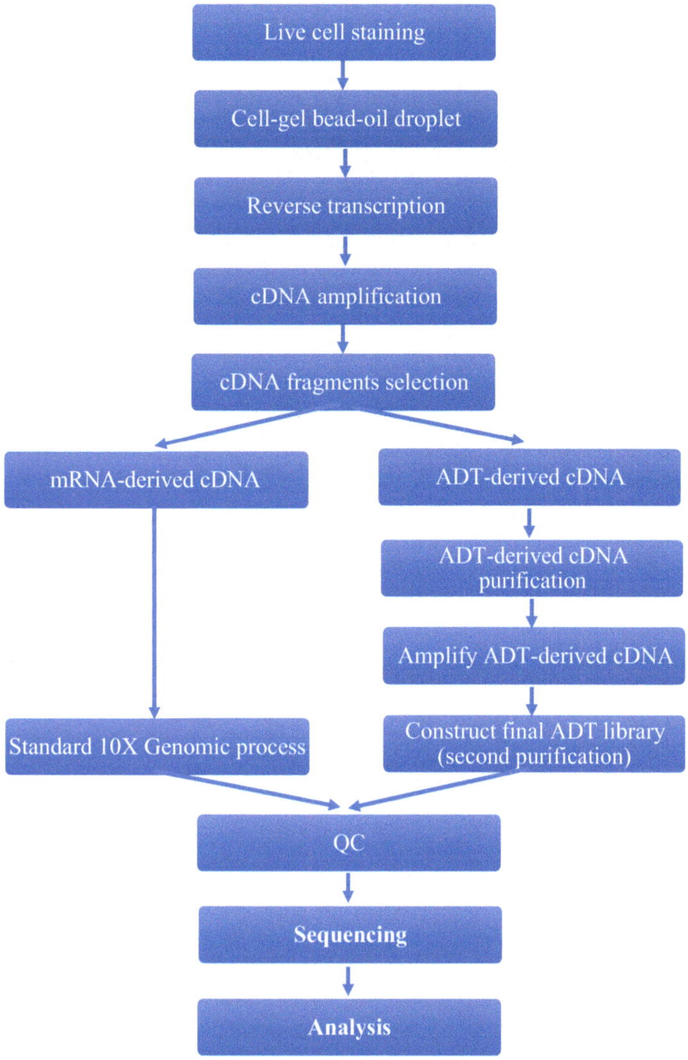

Fig. 1 Schematic of the workflow for CITE-seq. Live single cells are first stained with oligonucleotide-labeled antibodies which can bind to the target cell-surface proteins. Then, protein-labeled cells are captured by barcoded primer gel beads and encompass by oil to generate cell-gel bead-oil droplets. Cell lyses in the droplet, followed with the reverse transcription; mRNA generates cDNA indexed with unique barcode contained by gel bead, while the antibody-derived tags generate ADT-derived cDNA indexed with the same barcode as mRNA-derived cDNA. Amplify the mRNA-derived cDNA and ADT-derived cDNA mixture, and then separate the amplified mRNA-derived cDNA and ADT-derived cDNA by fragment size selection. Then, the two separate cDNA types can be further converted into sequencing libraries independently. mRNA-derived DNA library construction follows the standard 10× Genomics procedures, while ADT library construction is achieved after ADT-derived cDNA purification, amplification, and the second purification steps. Following with QC and sequencing, analysis of the sequencing data of these two libraries can integrate the gene expression level and cell-surface protein information into each individual cell

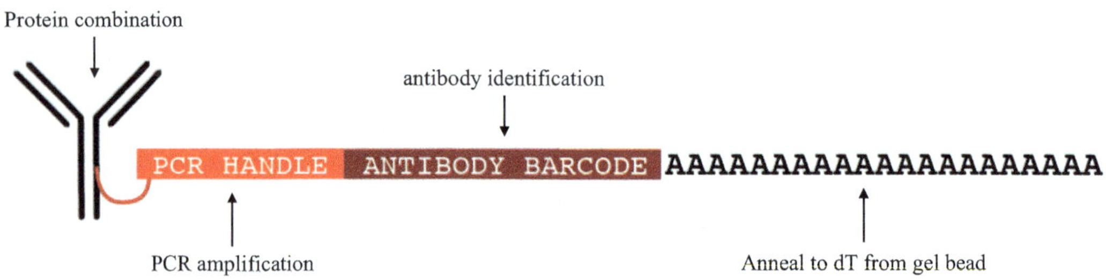

Fig. 2 Schematic of the design of oligonucleotide-labeled antibodies used in CITE-seq

2 Materials

2.1 Single-Cell RNA-seq Material

2.1.1 Reagent and Supply

1. Chromium™ Single-Cell 3′ Library and Gel Bead Kit v2 (10× Genomics, USA), store at −20 °C.
2. Chromium™ Single-Cell A Chip Kit (10× Genomics, USA), store at room temperature.
3. Chromium™ i7 Multiplex Kit (10× Genomics, USA), store at −20 °C.
4. DNA LoBind tubes, 1.5 ml.
5. TempAssure PCR 8-tube strip.
6. DynaBeads® MyOne™ Silane Beads (Thermo Fisher Scientific, USA), store at 4 °C.
7. Nuclease-free water.
8. Low TE buffer (10 mM Tris–HCl pH 8.0, 0.1 mM EDTA), store at room temperature.
9. Ethanol, pure (200 proof, anhydrous).
10. SPRIselect Reagent Kit (Beckman Coulter, USA), store at room temperature (*see* **Note 2**).
11. 10% Tween 20 (Bio-Rad, USA), store at room temperature.
12. Glycerin (glycerol), 50% (v/v) aqueous solution (Ricca Chemical Company, USA), store at room temperature.
13. Pipets (P2, P20, P200, P1000).
14. LoBind pipet tips (20 μl, 200 μl, 1000 μl).
15. Divided polystyrene reservoirs.
16. High sensitivity DNA kit (Agilent, USA), store at 4 °C and room temperature.
17. Cell staining buffer (BioLegend, USA), store at 4 °C.
18. ViaStain™ AOPI Staining Solution for cell counting (Nexcelom Bioscience, USA), store at 4 °C.
19. SD100 slides for cell counting (Nexcelom Bioscience, USA).
20. DynaMag™-2, working volume: 10–1500 μl (Thermo Fisher Scientific, USA).

2.1.2 Equipment

1. Chromium Controller and Accessory Kit (10× Genomics, USA).
2. Cellometer Auto 2000 Cell Viability Counter (Nexcelom Bioscience, USA).
3. Vortex mixer.
4. 2100 Bioanalyzer Laptop Bundle (Agilent, USA).
5. Mini-spin.
6. NanoDrop.
7. C1000 Touch™ thermal cycler with 96-deep well reaction module (Bio-Rad, USA).
8. ThermoMixer (Eppendorf, USA).

2.2 Other Materials for CITE-seq

1. Oligonucleotide-labeled antibodies (BioLegend, USA), store at 4 °C (*see* **Note 3**).
2. Human/mouse/rat Fc receptor blocking solution (BioLegend, USA), store at 4 °C.
3. KAPA HiFi HotStart ReadyMix (2×), store at −20 °C.
4. SI-PCR primer, stock concentration is 10 μM, store at −20 °C.

   5′AATGATACGGCGACCACCGAGATCTACACTCTTT CCCTACACGACGCTC
5. ADT cDNA PCR additive primer, stock concentration is 0.2 μM, store in −20 °C.

   5′CCTTGGCACCCGAGAATTCC
6. Illumina Small RNA RPI1 (or RPI2,3,4, etc.) primer, stock concentration is 10 μM, store in −20 °C (*see* **Note 4**).

   5′CAAGCAGAAGACGGCATACGAGATCGTGATGT GACTGGAGTTCCTTGGCACCCGAGAATTCCA
7. DynaMag™-2 (Invitrogen, USA).
8. Cell staining buffer: 2%BSA/0.01%Tween in PBS, store in 4 °C.
9. 40 μm cell strainer.

3 Methods

Carry out all procedures at suitable temperature as illustrated (*see* **Note 5**).

3.1 Single-Cell RNA-seq Prep

Single-cell RNA-seq library construction is performed using Single-Cell 3′ Reagent Kit v2 from 10× Genomics; details can be found at https://assets.ctfassets.net/an68im79xiti/UhAMGmlaEMmYMaA4A4Uwa/d65ff7b9bb5e88c2bb9e15e58f280e18/CG00052_SingleCell3_ReagentKitv2UserGuide_RevE.pdf

3.2 ADT Library Construction

3.2.1 Live Cell Staining

1. Count the ready-to-test single cells with a cellometer to ensure accurate cell population and viability. Cell viability >85% at least is recommended for next step (*see* **Note 6**).
2. Spin cells for 5 min in 400 × *g* at 4 °C, carefully remove the supernatant, and resuspend 1–2 million cells in 100 μl precold cell staining buffer on ice.
3. Add 10 μl of Fc receptor blocking solution, and mix it with cell solution by pipetting.
4. Incubate for 10 min at 4 °C.
5. During the incubation of Fc blocking, prepare antibody cocktail. Use 1 μg of each oligonucleotide-labeled antibody to stain individual cell sample like flow cytometry staining. For example, if we are going to stain one sample with 10 target cell-surface antibodies, add 1 μg antibody from each of the 10 antibodies into a new tube, respectively, and mix the antibody cocktail by pipetting. Store it on ice.
6. After the incubation of Fc blocking, add the antibody cocktail to the cell suspension. Mix the solution by pipetting.
7. Incubate for 30 min at 4 °C.
8. Directly add 1 ml precold cell staining buffer to the cell solution, spin for 5 min in 400 × *g* at 4 °C. Then, remove the supernatant and wash cells 2 more times with 1 ml precold cell staining buffer; spin for 5 min in 400 × *g* at 4 °C.
9. Resuspend cells in 1× PBS (calcium and magnesium free) containing 0.04% weight/volume BSA (400 μg/ml) at appropriate concentration (500–1000 cells/μl) for 10× Genomics (*see* **Note 7**).
10. Filter cells through 40 μm strainers.
11. Recount the cell number on cellometer after filtration; record the accurate cell concentration at this step.

3.2.2 Run 10× Genomics (Single Cell 3′ Reagent Kits v2) as Described in the Link at Subheading 3.1 Single-Cell RNA-seq Prep Until Before cDNA Amplification (See **Note 8**)

1. At cDNA amplification step: in order to increase yield of antibody-derived tags (ADTs), add "additive" primer (stock concentration is 0.2 μM) to 10× Genomics cDNA amplification PCR system. Specifically, replace the 10× Genomics cDNA amplification PCR system with the PCR system:

cDNA amplification PCR system	1× (μl)
Amplification master mix	50
cDNA additive	5
cDNA primer mix	2
Nuclease-free water	6
ADT additive primer (0.2 μM stock)	2

(continued)

cDNA amplification PCR system	1× (μl)
Purified GEM-RT product	35
Total volume	100

2. Continue the cDNA amplification according to the 10× Genomics instructions.

*3.2.3 ADT-Derived cDNA and mRNA-Derived cDNA Separation (See **Note 9**)*

1. After cDNA amplification, add 60 μl fully resuspended SPRI-select reagent (0.6×) to the sample in the tube strip, and pipette mix 15 times (pipette set to 150 μl) (*see* **Note 10**).
2. Incubate the tube strip at room temperature for 5 min, and place the tube strip in a 10× magnetic separator in the high position until the solution is clear.
3. Carefully transfer all the supernatant (about 150 μl solution) into a new 1.5 ml DNA LoBind tube. This solution contains the ADT-derived cDNAs (about 180 bp).
4. The remaining selection beads in the tube strip contain full length mRNA-derived cDNAs (>300 bp). Continue to proceed with 10× Genomics for final DNA library preparation.

3.2.4 Final ADT Library Construction

Purify ADT-Derived cDNAs

1. Add 1.4× SPRI (140 μl fully resuspended SPRIselect reagent, based on 100 μl sample volume) to supernatant which contains the ADT-derived cDNAs to obtain a final SPRI volume of 2× SPRI (*see* **Note 11**).
2. Incubate 10 min at room temperature.
3. Place tube on magnet and carefully remove and discard the supernatant after the solution is clear.
4. Add 400 μl fresh 80% ethanol to wash the beads without disturbing the bead pellet, and stand for 30 s. Then remove and discard the ethanol solution.
5. Centrifuge the tube briefly with a mini-spin and return it to magnet. Remove and discard any remaining ethanol.
6. Resuspend beads in 50 μl water.
7. Add 100 μl SPRI reagent directly to the resuspended beads (2× SPRI again). Mix by pipetting, and incubate for 10 min at room temperature.
8. Place tube on magnet and carefully remove and discard the supernatant after the solution is clear.
9. Add 200 μl fresh 80% ethanol to the tube without disturbing the pellet, and stand for 30 s (first ethanol wash). Carefully remove and discard the ethanol wash. And then repeat the wash (second ethanol wash).
10. Centrifuge tube briefly and return it to magnet. Remove and discard any remaining ethanol, and allow the beads to air dry for less than 2 min (*see* **Note 12**).

11. Resuspend beads in 45 μl water and incubate at room temperature for 5 min.
12. Place the tube back on magnet and transfer clear supernatant into a new PCR tube.

Amplify ADT-Derived cDNAs

1. Set up a 100 μl PCR reaction with purified ADT-derived cDNAs:

ADT-derived cDNA amplification PCR system	1× (μl)
Purified ADT-derived cDNAs	45
KAPA HiFi HotStart ReadyMix (2×)	50
Illumina small RNA RPIx primer (stock concentration 10 μM)	2.5
10× Genomics SI-PCR primer (stock concentration 10 μM)	2.5
Total volume	100

2. Run the PCR using the following conditions:

Cycles	Temperature (°C)	Time
1	95	3 min
6–10 cycles	95 60 72	20 s 30 s 20 s
1	72	5 min

Final ADT Library Construction

1. Add 160 μl SPRI reagent to PCR product from the last step (1.6× SPRI); mix by pipetting.
2. Incubate 5 min at room temperature. Then place the PCR tube on magnet and wait until solution is clear. Remove and discard the supernatant.
3. Add 200 μl fresh 80% ethanol to the tube without disturbing the pellet, and stand for 30 s (first ethanol wash). Carefully remove and discard the ethanol wash. Then repeat the wash (second ethanol wash).
4. Centrifuge tube briefly and return it to magnet. Remove and discard any remaining ethanol, and allow the beads to air dry for less than 2 min.
5. Resuspend beads in 20 μl water and incubate at room temperature for 5 min.
6. Place the PCR tube back on magnet and transfer clear supernatant into a new PCR tube. This supernatant is the final ADT library.

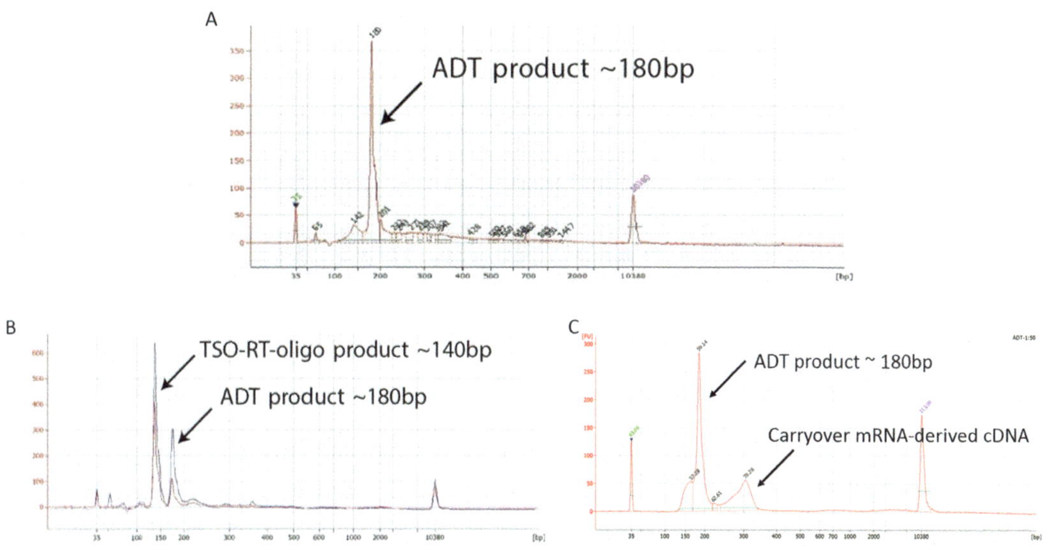

Fig. 3 QC for ADT library. (**a**) Normal ADT library will show an enriched peak at around 180 bp. (**b**) A TSO-RT-oligo product (~140 bp) showed up probably due to the carryover primers from cDNA amplification being amplified during the ADT PCR process. (**c**) Carryover mRNA-derived cDNA peak (around 300 bp) comes right behind the ADT product

QC for ADT Library

Quantify the ADT library by running the Agilent 2100 Bioanalyzer high sensitivity chip with 1 μl diluted ADT library (1:10 or 1:50 dilution with water). ADT library will be around 180 bp, Be sure to distinguish the ADT library peak from other peaks such as TSO-RT-oligo product, and carryover mRNA-derived cDNAs (Fig. 3) (*see* **Note 13**).

3.3 Sequencing

Final 10× DNA library and ADT library can be pooled together to be sequenced on Illumina Hiseqs. Generally sequencing ADT libraries in 5–10% of a lane and DNA library fraction at 90–95% of a lane (HiSeq2500 Rapid Run Mode Flowcell) is a sufficient read coverage for both libraries (*see* **Note 14**).

3.4 Analysis

Data were processed using Cell Ranger 3.0 using default parameters, and no further filtering was applied.

4 Notes

1. After cell lysis, both the antibody-derived tags and the cellular mRNA from the same single cell annealed to a unique barcode from gel bead in the droplet. This is the foundation to identify if the ADT data (protein information) and the genomics data (gene expression level) are from the same cell in the final analysis.

2. Selection beads from SPRIselect reagent in different SPRIselect reagent/sample ratios could bind to different sizes of DNA fragments. That is why ADT-derived cDNA can be separated from mRNA-derived cDNA by bead selection.
3. Please verify the species of the cells; make sure to use corresponding antibodies.
4. Illumina Small RNA RPI-x primer is used for ADT amplification. There are 48 different Small RNA RPI-x primers from Illumina which could be applied to different samples. ADT libraries from different samples could be pooled together for sequencing only when different Illumina Small RNA RPI-x primers are used to amplify each ADT library. In the protocol, we only provide the sequence of Small RNA RPI1 as an example. All the sequence information of other Small RNA RPI-x primers such as RPI2 and RPI3 can be obtained at https://support.illumina.com/content/dam/illumina-support/documents/documentation/chemistry_documentation/experiment-design/illumina-adapter-sequences-1000000002694-09.pdf, page 23–24.
5. Good cell viability is vital for high quality of the final libraries, so prepare the single-cell isolation steps and CITE-seq antibody staining steps on ice to get better cell viability.
6. If cell viability is low (<80%), dead cell removal kit is suggested to enrich live cells.
7. BSA is added, and Ca^{2+}/Mg^{2+} is excluded primarily to minimize cell losses and aggregation.
8. When super-loading cells to 10× Genomics platform, optional addition of RNAse inhibitor to single-cell RT buffer is recommended.
9. The ADT-derived and mRNA-derived cDNA separation is achieved by double-sided size selection of cDNA fragments. When 0.6× SPRI is applied, bigger sizes of cDNA fragments bind to the selection beads; while the smaller sizes of cDNA fragments remain in the supernatant, different sizes of cDNA fragments are then separated by separating the supernatant and beads. 0.6× SPRI is enough to separate ADT-derived cDNA (about 180 bp) and mRNA-derived cDNA (>300 bp).
10. 0.6× means the ratio of SPRIselect reagent volume/sample volume, in another word, 0.6× SPRIselect reagent concentration means 60 μl SPRIselect reagent in 100 μl sample.
11. All the supernatant (about 150 μl solution) containing the ADT-derived cDNAs has been collected, in which it is still containing about 60 μl SPRIselect reagent and about 100 μl sample solution from the previous step. Now plus the extra 140 μl SPRIselect reagent added into the solution, the solution

mix contains about 200 μl SPRIselect reagent, 100 μl sample solution, and some beads. Then, it turns out a final SPRI volume of 2× SPRI.

12. Overdry beads would lead to less library yield.
13. Carryover primers from cDNA amplification can be amplified during the ADT PCR, as a TSO-RT-oligo product (~140 bp) detected by high sensitivity chip. This product will interfere with quantification. It is impossible to remove the TSO-RT-oligo product by another clean-up to the ADT library, but sequential 2× SPRI purification of the ADT-derived cDNA after cDNA amplification can help reduce carryover of primers. Carryover mRNA-derived cDNA will not interfere with quantification, because it will not have the Illumina cluster-generating sequences appended; thus, it will not get amplified during a high-throughput sequencing run. It is not necessary to remove the carryover mRNA-derived cDNAs by DNA gel separation.
14. Final 10× DNA library and ADT library can be pooled together to be sequenced, because the 10× DNA library is recorded by Chromium i7 Sample Index, while ADT library is indexed with Illumina Small RNA RPI-x primer.

Acknowledgement

This work was supported by NIH grant R01HL137709.

References

1. Tanay A, Regev A (2017) Scaling single-cell genomics from phenomenology to mechanism. Nature 541(7637):331–338
2. Macosko EZ, Basu A, Satija R et al (2015) Highly parallel genome-wide expression profiling of individual cells using nanoliter droplets. Cell 161:1202–1214
3. Klein AM, Mazutis L, Akartuna I et al (2015) Droplet barcoding for single-cell transcriptomics applied to embryonic stem cells. Cell 161:1187–1201
4. Zheng GX, Terry JM, Belgrader P et al (2017) Massively parallel digital transcriptional profiling of single cells. Nat Commun 8:1–12
5. Kanter I, Kalisky T (2015) Single cell transcriptomics: methods and applications. Front Oncol 5:53
6. Liu S, Trapnell C (2016) Single-cell transcriptome sequencing: recent advances and remaining challenges. Version 1. F1000Res 5:F1000 Faculty Rev-182
7. Jaitin DA, Kenigsberg E, Keren-Shaul H et al (2014) Massively parallel single-cell RNA-seq for marker-free decomposition of tissues into cell types. Science 343:776–779
8. Bendall SC, Davis KL, Amir e-AD et al (2014) Single-cell trajectory detection uncovers progression and regulatory coordination in human B cell development. Cell 157:714–725
9. Krishnaswamy S, Spitzer MH, Mingueneau M et al (2014) Conditional density-based analysis of T cell signaling in single-cell data. Science 346:1250689
10. Patel AP, Tirosh I, Trombetta JJ et al (2014) Single-cell RNA-seq highlights intratumoral heterogeneity in primary glioblastoma. Science 344:1396–1401
11. Grün D, Lyubimova A, Kester L et al (2015) Single-cell messenger RNA sequencing reveals rare intestinal cell types. Nature 525:251–255

12. Pontén F, Gry M, Fagerberg L et al (2009) A global view of protein expression in human cells, tissues, and organs. Mol Syst Biol 5:1–9
13. Stoeckius M, Hafemeister C, Stephenson W et al (2017) Simultaneous epitope and transcriptome measurement in single cells. Nat Methods 14(9):865–868
14. Mimitou E, Cheng A, Montalbano A et al (2019) Expanding the CITE-seq tool-kit: Detection of proteins, transcriptomes, clonotypes and CRISPR perturbations with multiplexing, in a single assay. Nat Methods 16 (5):409–412

Chapter 4

Analysis of Transcriptional Profiling of Immune Cells at the Single-Cell Level

Annabel Ferguson and Kong Chen

Abstract

RNA sequencing has proven to be a key innovation for the study of biological processes by enabling scientists to measure differences in gene expression in different tissues.With recent advances in sequencing technology, researchers are able to measure gene transcription at the single-cell level, revealing previously unknown diversity and specificity of immune cells. The single-cell sequencing method now enables profiling of the T-cell receptor (TCR) genes resulting from V(D)J recombination.Here we describe how to adapt single-cell RNA sequencing data generated using the 10× genomics 5′V(D)J immune cell profiling workflow for integration into the R analysis pipeline.We will start with the data matrix files generated from the 10× genomics Cell Ranger alignment software and detail how to format this data as input for the R analysis package called Seurat such that data from both the overall cell transcript abundance and the targeted V(D)J transcript abundance data can be visualized on the same plots.

Key words 10× genomics V(D)J, Drop-seq, Single-cell RNA sequencing, T-cell receptor repertoire profiling

1 Introduction

1.1 T-Cell Receptors and V(D)J Recombination

The polyclonal nature of T cells derives from their T-cell receptor (TCR) structure, which is one component of the adaptive immune system that allows for specific recognition of diverse foreign antigens and enables the immune system to fight off a vast breadth of pathogens. The diversity of the T-cell receptor repertoire arises from somatic recombination of genes encoding the TCR, including gene segments for the variable (V), diverse (D), and joining (J) regions; this process is referred to as V(D)J recombination.V (D)J recombination followed by a positive and negative selection process for TCR containing T cells gives rise to an estimated 10^6–10^{10} different T-cell clones each with a different TCR sequence [1]. In a healthy, uninfected individual, each T-cell clone has a low frequency of approximately 100 cells; however, upon activation of the immune system usually in recognition of an invading pathogen,

Chaohong Liu (ed.), *T-Cell Receptor Signaling: Methods and Protocols*, Methods in Molecular Biology, vol. 2111, https://doi.org/10.1007/978-1-0716-0266-9_4,

a particular T-cell clone may be triggered to proliferate and undergo clonal expansion [2].Until very recently, it has not been possible to obtain comprehensive information about a population of T cells, with respect to TCR sequence and clonal frequency.With the advent of single-cell RNA sequencing, this is now attainable.

1.2 Transcriptional Profiling at the Single-Cell Level

RNA sequencing at bulk tissue level typically involves homogenization and lysis of the cells, followed by isolation of mRNA, cDNA synthesis, and further processing such that the samples may be sequenced on a high-throughput sequencer with short (75 to 150 basepair) read lengths.Through the addition of 6 to 8 nucleotide barcodes, multiple biological samples may be run on the same sequencing flow cell, and the data can be parsed out bioinformatically due to the sample-identifying indices being linked to the molecules coming from that sample.In bulk RNA-seq, information about each cell cannot be parsed out because in the first step of the process, all cells from one sample are mixed and lysed in one tube. The 10× genomics system isolates individual cells in a method known as drop-seq and uses barcoding technology to encapsulate single cells along with the reagents necessary to uniquely barcode each cell.Through the use of capillary tubes, this technology furthermore allows for the isolation of 2000 to 3000 cells with a less than 2% doublet rate.

1.3 Immune Cell Profiling with 10× Genomics 5 Prime V(D)J Preparation

In order to profile the V(D)J region of T cells or B cells, 10× genomics designed a single-cell RNA seq kit in which mRNA sequences are sequenced starting from the 5 prime end of the molecule, ensuring better read accuracy in the 5 prime end of the cDNA strand, which is where the product of V(D)J recombination is located. The resulting cell-indexed cDNA may then be processed to generate multiple sequencing libraries: one in which a profile of all transcripts in the sample is captured and another in which the T-cell or B-cell receptors are enriched through PCR amplification using targeting primers.Importantly, with this method, sequencing reads resulting from either the V(D)J targeted library or the entire transcriptome library will be linked to the cell that they originated from.Therefore, both the V(D)J sequence and the background transcriptome may be measured in the same cells simultaneously.

1.4 Tools for Visualization Analysis of Single-Cell TCR Sequencing Data

Once the immune cells have been isolated, barcoded, prepared as libraries, and sequenced, the next step is to align and count the reads.For 10× genomics, this step may be done using a Linux computational cluster with the software package called Cell Ranger, which automatically produces files that may be used for analysis and visualization.Numerous tools for analyzing single-cell RNA seq alignment data exist, including the convenient point-and-click software developed by 10× genomics called vloupe or cloupe, for analyzing V(D)J data or expression count data, respectively.While useful for obtaining a first glance at the data, it is not as customizable or adaptable as R language pipelines.The popular R language pipeline called Seurat additionally provides the desirable feature of canonical correlation analysis (CCA) normalization, which is often helpful for comparison of multiple libraries of samples with different conditions[3, 4].There are extensive tutorials and manuals describing the process of aligning and counting the single-cell data offered by 10×genomics[5].In addition, the Seurat R manual and vignettes are comprehensive [3, 4].Therefore, in this chapter, we will only demonstrate the process of adapting the Cell Ranger VDJ data output such that it can be analyzed in the Seurat R pipeline (*see* **Note 1**).

2 Materials

2.1 Computer

A computer is capable of running R Seurat analysis pipeline, with at least 16 gb random access memory.

Software: R, R studio

R packages: Seurat, Matrix, dplyr, cowplot, plyr, reshape2

2.2 Data Matrix Files

1. Single-cell 5 prime T-cell V(D)J dataset. This is the output from running the `cellrangervdj` option command.The file is located in the following automatically generated "outs" directory and will have the name "filtered_contig_annotations.csv."This file contains information about the TCR V(D)J components as they have mapped to the reference file.It is formatted as a table with the following structure; each row represents a unique assembled contig for each cell, and each column represents specific information about that contig.A full description of the columns of this table may be found at the 10× genomics support site[5, 6].The columns of interest are described in Table 1.
2. Single-cell RNA-seq5 prime gene expression dataset [7]. This is the set of output files from running `cellranger count` that are required as input for the Seurat single-cell analysis pipeline [3].

3 Methods

3.1 Formatting V(D)J Data Matrices to Use for Annotating the Expression Data

1. Open R studio, and load the following packages.

```
library(Seurat)
library(Matrix)
library(dplyr)
library(cowplot)
library(plyr)
```

Table 1
Column names and descriptions of columns of interest in the dataset resulting from running Cell Ranger VDJ

Column name	Column description
contig_ID	This contains an identifier that has two parts: first is a unique nucleic acid sequence that barcodes the cell and the second is an identifier for the assembled V(D)J contig
Umis	Number of unique molecular identifiers (UMIs)
v-gene	Annotated V gene name
d-gene	Annotated D gene name
j-gene	Annotated J gene name
c-gene	Annotated C gene name
productive	Describes whether the contig is predictive of whether the transcript translates to a protein with a CDR3 region. The values are TRUE, FALSE, or none

2. Load the V(D)J dataset, and subset the dataset to include only productive T cells.

```
healthy_VDJ <-
read.csv("./vdj_v1_hs_pbmc_t_filtered_contig_annotations.
csv",
sep = ",", header = TRUE)
healthy_VDJ <- subset(healthy_VDJ, productive == "True")
```

3. Create a diagonal matrix of the UMIs, and label each row according to the contig.

```
umi <- healthy_VDJ[,16]
umi_diag <- diag(umi, length(umi), length(umi))
colnames(umi_diag)<- healthy_VDJ[,3]
```

4. Add more information to the matrix that we generated.

```
umi_2 <- cbind(healthy_VDJ[, c(7, 8, 9, 10)], umi_diag)
```

5. Now aggregate the data so that we tabulate the number of UMIs for each cell by V-genes.

```
umi_v_gene <- aggregate(umi_2[,c(5:ncol(umi_2))],
list(umi_2[,1]), FUN = sum)
rownames(umi_v_gene) <- umi_v_gene[,1]
umi_v_gene <- umi_v_gene[,-1]
```

6. Transpose the matrix, and modify the cell barcode labels so that they will match the barcode labels for the GEX matrix.

```
umi_v_gene_t <- t(umi_v_gene)
umi_rownames <- rownames(umi_v_gene_t)
umi_barcode <-substring(umi_rownames, 1, 16)
umi_v_gene_t <- cbind(umi_barcode, umi_v_gene_t)
```

7. Aggregate the data so that each row represents a unique cell barcode.

```
library(dplyr)
umi_v_gene_1_2 <- as_tibble(umi_v_gene_t)
umi_v_gene_1_2 <- mutate_at(umi_v_gene_1_2,
(2:ncol(umi_v_gene_1_2)), funs(as.numeric))
umi_v_gene_t_2 <- umi_v_gene_1_2 %>%
 group_by(umi_barcode) %>%
 summarise_all(sum)
```

8. Change the rownames to the cell barcodes.

```
umi_v_gene_t_2 <- as.data.frame(umi_v_gene_t_2)
rownames(umi_v_gene_t_2) <- umi_v_gene_t_2[,1]
umi_v_gene_t_2 <- umi_v_gene_t_2[,-1]
```

9. Finally, transform the dataframe again to make the columns barcodes and the rows genes. The resulting object will be used in the next steps.

```
umi_v_gene <- t(umi_v_gene_t_2)
```

3.2 Loading the 5′ Expression Dataset and the V(D)J Matrix to R as a Seurat Object

1. Take the object created in the previous step, and format a Seurat object from it.

```
VDJ_seurat <- CreateSeuratObject(umi_v_gene, min.genes = 1)
```

2. Read-in the 5 prime gene expression matrix, and convert it into a Seurat object.

```
GEX_matrix <- Read10X(data.dir = "./raw_gene_bc_matrice-
s_healthy_150_150/GRCh38/")GEX_seurat <- CreateSeuratObject
(GEX_matrix, min.genes = 200)
```

3.3 Combining the Gene Expression and V(D)J Seurat Objects

1. First create a dataframe consisting of the raw UMI data from each Seurat object previously generated. Convert the rownames into a column in the gene expression dataframe.

```
GEX_df <- as.data.frame(as.matrix(GEX_seurat@raw.data))
VDJ_df <- as.data.frame(as.matrix(VDJ_seurat@raw.data))
GEX_df <- cbind(rownames(GEX_df), GEX_df)
colnames(GEX_df)[1]<- "rownames"
```

2. Label the VDJ genes from the VDJ dataset so we can tell these from the genes in the other dataset. Add the rownames of the VDJ dataframe as a new column.

```
rownames(VDJ_df) <- paste(rownames(VDJ_df), "VDJ", sep = "-")
VDJ_df <- cbind(rownames(VDJ_df), VDJ_df)
colnames(VDJ_df)[1]<- "rownames"
```

3. Combine the gene expression and VDJ dataframes using the rbind.fill function. This will add the VDJ-labeled genes originating from the VDJ dataset to the gene expression dataframe, and fill in any columns that are not present in the VDJ dataframe with NA values. NA values will be replaced with zeros since these represent cells that were captured but did not have measurable T-cell receptor genes and were therefore not enriched during the VDJ library preparation.

```
library(plyr)
GEX_VDJ <- rbind.fill(GEX_df, VDJ_df)
GEX_VDJ[is.na(GEX_VDJ)] <- 0
rownames(GEX_VDJ)<- GEX_VDJ[,1]
GEX_VDJ<-GEX_VDJ[,-1]
```

4. Remove files in the R workspace that are no longer being used (*see* **Note 2**).

```
rm(healthy_VDJ, umi_2, umi_diag, GEX_matrix, GEX_df, GEX_seur-
at,
umi_v_gene, umi_v_gene_1_2, umi_v_gene_t, umi_v_gene_t,
VDJ_seurat,
umi_v_gene_t_2, VDJ_df, umi, umi_barcode, umi_rownames)
gc()
```

5. Create a Seurat object from the combined dataframe.

```
GEX_VDJ <- CreateSeuratObject(GEX_VDJ, min.cells = 1)
```

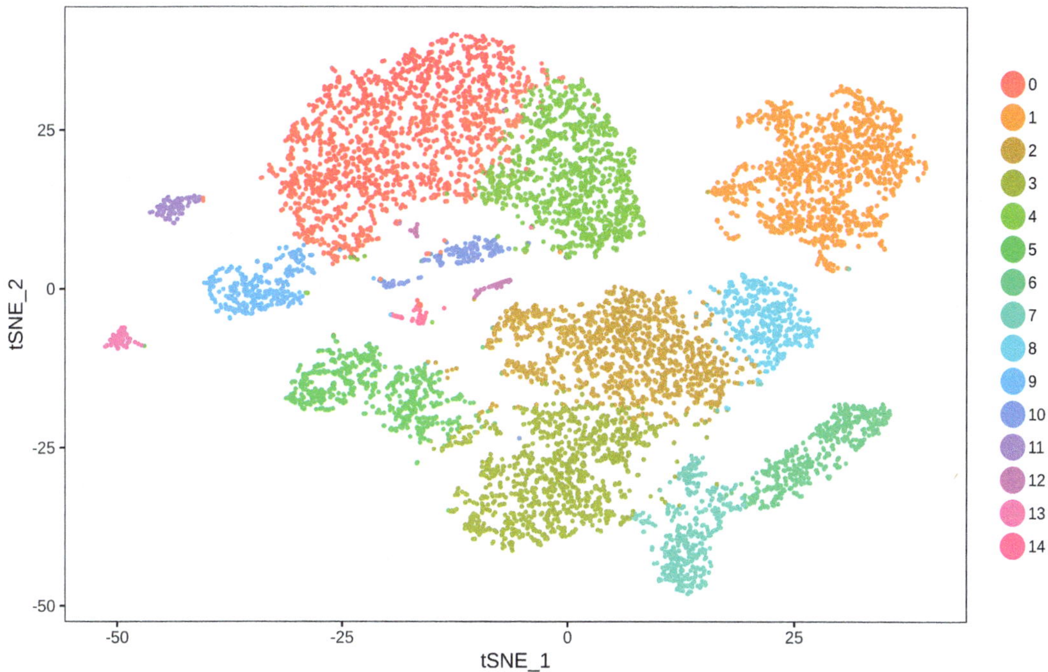

Fig. 1 TSNE plot for all cells in the dataset. Colors represent the clusters determined by running the FindClusters function

6. Add a column to the metadata dataframe in the Seurat object to annotate which cells are from the VDJ dataset. To do this, it is necessary to construct a matrix in which each column matches the cell barcodes of the Seurat object and each row contains the metadata. In this case, the meatadata is 1 or 0, where 1 indicates that the cell has VDJ expression.

```
VDJ.genes <- grep(pattern = "-VDJ$", x=rownames(x=GEX_VDJ@da-
ta), value=TRUE)
VDJ_matrix <- as.matrix(GEX_VDJ@raw.data[VDJ.genes,])
VDJ_matrix_sum <- Matrix::colSums(VDJ_matrix)
is_VDJ <- VDJ_matrix_sum
is_VDJ[is_VDJ <0] <- 0
is_VDJ[is_VDJ > 0] <- 1
GEX_VDJ <- AddMetaData(object = GEX_VDJ, metadata = is_VDJ,
col.name = "is_VDJ")
```

7. Run the Seurat pipeline commands to obtain a TSNE plot for the merged dataset. The resulting TSNE plot is shown in Fig. 1.

```
GEX_VDJ <- NormalizeData(GEX_VDJ, normalization.method = "Log-
Normalize", scale.factor = 10000)
GEX_VDJ <- FindVariableGenes(GEX_VDJ, do.plot = FALSE, y.cut-
off = .5)
```

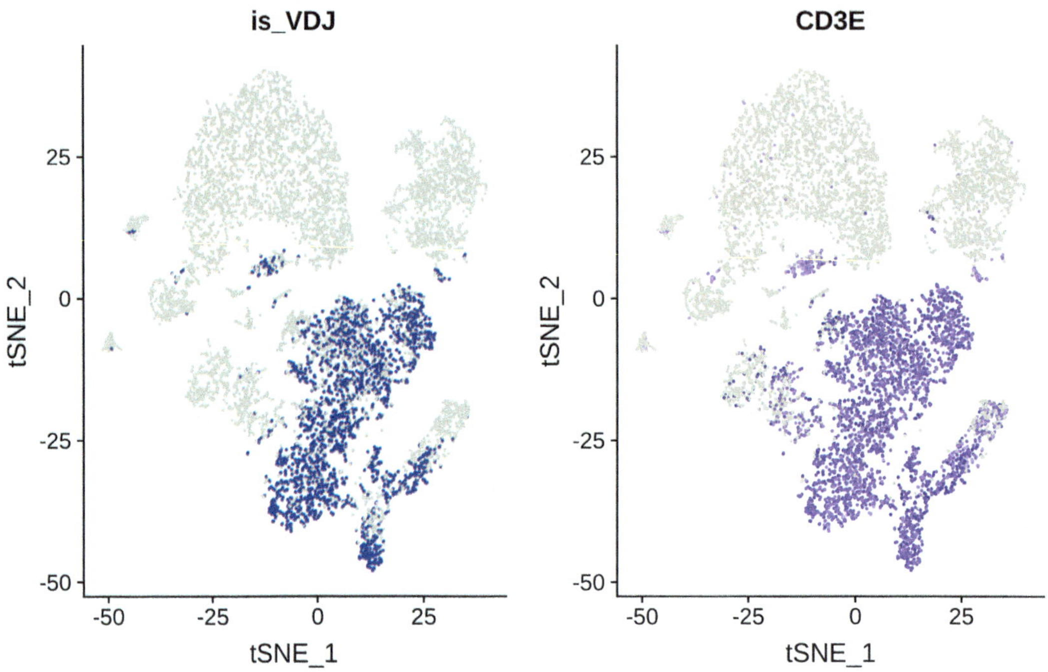

Fig. 2 TSNE plots of T-cell subset of the dataset. The plot on the left depicts blue color to represent cells that have a VDJ library barcode. The plot on the right depicts lavender color, which indicates cells that have CD3E gene expression in the gene expression library dataset

```
GEX_VDJ <- ScaleData(GEX_VDJ, display.progress = TRUE)
GEX_VDJ <- RunPCA(GEX_VDJ, pcs.compute = 15, pcs.print = 0)
GEX_VDJ <- FindClusters(GEX_VDJ, dims.use = 1:12, print.output
= TRUE, resolution=.5)
GEX_VDJ <- RunTSNE(GEX_VDJ, dims.use = 1:12, reduction.use =
"pca")
TSNEPlot(GEX_VDJ, do.label = FALSE, pt.size = 0.5)
```

8. Next, to demonstrate that the VDJ-labeled genes overlay the T cells from the gene expression dataset, generate the following plots, which are shown in Fig. 2.

```
FeaturePlot(GEX_VDJ, features.plot = c("is_VDJ", "CD3E"),
 cols.use = c("lightgrey", "blue"), pt.size = 0.5)
FeaturePlot(GEX_VDJ, features.plot = c("TRAV1.2-VDJ"),
 cols.use = c("lightgrey", "red"), pt.size = 0.5)
```

3.4 Annotating Each VDJ Cell According to Which TRAV Gene ItExpresses

1. The next commands will label each VDJ cell according to which TRAV gene it expresses. This is done by generating a matrix that can be added to the Seurat object metadata. The contents of this matrix are columns corresponding to the cell barcodes and one row with either the name of the T-cell receptor alpha variable (TRAV) gene that the cell expresses, or “NA” if the cell does not express a TRAV gene.

```
library(dplyr)
library(reshape2)
VDJ_TRAV.genes <- grep(pattern = "^TRAV", x=VDJ.genes, value=-
TRUE)
TRAV_cells <- GEX_VDJ@raw.data[VDJ_TRAV.genes, ]
TRAV_cells <- cbind(rownames(TRAV_cells), TRAV_cells)
colnames(TRAV_cells)[1] <- "Genes"
TRAV_cells <- as_tibble(TRAV_cells)
TRAV_cells_melt <- melt(TRAV_cells, id="Genes")
TRAV_cells_melt <- as_tibble(TRAV_cells_melt)
TRAV_cells_melt <- TRAV_cells_melt %>%
 mutate(VDJ_expressed= if_else(value>0, 1, 0))
TRAV_cells_melt_expressed <- TRAV_cells_melt %>%
 filter(VDJ_expressed>0) %>%
 add_count(variable)
```

2. Remove any cells from the dataframe that express more than one TRAV gene.

```
TRAV_cells_melt_expressed <- TRAV_cells_melt_expressed %>%
 filter(n==1)
```

3. Select only the unique cell barcodes; note, this will remove any cell that expresses more than oneTRAV gene.

```
barcodes <- TRAV_cells_melt %>%
 select(variable) %>%
 distinct(variable)
```

4. Combine the cell barcodes with TRAV gene labels with the other cells, and format the combined dataframe so that it can be added as metadata to the Seurat object.

```
TRAV_annotated  <-  left_join(barcodes,  TRAV_cells_melt_ex-
pressed) %>%
 select(variable, TRAV_gene)
TRAV_annotated <- as.data.frame(as.matrix(TRAV_annotated))
GEX_VDJ_meta <- GEX_VDJ@meta.data
GEX_VDJ_meta["TRAV"] <- TRAV_annotated$TRAV_gene
TRAV_meta <- subset(GEX_VDJ_meta, select = c("TRAV"))
```

5. Add the TRAV metadata table to the Seurat object.

```
GEX_VDJ <- AddMetaData(object = GEX_VDJ, metadata = TRAV_meta,
col.name = "TRAV")
```

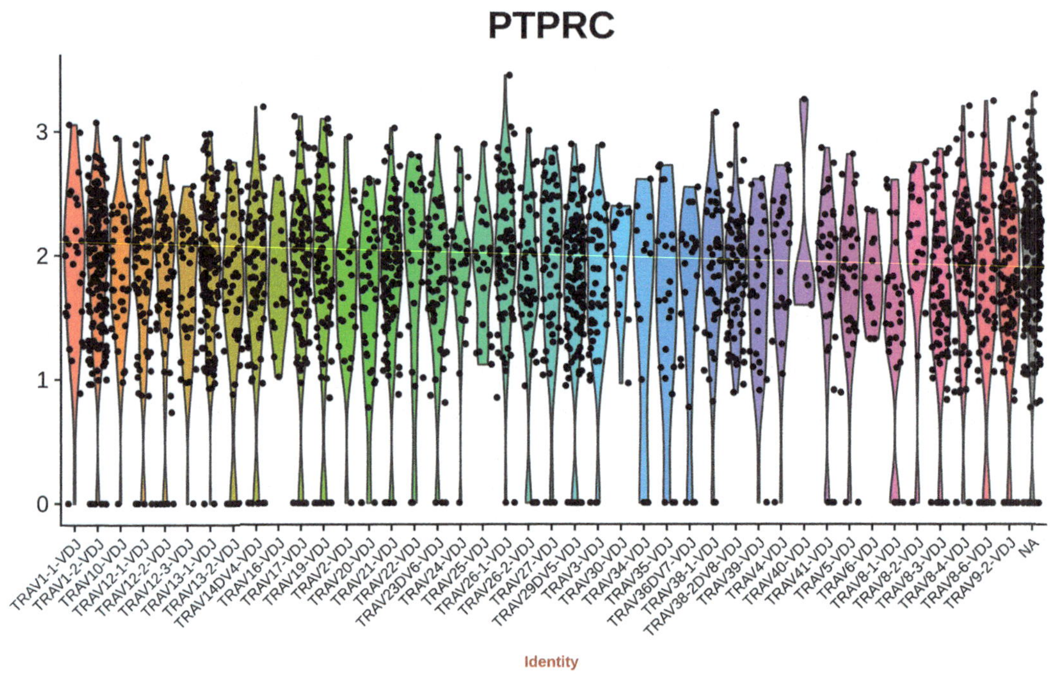

Fig. 3 Violin plot of normalized PTPRC gene expression values for each cell versus TRAV identity

6. Finally, generate a violin plot to show the expression of PTPRC for each TRAV cell. The resulting plot is shown in Fig. 3.

```
VlnPlot(object = GEX_VDJ, features.plot = "PTPRC", use.raw =
FALSE, x.lab.rot =T, group.by="TRAV")
```

4 Notes

1. There are many steps involved from obtaining the tissue sample to getting to a read count cell barcode matrix that is used as the starting point for the methods described here.10× genomics provides in-depth detailed instructions for reaching this point. In addition, the company has provided a freely available by download example dataset, which profiles both B cells and T cells for V(D)J sequences from PBMCs from a healthy volunteer.Files for each step of the data processing pipeline are available for download, from demultiplexed fastq files to the count matrix.
2. This step is necessary to free up RAM space.The next steps will require a large amount of RAM, and if the previously generated R objects are not removed, the next steps may cause the program to crash.

Acknowledgement

This work was supported by NIH grant R01HL137709.

References

1. Lythe G, Callard R, Hoare RL, Molina-Paris C (2016) How many TCR clonotypes does a body maintain? J TheorBiol 389:214–224
2. Murphy K, Weaver C (eds) (2017) Janeway'simmunobiology, 9th edn. Garland Science, Taylor & Francis Group, New York, p 499
3. R toolkit Seurat. https://satijalab.org/seurat/pbmc3k_tutorial_v3.html. Accessed 22 Mar 2019
4. Butler A, Hoffman P, Smibert P, Papalexi E, Satija R (2018) Integrating single-cell transcriptomic data across different conditions, technologies, and species. Nat Biotechnol 36:411–420
5. 10x Genomics Support.https://support.10xgenomics.com/. Accessed 22 Mar 2019
6. 10x Genomics Datasets.https://support.10xgenomics.com/single-cell-vdj/datasets/2.2.0/vdj_v1_hs_pbmc_t. Accessed 22 Mar 2019
7. 10x Genomics Datasets.https://support.10xgenomics.com/single-cell-vdj/datasets/2.2.0/vdj_v1_hs_pbmc_5gex. Accessed 22 Mar2019

Chapter 5

CRISPR/Cas9-Based Genetic Screening to Study T-Cell Function

Wanjing Shang, Fei Wang, Qi Zhu, Liangyu Wang, and Haopeng Wang

Abstract

T-cell-based cancer immunotherapies have emerged as a promising approach for cancer treatment, highlighting the importance of understanding the regulation of T-cell function. However, the molecular mechanisms underlying T-cell activation are not fully understood. The CRISPR/Cas9 system can serve as a robust method to systematically study signaling pathways. In this chapter, we describe details of using the CRISPR screen to identify regulators in TCR signaling, from the sgRNA library construction to genomic DNA sequencing. We also add some notes to further help readers performing the CRISPR screen. This approach can be readily adapted to study the activation of other immune cells, including B cells and dendritic cells.

Key words gRNA library, T-cell activation, Lentivirus production and titer, Lentiviral transduction, Cell sorting, Genomic DNA extraction

1 Introduction

T cells are a vital component of the adaptive immune system. Upon antigen engagement, Src family kinases (e.g., Lck and Fyn) are activated and phosphorylate the ITAM motifs within the TCR: CD3 complex, which recruits kinase ZAP70. ZAP70 is further phosphorylated by Lck. Activated ZAP70 recruits adaptors such as LAT and SLP-76 [1, 2]. These events lead to the activation of downstream signaling events including the Ras-MAPK pathway, calcium mobilization, and cytoskeletal reorganization. While most of these known TCR signaling molecules have been identified and characterized by traditional genetic and biochemical approaches, it remains a challenge to uncover the novel signaling molecules in T cells, which could be potential therapeutic targets of cancer immunotherapy. In the context of the current T-cell-based therapies, fully understanding T-cell signaling pathway is critical for cancer immunotherapy and autoimmune disease treatment.

Chaohong Liu (ed.), *T-Cell Receptor Signaling: Methods and Protocols*, Methods in Molecular Biology, vol. 2111,
https://doi.org/10.1007/978-1-0716-0266-9_5,

With the completion of the Human Genome Project in 2003, annotating gene functions has become one of the primary goals in the post-genomic era. Conventional gene interpretation methods include RNA interference (RNAi) screens, cDNA overexpression, and insertional mutagenesis or chemical mutagenesis such as EMS or ENU experiments. The former method (RNAi screen) is based on mRNA knockdown of the target gene. This method has enabled functional screenings in various species and cell types. One of the examples of RNAi screening in T cells was published in 2014 [3]. The authors performed an shRNA screen in T cells in vivo and screened for regulators controlling T-cell function within the tumor microenvironment. However, this RNAi method often results in hypomorphic phenotypes, instead of complete loss-of-function phenotypes, which may mask the importance of the target gene. Therefore, new tools for genetic screening are keenly desired to uncover new regulators of T-cell activation [4].

CRISPR/Cas9-based screens are effective and powerful in dissecting gene functions. The principle behind Cas9 involves utilizing the single-guide RNA (sgRNA) that is complementary to a target sequence to recruit Cas9 protein to the target site and induce double-strand breaks (DSBs). The DSBs are mainly repaired by nonhomologous end-joining (NHEJ) pathways in mammalian cells, leading to frameshift mutations in target genes. CRISPR-based screens include CRISPR, CRISPR interference (CRISPRi), and CRISPR activation (CRISPRa). Recently, our lab [5] and Marson lab [6] have performed genome-wide CRISPR screens in Jurkat T cells and human primary T cells, respectively. Specifically, our lab used an sgRNA library targeting 24 K genes in the human genome with approximately 12 sgRNAs per gene. In our screen, we stimulated T cells using anti-TCR antibody and assessed the T-cell activation status by measuring upregulation of CD69, an early T-cell activation marker. Our screen not only confirmed many of the known signaling molecules in proximal T-cell signaling but also identified a list of novel TCR signaling regulators. For example, we found that FAM49B, a previously uncharacterized gene, acts as a negative regulator in T-cell activation through repressing Rac activity and modulating cytoskeleton reorganization [5].

In this chapter, we describe how to perform a CRISPR screen in T cells. Our protocol contains the following four parts: generation of a Jurkat T-cell line expressing functional Cas9, lentiviral packaging of sgRNA libraries, T-cell activation-based screening, and genomic DNA extraction for high-throughput sequencing. Although we specifically focus on T lymphocytes, this protocol can also be applied to study of the activation of other immune cells, including B cells and dendritic cells. Some notes are included which may be helpful for completing the screen.

2 Materials

2.1 Generating a Jurkat T-Cell Line Expressing Functional Cas9

1. NucleoBond® Xtra Midi EF Prep Kit (Macherey-Nagel).
2. Plasmid: pCMVdR8.91 (Addgene, #8455), pMD2.G (Addgene, #12259), and Cas9-T2A-BFP [7].
3. Cells: Lenti-X™ 293T cells (Takara #632180) for lentiviral packaging; Jurkat T cells.
4. TransIT-LT1 Transfection Reagent (Mirusbio).
5. Opti-MEM I reduced serum medium (Thermo Scientific).
6. Complete RPMI 1640 medium: 500 mL RPMI 1640 (Gibco), 5% fetal bovine serum (FBS, Gibco), 100 U/mL penicillin, 100 μg/mL streptomycin, and 292 μg/mL glutamine.
7. Trypsin 0.25% EDTA (Life).
8. PBS (PH = 7.2) (Gibco).
9. 6-well plate; 24-well plate; 96-well round-bottom plate (Corning).
10. Complete DMEM medium: 500 mL DMEM (Gibco), 10% fetal bovine serum (FBS, Gibco), 100 U/mL penicillin, 100 μg/mL streptomycin, and 300 μg/mL glutamine.
11. Anti-human CD28-PE (Tonbo, clone: CD28.2).
12. Anti-human CD3-APC (Tonbo, clone: UCHT1).
13. Anti-human HLA-PE-Cy7 (BD Pharmingen™, clone: G46-2.6).
14. DAPI readymade solution (1 mg/mL, Sigma).
15. Cell staining buffer (4A Biotech).
16. Nalgene™ General Long-Term Storage Cryogenic Tubes (Thermo Scientific).
17. sgRNA sequence against human B2M (5′ to 3′): GGCCGAGATGTCTCGCTCCG.

2.2 Lentiviral Packaging of sgRNA Library

1. Ampicillin; LB broth (Sangon biotech).
2. 1.5 mL low-binding tubes (Eppendorf).
3. HST08 electrocompetent cell (Takara).
4. 0.1 cm electroporation cuvette (Biorad).
5. Cells: Lenti-X™ 293T cells (Takara #632180) for lentiviral packaging; Jurkat T cells.
6. Complete DMEM medium: 500 mL DMEM (Gibco), 10% fetal bovine serum (FBS, Gibco), 100 U/mL penicillin, 100 μg/mL streptomycin, and 300 μg/mL glutamine.
7. Puromycin (Invivogen).

2.3 T-Cell Activation-Based Screening

1. Complete RPMI 1640 medium: 500 mL RPMI 1640 (Gibco), 5% fetal bovine serum (FBS, Gibco), 100 U/mL penicillin, 100 μg/mL streptomycin, and 300 μg/mL glutamine.
2. Anti-TCR antibody C305 (Millipore).
3. Anti-human CD69-APC (Biolegend).
4. CD69 MicroBead Kit II (Miltenyi Biotec).

2.4 Genomic DNA Extraction

1. QIAamp DNA mini Kit (Qiagen).
2. PBS (PH = 7.2) (Gibco).
3. Proteinase K (Sangon biotech).
4. Syringe needles gauge 21, gauge 23, gauge 25, and gauge 27 (Sigma).
5. Phenol:chloro:isoamyl alcohol = 25:24:1 (UltraPure).
6. 3 M sodium acetate (Corning).
7. Titanium Taq (Clontech).

3 Methods

3.1 Generating a Jurkat T-Cell Line Expressing Functional Cas9

In the immunology field, the Jurkat T-cell line is widely used to study T-cell function. Therefore, it is useful to generate a Jurkat cell line stably expressing functional Cas9 optimized for large-scale genetic screens. This $Cas9^{+}$ Jurkat cell line can be generated by transducing wild-type Jurkat cells with the lentivirus constitutively expressing Cas9 (Cas9 lentivirus).

3.1.1 Packaging of the Cas9 Lentivirus

The lentiviral construct containing the Cas9 gene and a BFP reporter gene was described previously [7]. To ensure the maximal viral production, we recommend to prepare plasmids by NucleoBond® Xtra Midi EF Prep Kit (*see* **Note 1**).

1. (Day 0) Seed 0.55 million Lenti-X™ 293T cells in one well of the 6-well plate containing 2.5 mL complete DMEM medium without Penicillin-Streptomycin (P/S). The cell density will be at 60–70% on the next day for transfection.
2. (Day 1) To transfect the Lenti-X™ 293T cells, a mixture of 500 ng pCMV dR8.91, 50 ng pMD2.G, and 500 ng Cas9-T2A-BFP plasmids is added to 37 μL Opti-MEM I reduced serum medium. Dilute 3 μL MirusTransIT LT1 in 15 μL Opti-MEM I reduced serum medium. Then add diluted TransIT-LT1 to the DNA and mix gently. After incubating for 30 min at room temperature, add the transfection mixture to the Lenti-X™ 293 T cells, and incubate at 37 °C in cell incubator.
3. (Day 2) 18 h posttransfection, replace the original transfection reagent medium with 2.5 mL complete DMEM medium.

4. (Day 4) Approximately 72 h after transfection, harvest the supernatants and spin at 800 × *g* for 5 min to pellet any Lenti-X™ 293T cells that were collected during harvesting. Virus could be stored at 4 °C for a few days, but should be frozen at −80 °C for long-term storage.

3.1.2 Transduction of Jurkat T Cells with Cas9 Lentivirus

1. Culture Jurkat cells in complete RPMI 1640 medium at the density of 0.5 million/mL.
2. Once Jurkat cells enter the exponential growth, replace half of the culture medium with freshly harvested Cas9 virus supernatant generated in Subheading 3.1.1.
3. Continue to culture Jurkat T cells for 6–7 days.
4. Harvest the transduced Jurkat T cells and resuspend cells in cell staining buffer at the concentration of ten million cells/mL. Add final concentration at 1:500 anti-CD28-PE (Tonbo, clone: CD28.2) and final concentration at 1:500 anti-CD3-APC (Tonbo, clone: UCHT1) into the buffer, gently mix, and incubate the mixture at 4 °C for 30 min in dark (*see* **Note 2**).
5. After incubation, wash the cells twice with cell staining buffer, and spin down the cells at 500 × *g* for 5 min at 4 °C.
6. Resuspend in cell staining buffer with final concentration at 1 μg/mL DAPI, and then transfer to sorting tube. Keep cells on ice and in dark. Proceed to the sorting step.
7. Sort individual $BFP^+CD28^+CD3^+$ Jurkat cells into a 96-well round-bottom plate with RPMI 1640 medium through single-cell sorting using BD FACSAria II flow cytometry (*see* **Note 3**).
8. Continue to grow these single-cell clones for 4 weeks (*see* **Note 4**).

3.1.3 Testing the Function Activity of Transduced Cas9 in Jurkat Cells

We often find that some cell clones expressing BFP completely lose the genome-editing ability of Cas9 protein for unknown reason. Therefore, it is important to choose a single-cell clone of Jurkat T cell expressing functional Cas9. To test whether Cas9 protein in Jurkat cells had any genome-editing function, we designed an sgRNA to specifically target the beta-2 microglobulin (B2M) gene (sgRNA sequence: GGCCGAGATGTCTCGCTCCG). B2M is a subunit of MHC class I molecules, and knockout of the B2M gene results in ablating MHC class I surface expression [7]. MHC class I molecules are highly expressed in all nucleated cells, including human T cells. Therefore, the loss of MHC class I expression after expressing sgRNA targeting B2M serves as a functional readout of Cas9 activity of different clones.

1. Select 10–20 clones with high-level expression of TCR and CD28 receptors, and expand them in 24-well plates.

2. Take out about 20,000 cells per clone. Transduce these individual cells with lentivirus expressing sgRNA targeting B2M.
3. After 4–6 days, harvest transduced Jurkat T cells and stain with anti-HLA-PE-Cy7 (clone: G46-2.6). Measure the expression level of MHC class I by FACS. The higher the activity of Cas9 within transduced cells, the higher the percentage of cells that lose expression of MHC class I molecules. We normally select these clones in which over 70% of cells have lost MHC class I expression after expressing sgRNA targeting B2M.
4. Expand the selected clones and freeze them for long-term storage.

3.2 Packaging and Transducing Lentiviruses Encoding the sgRNA Library

There are multiple sgRNA libraries existing in the field, including both unbiased genome-wide sgRNA libraries and selected sgRNA pool targeting a specific list of genes [8]. After choosing an sgRNA library according to your experimental purpose, the selected sgRNA library DNA should be amplified and packaged into lentiviruses. Our lab uses Lenti-X™ 293T cells to generate lentivirus. Sometimes, the lentivirus has to be concentrated to get a high viral titer. $Cas9^+$ Jurkat T cells generated in Subheading 3.1 are infected at a low MOI, normally 0.3 to 0.5, to ensure that one cell only receives one sgRNA construct.

3.2.1 Amplification of sgRNA Library

1. Mix 100 ng DNA of sgRNA library with 100 μL HST08 electrocompetent cell. Caution: ON ICE.
2. Add DNA/HST08 mix into two prechilled 0.1 cm electroporation cuvettes. Caution: ON ICE.
3. Electroporate both cuvettes at 2 kV, 200 Ohm, 25 μF, and recover immediately in a total of 5 mL recovery media by shaking for 1 h at 37 °C.
4. Take 1 μL of transformation mixture and dilute with 99 μL LB. Vortex briefly and take 1 μL of this dilution to add to 99 μL LB. Vortex briefly again and plate a total of 100 μL on a petri dish with LB^{amp} media (100 μg/mL) at 37 °C overnight.
5. Add the remaining recovered cells to 200 mL (or more) of LB^{amp} (100 μg/mL), and grow the bacteria at 37 °C with 200 RPM shaking for 6–16 h.
6. Count the colonies on the LB^{amp} plate 1 day later. The coverage can be calculated with the following formula:

$$\text{Coverage} = \frac{n_{\text{colony}} \times V_{\text{transformation mix}(\mu\text{L})} \times 100}{n_{\text{sgRNA}}}.$$

7. If the coverage is more than 100, proceed to the next step; otherwise repeat Subheading 3.2.1 until the desired coverage is reached.

8. Use NucleoBond® Xtra Midi EF Prep Kit to isolate library plasmid from liquid culture.

3.2.2 Production and Titering of Lentiviruses Encoding the sgRNA Library

Lentiviral titers can be measured by flow cytometry if the sgRNA construct contains a florescence protein. Ideally, the lentiviral titer should be determined in the cell line used for the screen. Therefore, we recommend to infect the $Cas9^+$ Jurkat cell line generated in Subheading 3.1 to determine the titer.

1. Produce the sgRNA lentivirus in accordance with Subheading 3.1.1. Take out 100 μL of supernatant to determine the titer. Aliquot and freeze the rest of the virus at −80 °C for long-term storage.
2. Culture the $Cas9^+$ Jurkat cells in complete RPMI 1640 medium at the concentration of 0.5 million/mL.
3. Once Jurkat cells enter exponential growth, harvest and count the cells.
4. In 8 wells of a 96-well plate, perform the following twofold serial dilutions: add 50 μL of complete media to all eight wells. In the first well, add 50 μL of media with no virus (negative control). In the second well, add 50 μL of undiluted virus. In the third well, add 50 μL of the mix from the second well. In the fourth well, add 50 μL from the third well. In the fifth well, add 50 μL from the fourth well. In the sixth well, add 50 μL from the fifth well. In the seventh well, add 50 μL from the sixth well. In the eighth well, add 50 μL from the seventh well. Finally, take out 50 μL out of the eighth well to equalize the volume of all eight wells.
5. Seed 15,000 cells in each well of 8 wells; adjust the volume of culture medium to 150 μL per well. Mix each well thoroughly by pipetting up and down.
6. Culture these cells for 72 h.
7. Harvest the cells and analyze the expression of fluorescent reporter proteins using the flow cytometer. Viral particles per mL are quantitated by the percent of fluorescent cells per well with the following formula:

$$\text{Titer} = \frac{15,000 \times \%\text{of reporter positive cells} \times 1000}{\mu\text{l of virus added to well}}.$$

3.2.3 Infection of $Cas9^+$ Jurkat Cells with Lentiviruses Encoding the sgRNA Library

To ensure most cells receive only one viral particle, the MOI should be around 0.3 (*see* **Note 5**). The coverage of Jurkat cells per sgRNA should be at least 1000-fold. For example, if a library of 6000 sgRNA is used, transduce 20 million $Cas9^+$ Jurkat cells with six million lentiviral particles at an MOI of 0.3.

1. Culture $Cas9^+$ Jurkat cells in complete RPMI 1640 medium at the concentration of 0.5 million/mL. Expand the $Cas9^+$ Jurkat cells to a sufficient amount for screening.
2. Infected cultured cells with viruses are generated in Subheading 3.2.1 with the MOI of 0.3.
3. About 2–3 days after transduction, sgRNA and the selection marker (puroR) start to express.
4. Culture these transduced cells in the presence of 2 μg/mL puromycin for 5–7 days. The cells are ready for screening process.

3.3 T-Cell Activation-Based Screening

One of the most important factors of designing the screening is to determine how to apply the selection pressure. There are three types of selection strategies that have already been applied for the CRISPR library screen. First, *positive selection screens* are designed to identify novel genes resistant to a drug or toxin. Cell populations transduced with sgRNAs targeting drug- or toxin-resistant genes can be enriched after selection. Second, *negative selection screens* rely on the depletion of sgRNAs and a selection phenotype as a result of cell death. Normally, the negative screens are used to identify the essential genes that are critical for cell survival. T-cell activation-based screenings often rely on the third type of selection strategy, which is dependent on a *FACS-based assay.* T-cell activation results in the upregulation of a series of activation markers, including CD69, CD25, PD-1, and CD44, and the production of multiple cytokines (e.g., IL-2 and IFN-γ), as well as cell proliferation. All these events can be assessed and quantitated by FACS technology and therefore used as a selection method for the T-cell activation screen. Taking the upregulation of CD69 as an example, the whole T-cell population could be sorted into two populations according to the expression of CD69 (the strong activated population, $CD69^{high}$, and the weak activated population, $CD69^{low}$). sgRNAs targeting positive regulators of T-cell activation are presumably enriched in the $CD69^{low}$ population, whereas sgRNAs against negative regulators are depleted in the $CD69^{high}$ population.

3.3.1 T-Cell Activation

1. Culture sgRNA transduced $Cas9^+$ Jurkat cells in complete RPMI 1640 medium at the concentration of one million/mL.
2. Add anti-TCR antibody (clone: C305) at the final concentration of 6.8 ng/mL to transduced Jurkat T cells (*see* **Note 6**).
3. Continue to culture the cells for 13 h.
4. Harvest the cells and wash the cells with PBS once.

3.3.2 CD69 Staining and FACS Sample Preparation

1. Resuspend the T cell in cell staining buffer at the concentration of ten million cells/mL.
2. Add anti-hCD69-APC antibody (1:500) to cells and incubate the mix at 4 °C for 30 min in dark.
3. After incubation, wash the cells twice with cell staining buffer and spin down the cells at 500 × *g* for 5 min at 4 °C.
4. Resuspend cells in cell staining buffer with final concentration at 1 μg/mL DAPI; then transfer to sorting tube. Keep cells on ice and in dark.
5. Sort two populations according to expression of CD69 (strong activated population, $CD69^{high}$, and weak activated population, $CD69^{low}$) using BD Aria II (Fig. 1, *see* **Note 7**).
6. Harvest the sorted cells, and proceed to genomic DNA exaction step (*see* **Note 8**).

3.4 Genomic DNA Extraction for High-Throughput Sequencing

It is important to harvest enough cells for genomic DNA extraction. We normally maintain a coverage of >250×. For example, for a library size of 60,000 sgRNAs, sort and harvest about 15 million cells for genomic DNA extraction. For small numbers of cells (below ten million cells), we recommend to use QIAamp DNA mini Kit (Qiagen). For larger numbers of cells, the genomic DNA can also be extracted with the following protocol:

1. Centrifuge the cells at 500 × *g* for 5 min and discard supernatant. Wash once with 30 mL PBS, discard supernatant, and resuspend pellet in 400 μL Qiagen P1 buffer with RNase A (0.1 mg/mL). Then add 40 μL 10% SDS and incubate this mixture at room temperature for 15 min.
2. Add proteinase K to a final concentration of 100 μg/mL and incubate for 30 min at 55 °C. Using syringe needles gauge 21, gauge 23, gauge 25, and gauge 27, shear sample for 10 times/needle. Then transfer the DNA to a new 1.5 mL tube.
3. Add 400 μL phenol:chloro:isoamyl alcohol = 25:24:1 to the DNA, shake the mixture, and centrifuge at 13,000 × *g* for 10 min at 20 °C.
4. Take the supernatant to a new ultracentrifuge tube; add 0.1-fold of supernatant volume of 3 M sodium acetate and 0.8-fold of isopropanol. Centrifuge at 12,000 × *g* for 30 min at 4 °C and discard the supernatant.
5. Wash the DNA pellet with 1 mL 70% ethanol and centrifuge at 13,000 × *g* for 15 min.
6. Discard the supernatant; then dry DNA at room temperature for 30 min.
7. Resuspend the DNA pellet in 50 μL ddH_2O and store at 4 °C.

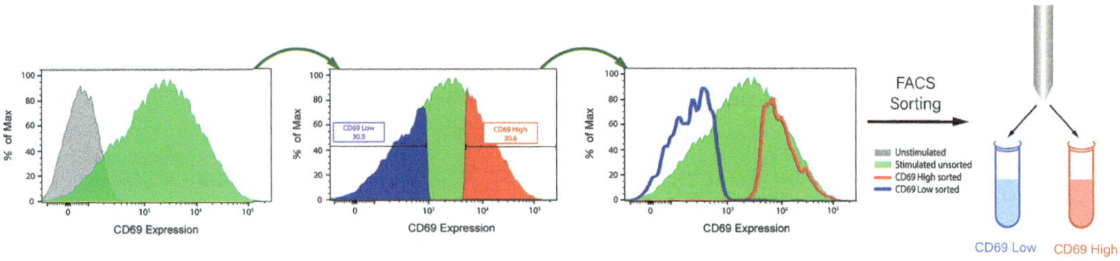

Fig. 1 Sorting strategy for the T-cell activation-based screening. Histogram of the CD69 expression in activated Jurkat cells. Resting and activated Jurkat cell populations are in gray and green, respectively. Gated CD69^{low} and CD69^{high} populations are in blue and red, respectively

8. According to the sgRNA library you used, choose the suggested primers for PCR amplification of the sgRNA region from the genomic DNA for NGS analysis (*see* **Note 9**).

4 Notes

1. Generation of high-quality lentiviruses is crucial for the CRISPR screen. Here are some hints to generate high-quality lentiviruses. Firstly, maintaining 293T cells at an appropriate density is critical for lentivirus production. Either cell overgrowth or a low cell density will decrease viral titer. Secondly, after digesting Lenti-X™ 293T cells, the trypsin/EDTA should be neutralized with serum-containing medium and completely removed by centrifuge. Thirdly, 1 day before harvesting viruses, we strongly recommend changing the medium to remove the transfection reagent to reduce toxicity for cells of interest. Lastly, an endotoxin-free maxi/midi prep kits should be used to prepare the DNAs for viral production.
2. During the antibody staining and sorting steps, it is important to keep the cell sample sterile. The staining antibody solution should be filtered using 0.22 μm filters. All the procedures should be completed in the tissue culture hood. We do not add sodium azide in the FACS buffer or staining buffer since it is toxic to cells.
3. One recent study reported that clonal heterogeneity may affect screening sensitivity [9]. We also observed that T-cell activity is lost in some subclones of Jurkat cells, likely due to the downregulation of surface TCR. Therefore, it is important to sort TCR^{high} and CD28^{high} cell populations to ensure proper sensitivity toward T-cell stimulation.
4. Here are some tips for single-cell clone generation: (1) To reduce evaporation during long-term culture, the 96 wells should be wrapped with aluminum foil. (2) Monitor cell

growth weekly. Once the cells have expanded but before they become overconfluent, transfer colonies into a 24-well plate. (3) In addition to FACS sorting, single-cell clones of cells can also be generated by serial dilution method.

5. The lentiviral sgRNA library should be infected at an MOI < 0.3 to ensure that the majority of cells harbors only one sgRNA. Transducing at a higher MOI may increase the false-positive discovery rate. However, it is difficult to control the MOI when primary T cells are used for transduction, or when an in vivo screen is performed. To address this point, the iBAR approach was recently developed [10]. In this iBAR method, each sgRNA is assigned with different internal barcodes, and the performance of each sgRNA can be traced multiple times within the same experiments, which improves the accuracy and efficiency of a screen with high MOI infection.
6. There are several ways to activate Jurkat T cells. Jurkat cells could be activated by using anti-CD3 antibody or anti-TCR antibody. The activation could be further promoted through adding costimulatory anti-CD28 antibody in the culture. In addition, Jurkat cells could be fully activated by using superantigens presented by APCs [11].
7. If the cell population is large (>100 million), we often choose to use MACS (e.g., CD69 MicroBead Kit II from Miltenyi Biotec) instead of FACS sorting, according to the manufacturer's directions. If it is a focused library with a limited cell number, we normally choose to use flow cytometry for sorting.
8. Sorted cells can be directly processed to genomic DNA extraction step or snap frozen at −80 °C for storage until next step. When thawing cells, put the cells in a 37 °C water bath and add an appropriate amount of PBS, usually 30 mL PBS in a 50 mL centrifugation tube.
9. **Note** 105: We used a high-fidelity polymerase (Titanium Taq from Clontech) in this step, in order to achieve a faithful amplification of the sgRNA from genomic DNA.

Acknowledgments

The authors thank F. Wang, Z. Lin (ShanghaiTech University), and J. M. Shen (The Semiconductor Manufacturing International Corporation Private School) for their critical reading of the manuscript. H.W. is funded by National Natural Science Foundation of China Grant 31670919 as well as the 1,000-Youth Elite Program of China.

References

1. Courtney AH, Lo WL, Weiss A (2018) TCR signaling: mechanisms of initiation and propagation. Trends Biochem Sci 43(2):108–123
2. Wang H et al (2010) ZAP-70: an essential kinase in T-cell signaling. Cold Spring Harb Perspect Biol 2(5):a002279
3. Zhou P et al (2014) In vivo discovery of immunotherapy targets in the tumour microenvironment. Nature 506(7486):52–57
4. Boettcher M, McManus MT (2015) Choosing the right tool for the job: RNAi, TALEN, or CRISPR. Mol Cell 58(4):575–585
5. Shang W et al (2018) Genome-wide CRISPR screen identifies FAM49B as a key regulator of actin dynamics and T cell activation. Proc Natl Acad Sci U S A 115(17):E4051–E4060
6. Shifrut E et al (2018) Genome-wide CRISPR screens in primary human T cells reveal key regulators of immune function. Cell 175 (7):1958–1971.e1915
7. Chi S, Weiss A, Wang H (2016) A CRISPR-based toolbox for studying T cell signal transduction. Biomed Res Int 2016:5052369
8. Shang W, Wang F, Fan G, Wang H (2017) Key elements for designing and performing a CRISPR/Cas9-based genetic screen. J Genet Genomics 44(9):439–449
9. Boettcher M et al (2019) Tracing cellular heterogeneity in pooled genetic screens via multi-level barcoding. BMC Genomics 20(1):107
10. Zhu S et al (2019) Guide RNAs with embedded barcodes boost CRISPR-pooled screens. Genome Biol 20(1):20
11. Tian R et al (2015) Combinatorial proteomic analysis of intercellular signaling applied to the CD28 T-cell costimulatory receptor. Proc Natl Acad Sci U S A 112(13):E1594–E1603

Chapter 6

Preferential Expansion of $CD4^+Foxp3^+$ Regulatory T Cells (Tregs) In Vitro by Tumor Necrosis Factor

Chon-Kit Chou and Xin Chen

Abstract

$CD4^+Foxp3^+$ regulatory T cells (Tregs) are a distinct subset of CD4 T cells that play indispensable role in the maintenance of immune homeostasis and prevention of deleterious immune responses to self-antigens. Tumor necrosis factor (TNF) is a key cytokine in the autoimmune inflammatory responses. The effect of TNF on Treg activity was extensively studied in the past decade. We for the first time reported that TNF through TNFR2 preferentially activates and expands Tregs. Our discovery is increasingly supported by the research community; however, some controversial results were reported. The differential results are likely caused by different experimental condition. A standard experiment protocol can help researchers to obtain more consistent results. In this chapter, we detail methods used to examine in vitro effect of exogenous TNF on the proliferative expansion of Tregs in unfractionated mouse $CD4^+$ T cells. The related technic issues are analyzed and discussed.

Key words $CD4^+FOXP3^+$ regulatory T cells (Tregs), Tumor necrosis factor (TNF), TNF receptor 2 (TNFR2), Selective expansion, CFSE cell division assay, Flow cytometry

1 Introduction

Tumor necrosis factor (TNF) is a pleiotropic cytokine that has both proinflammatory and immunosuppressive functions [1]. Biological activity of TNF is mediated by two functionally distinct receptors, TNFR1 (p55) and TNFR2 (p75) [2]. Unlike the ubiquitous expression of TNFR1, TNFR2 is primarily expressed by Tregs in normal mice [3]. We for the first time reported that the expression of TNFR2 on Tregs has important functional implications, as evidenced by the activation and expansion of Tregs after TNF-TNFR2 interaction [4]. In fact, the expression of TNFR2 itself identifies the most suppressive subset of Tregs [3, 5]. Furthermore, we found that TNFR2 signal is required for the sustained foxp3 expression in purified Tregs in both in vitro and in vivo settings and required for the in vivo function of Tregs [6, 7]. Intriguingly, TNF preferentially upregulates TNFR2 expression on Tregs [8]. We also reported that

Chaohong Liu (ed.), *T-Cell Receptor Signaling: Methods and Protocols*, Methods in Molecular Biology, vol. 2111,
https://doi.org/10.1007/978-1-0716-0266-9_6,

the number of Tregs was markedly decreased in the thymus and peripheral lymphoid tissues of mouse deficient in TNFR2 or its ligands [6] or its major signaling component [9]. Consequently, we proposed that up- or downregulation of Treg activity by targeting of TNFR2 could be therapeutically harnessed for the treatment of major human diseases [10–13]. Although counterintuitive, the concept established by us that TNF-TNFR2 interaction plays a decisive role in the activation, expansion, and phenotypical stability of functional Tregs in an immune response and in the inflammatory environment has been supported and substantiated by increasing studies from other groups [14–25]. Nevertheless, some studies reported the opposite effect of TNF on Tregs. For example, Valencia et al. reported that stimulation of human circulating $CD4^{+}CD25^{high}$ Tregs with TNF resulted in the loss of their immunosuppressive functions [26]. The controversial reports of the effect of TNF-TNFR2 signaling on Tregs are most likely due to different protocols used by investigators. Therefore, a standard protocol is likely to be helpful for the research community to obtain more consistent results. In this article, the details of protocol we used to show the selective expansion of mouse Tregs in the in vitro cultured $CD4^{+}$ T cells induced by TNF are introduced. The relevant technic issues are analyzed and discussed.

2 Materials

2.1 Single Cell Suspension Preparation and $CD4^{+}$ T-Cell Enrichment

1. Dissection tools: sterile scissor, forceps, and stainless steel mesh screen.
2. Phosphate buffered saline (PBS) supplemented with 2% heat-inactivated fetal bovine serum (FBS).
3. Red blood cell (RBC) lysis buffer: 10× RBC lysis buffer (BioLegend) needs to be diluted to 1× working concentration with deionized water.
4. Falcon cell strainer with 70 μm nylon mesh (BD Biosciences).
5. Mouse $CD4^{+}$ T-Cell Isolation Kit (Miltenyi Biotec) contains MicroBeads conjugated to CD4 (L3T4) monoclonal antibodies for magnetic labeling of mouse CD4-expressing cells.
6. MACS buffer: PBS supplemented with 2% heat-inactivated FBS and 2 mM EDTA.

2.2 Culture of Unfractionated $CD4^{+}$ T Cells with TNF

1. Complete RPMI 1640 medium: RPMI 1640 (Lonza BioWhittaker) supplemented with 10% heat-inactivated FBS, 2 mM glutamine, 100 IU/ml penicillin, 100 mg/ml streptomycin, 10 mM HEPES, 1 mM sodium pyruvate, 0.1 mM nonessential amino acids, and 50 mM 2-mercaptoethanol.
2. Round-bottom 96-well tissue culture plate (Thermo Fisher Scientific).

3. Recombinant mouse IL-2 (BD Pharmingen).
4. Recombinant mouse TNF (BD Pharmingen).
5. CD4 Monoclonal Antibody (RM4-5), PerCP-Cyanine5.5 (eBioscience).
6. Rat IgG2a kappa Isotype Control (eBR2a), PerCP-Cyanine5.5 (eBioscience).
7. FOXP3 Monoclonal Antibody (FJK-16s), APC (eBioscience).
8. Rat IgG2a kappa Isotype Control (eBR2a), APC (eBioscience).
9. PE Hamster Anti-Mouse CD120b (TR75-89) (BD Biosciences).
10. PE Hamster IgG1, λ1 Isotype Control (BD Biosciences).
11. Fc blocking solution: anti-mouse CD16/CD32 antibodies (BD Biosciences) in PBS at a final concentration of 5 ng/μl.
12. Fixation/permeabilization buffer mixture: Add one part of concentrate (eBioscience Cat. 00-5123) to three parts of diluent (eBioscience Cat. 00-5223) before use.
13. 1× permeabilization/wash buffer: Dilute permeabilization buffer (10×) (eBioscience Cat. 00-8333) by adding one part of the concentrate to nine part of distilled water.

3 Methods

3.1 Collection of Spleen and Lymph Nodes and Single-Cell Suspension Preparation

1. Normal C57BL/6 mice or Balb/c mice, or other strains of mice, male or female, 8–12 weeks old.
2. Mice will be sacrificed using CO_2 asphyxiation.
3. Sterilize a set of dissection tools (sterile scissor, forceps, and stainless steel mesh screen.) by autoclave before dissection. The tools should be dipped in 70% ethanol and then passed through a flame between the dissection of each mouse.
4. Mice should be placed face-up on absorbent paper. Wet the fur on the abdominal area with 70% ethanol to minimize fur entering the peritoneum. From the ventral view, spread the limbs of the mouse and make a cut through the skin latitudinally around the umbilicus, and then extend incision outward in an opposite direction until the abdominal region is fully exposed while being careful not to puncture the underlying tissue.
5. Pin the skin onto a pedestal to allow comfortable access to brachial, axillary, and inguinal lymph nodes. Grab the accessible lymph node with forceps while trimming it out with scissor. Place lymph nodes in collection dish.
6. Carefully open the abdominal wall with sterile scissor to avoid damaging the underlying organs. Harvest spleen and

mesenteric lymph nodes by using forceps and scissors. External lymph nodes such as in the inguinal and axillary regions can also be collected. Try to exclude the connective tissue and adipose tissue surrounding the spleen and lymph nodes. Place spleen and mesenteric lymph nodes in separate collection dish.

7. Collected spleen and lymph nodes are placed on ice in 60 mm dish containing 3 ml PBS with 2% FBS.
8. Disaggregate the tissues (spleen or lymph nodes) by grinding them on stainless steel mesh screen. Pipette the suspension up and down several times until no large clumps remained. Filter the suspension through a Falcon 70 μm cell strainer into the labeled 15 ml conical tubes to remove the remaining debris. This step is necessary for dissociation of tissues to form single-cell suspensions.
9. Centrifuge the cell suspension at 300 × *g* for 5 min and discard the supernatant.
10. Resuspend the pellet in 1 ml of 1× RBC lysis buffer for every 10^8 cells. Incubate the lysis reaction for 5 min at room temperature with occasional inversion.
11. Stop the RBC lysis reaction by directly adding 10 ml of PBS containing 2% FBS.
12. Centrifuge cell suspension at 300 × *g* for 5 min, discard the supernatant, and wash with and resuspend pellet in 5 ml of cold PBS containing 2% FBS for viable cell counting.
13. Count the living cells using trypan blue and hemocytometer. Then resuspend cells in 90 μl of MACS buffer per 10^7 total cells (will probably be around 1×10^8 cells/ml).

3.2 The Enrichment of CD4$^+$ T Cells

1. Add 10 μl of CD4 MicroBeads per 10^7 total cells into cell suspension. Transfer obtained single-cell suspension into a 15 ml conical tube. Invert the tube several times and incubate for 20 min in the refrigerator (2–8 °C).
2. Wash the cell-bead mixture with 10 ml of MACS buffer by centrifugation at 300 × *g* for 10 min. Resuspend cell-bead mixture in 1 ml of MACS buffer by gently pipetting up and down.
3. Place the MS column in the magnetic field of MACS Separator. Then mount MACS preseparation filters on top of MS columns, and rinse the MS column with 3 ml of MACS buffer.
4. Apply the entire cell-bead mixture suspension through MACS preseparation filter-MS column, and allow it to drain through the column by gravity flow. Only the magnetically labeled CD4$^+$ T cells can be retained by the column, while the unlabeled CD4$^-$ cells are passed through that column.

5. Wash the column three times with 0.5 ml MACS buffer and leave the reservoir to empty completely.
6. Gently remove the column from the magnetic field and then place it on a new 15 ml collection tube.
7. Add 1 ml MACS buffer into MS column. Immediately flush out the magnetically labeled CD4$^+$ T cells with 1 ml MACS buffer. The flow-through should be flushed through the column with pressure by using the syringe plunger. Repeat this procedure if you need to obtain higher purity of CD4$^+$ cells (*see* **Note 1**).
8. Count the number of alive CD4$^+$ T cells using trypan blue and hemocytometer. Then validate the purity of collected CD4$^+$ T cells by staining a small aliquot of collected CD4$^+$ T cells with fluorophore-labeled anti-mouse CD4$^+$ antibodies and running flow cytometry.

3.3 CFSE Division Assay on CD4$^+$ T Cells Following Stimulation with TNF

1. Dilute recombinant mouse IL-2 (or other members of the common gamma chain cytokine family, to maintain survival of in vitro cultured CD4$^+$ T cells) with or without TNF in complete RPMI 1640 medium to a final concentration of 20 ng/ml (*see* **Notes 2** and **3**). For CFSE cell division assay, dilute CFSE stock solution in PBS to give a final concentration of 5 μM.
2. Adjust concentration of CD4$^+$ T cells to 2×10^6 cells/ml. Centrifuge cells at $300 \times g$ for 5 min, and carefully discard the supernatant.
3. To label cells with CFSE, resuspend 2×10^6 CD4$^+$ T cells in 1 ml of CFSE-containing PBS, and then incubate at 37 °C for 20 min.
4. Add 2×10^5 CFSE-labeled CD4$^+$ T cells into each well of U-bottom 96-well plate.
5. Add 100 μl of RPMI 1640 medium (in the presence of IL-2 and/or TNF) into their respective wells. Then incubate the cells at humidified 37 °C, 5% CO_2 incubator for 3 days.
6. On day 3 after stimulation, collect the cells by pipetting up and down in each well, and transfer into 15 ml BD Falcon round-bottom tubes.
7. Pellet the cells by centrifugation at $300 \times g$ for 5 min and discard supernatant.
8. Wash cells twice by 1 ml PBS by centrifugation at $300 \times g$ for 5 min.
9. Fix the cells by resuspending the pellet in 300 μl of fixation/permeabilization buffer for 12–18 h at 4 °C.
10. Wash and resuspend the fixed cells in 500 μl of 1× permeabilization/wash buffer.

11. Pellet the fixed cells by centrifugation at 300 × *g* for 5 min at 4 °C and discard supernatant.
12. Wash pellet two times in 1 ml of PBS. Centrifuge again at 300 × *g* for 5 min and discard supernatant.
13. The cells are ready for immunostaining for flow cytometry-based phenotypic analysis (*see* **Note 4**).

3.4 Immunostaining and Flow Cytometry Analysis of TNF-Expanded Tregs

1. Discard all of the supernatant from tube, and then resuspend cells in 50 μl of Fc blocking buffer (anti-mouse CD16/CD32 antibodies) per tube to block Fc receptor for 10 min at 4 °C.
2. Fluorescently labeled antibodies are prepared in permeabilization/wash buffer while the samples are incubating with Fc blocking buffer (protect from light), for example, anti-CD4-PerCP-Cyanine5.5 (RM4-5), anti-Foxp3-APC (FJK-16s), as well as their corresponding isotype controls. Add fluorescently labeled antibodies at an appropriate concentration (determined by a preliminary study) for 20–30 min at 4 °C.
3. Flow data are acquired with BD FACSCanto II flow cytometer and analyzed using FlowJo software (v10; Tree Star). The proliferation of Tregs and effector T cells (Teffs) is assessed by CFSE dilution using flow cytometry, by gating on $CD4^+Foxp3^+$ cells or $CD4^+Foxp3^-$ cells (*see* **Note 4**).

4 Notes

1. It is more desirable to isolate mouse $CD4^+$ T cells with negative selection methods, such as R&D System Mouse $CD4^+$ T-Cell Enrichment Column (Cat #: MCD4C-1000 or MCD43) or Miltenyi Biotec Mouse CD4+ T-Cell Isolation Kit (Cat #:130-104-454). Resultant cells from negative selection are antibody-free and bead-free and thus may reduce experiment variations. However, the cost of experiment may be increased substantially, and CD4 T-cell yield may be less as compared with positive selection described in this protocol.
2. In vitro cultured T cells, including Treg cells, need cytokines like IL-2 to maintain their survival. To mitigate the potential role of IL-2 in the expansion of Tregs, other common gamma chain cytokines may be used to replace IL-2. For example, in the presence of IL-7, TNF could also preferentially stimulate the proliferation of Tregs [8].
3. Since $CD4^+Foxp3^-$ effector T cells (Teffs) are much more responsive to in vitro TCR stimulation than Tregs [27], and activated Teffs can produce both IL-2 and TNF (and other cytokines); it is not recommended to examine the preferential stimulatory effect of TNF on Tregs in a culture system containing anti-CD3 and anti-CD28 antibodies. Nevertheless, TNF is

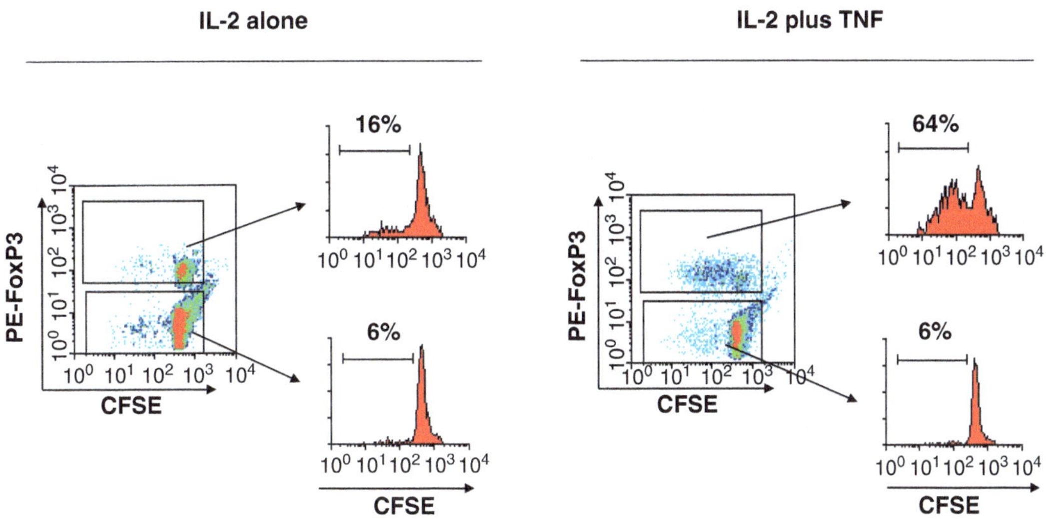

Fig. 1 $CD4^+$ T cells were purified from lymphocytes by using CD4 (L3T4) MicroBeads (Miltenyi Biotec) and MS column (Miltenyi Biotec). MACS-purified $CD4^+$ cells were labeled with CFSE, and cells (2×10^5 cells/well) were cultured in a 96-well round-bottom plate and then stimulated with IL-2 or IL-2 plus TNF for 3 days. Proliferation of Tregs was assessed by CFSE dilution assay with FACS, by gating on $Foxp3^+$ cells

able to break hyporesponsive state of Tregs to in vitro TCR stimulation, and this effect of TNF can be shown in purified Treg culture system [4].

4. In the aforementioned experiment system, Tregs and Teffs contained in unfractionated CD4 T cells are stimulated with TNF (and common gamma chain cytokine) at completely identical manner; thus the unbiased results can be observed. In this system, it can be anticipated that TNF preferentially stimulates the proliferative expansion of Tregs, as evidenced by the replication of Tregs (as shown by CFSE dilution in Fig. 1) and market increase of the proportion of Foxp3-expressing Tregs in CD4 T cells [4].

Acknowledgments

We thank Ms. Tianzhen He, Mr. Shuoyang Liu, and Ms. Jingbin Zheng at Institute of Chinese Medical Sciences, University of Macau, for the help in the preparation of manuscript.

Conflict of Interest Statement: The authors declare that the research was conducted in the absence of any commercial or financial relationships that could be construed as a potential conflict of interest. *Funding*: This work was supported by University of Macau under Grants MYRG2016-00023-ICMS-QRCM and MYRG2017-00120-ICMS and the Science and Technology Development Fund of Macao S.A.R. (FDCT) under grant 201/2017/A3 and 0056/2019/AFJ.

References

1. Chen X, Oppenheim JJ (2010) TNF-alpha: an activator of CD4+FoxP3+TNFR2+ regulatory T cells. Curr Dir Autoimmun 11:119–134
2. Vandenabeele P et al (1995) Two tumour necrosis factor receptors: structure and function. Trends Cell Biol 5(10):392–399
3. Chen X et al (2008) Cutting edge: expression of TNFR2 defines a maximally suppressive subset of mouse CD4+CD25+FoxP3+ T regulatory cells: applicability to tumor-infiltrating T regulatory cells. J Immunol 180 (10):6467–6471
4. Chen X et al (2007) Interaction of TNF with TNF receptor type 2 promotes expansion and function of mouse CD4+CD25+ T regulatory cells. J Immunol 179(1):154–161
5. Chen X et al (2010) Co-expression of TNFR2 and CD25 identifies more of the functional CD4+FOXP3+ regulatory T cells in human peripheral blood. Eur J Immunol 40 (4):1099–1106
6. Chen X et al (2013) TNFR2 is critical for the stabilization of the CD4+Foxp3+ regulatory T. cell phenotype in the inflammatory environment. J Immunol 190(3):1076–1084
7. He T et al (2018) The p38 MAPK inhibitor SB203580 abrogates tumor necrosis factor-induced proliferative expansion of mouse CD4(+)Foxp3(+) regulatory T cells. Front Immunol 9:1556
8. Hamano R et al (2011) TNF optimally activatives regulatory T cells by inducing TNF receptor superfamily members TNFR2, 4-1BB and OX40. Eur J Immunol 41(7):2010–2020
9. Chen X et al (2015) IKKalpha is required for the homeostasis of regulatory T cells and for the expansion of both regulatory and effector CD4 T cells. FASEB J 29(2):443–454
10. Zou H et al (2018) Modulation of regulatory T cell activity by TNF receptor type II-targeting pharmacological agents. Front Immunol 9:594
11. Shaikh F et al (2018) TNF receptor type II as an emerging drug target for the treatment of cancer, autoimmune diseases, and graft-versus-host disease: current perspectives and in silico search for small molecule binders. Front Immunol 9:1382
12. Nie Y et al (2018) Blockade of TNFR2 signaling enhances the immunotherapeutic effect of CpG ODN in a mouse model of colon cancer. Sci Signal 11(511):eaan0790
13. Chen X, Oppenheim JJ (2017) Targeting TNFR2, an immune checkpoint stimulator and oncoprotein, is a promising treatment for cancer. Sci Signal 10(462):eaal2328
14. Okubo Y et al (2013) Homogeneous expansion of human T-regulatory cells via tumor necrosis factor receptor 2. Sci Rep 3:3153
15. Nguyen DX, Ehrenstein MR (2016) Anti-TNF drives regulatory T cell expansion by paradoxically promoting membrane TNF-TNF-RII binding in rheumatoid arthritis. J Exp Med 213(7):1241–1253
16. Torrey H et al (2017) Targeting TNFR2 with antagonistic antibodies inhibits proliferation of ovarian cancer cells and tumor-associated Tregs. Sci Signal 10(462):eaaf8608
17. Grinberg-Bleyer Y et al (2010) Pathogenic T cells have a paradoxical protective effect in murine autoimmune diabetes by boosting Tregs. J Clin Invest 120(12):4558–4568
18. Housley WJ et al (2011) Natural but not inducible regulatory T cells require TNF-alpha signaling for in vivo function. J Immunol 186(12):6779–6787
19. Chopra M et al (2016) Exogenous TNFR2 activation protects from acute GvHD via host T reg cell expansion. J Exp Med 213 (9):1881–1900
20. Chopra M et al (2013) Tumor necrosis factor receptor 2-dependent homeostasis of regulatory T cells as a player in TNF-induced experimental metastasis. Carcinogenesis 34 (6):1296–1303
21. Mahmud SA et al (2014) Costimulation via the tumor-necrosis factor receptor superfamily couples TCR signal strength to the thymic differentiation of regulatory T cells. Nat Immunol 15(5):473–481
22. Leclerc M et al (2016) Control of GVHD by regulatory T cells depends on TNF produced by T cells and TNFR2 expressed by regulatory T cells. Blood 128(12):1651–1659
23. Pierini A et al (2016) TNF-alpha priming enhances CD4+FoxP3+ regulatory T-cell suppressive function in murine GVHD prevention and treatment. Blood 128(6):866–871
24. Atretkhany KN et al (2018) Intrinsic TNFR2 signaling in T regulatory cells provides protection in CNS autoimmunity. Proc Natl Acad Sci U S A 115(51):13051–13056
25. Zaragoza B et al (2016) Suppressive activity of human regulatory T cells is maintained in the presence of TNF. Nat Med 22(1):16–17
26. Valencia X et al (2006) TNF downmodulates the function of human CD4+CD25hi T-regulatory cells. Blood 108(1):253–261
27. Itoh M et al (1999) Thymus and autoimmunity: production of CD25+CD4+ naturally anergic and suppressive T cells as a key function of the thymus in maintaining immunologic self-tolerance. J Immunol 162(9):5317–5326

Chapter 7

In Vitro Differentiation of CD4$^+$ T Cell Effector and Regulatory Subsets

Jaclyn R. Espinosa, Joshua D. Wheaton, and Maria Ciofani

Abstract

In vitro differentiation of naïve CD4$^+$ T cells into effector and regulatory subsets offers a means to acquire large numbers of relatively homogeneous cell populations for experimentation. However, culture systems for T cell differentiation described in the literature vary widely in efficiency and complexity, limiting their comparison across studies. Here, we present a standardized and robust method for the isolation and in vitro differentiation of six CD4$^+$ T cell subsets from mouse naïve T cells.

Key words CD4 T cells, In vitro differentiation, Polarization, Helper T cells, Cytokines, Th1, Th2, Th17, iTreg

1 Introduction

Mammals are exposed to an extreme diversity of pathogenic microbes, necessitating a variety of distinct immunological strategies for effective host defense. The quality and magnitude of the adaptive immune response are controlled by CD4-expressing T lymphocytes (CD4$^+$ T cells), which differentiate into effector subsets specialized for particular classes of pathogen. Differentiation begins when a naïve CD4$^+$ T cell becomes activated by exposure to its cognate antigen on an antigen-presenting cell [1]. Subsequent exposure to environmental cues—such as inflammatory cytokines—induces broad transcriptional reprogramming to one of several distinct effector subsets characterized by the expression of both a specific lineage-defining transcription factor and signature cytokine(s) [2, 3].

Naïve CD4$^+$ T cells differentiate into specialized T-helper subsets, four of which have been particularly well-defined. T-helper type-1 (Th1) cells express the lineage-defining transcription factor

Jaclyn R. Espinosa and Joshua D. Wheaton contributed equally to this work.

Chaohong Liu (ed.), *T-Cell Receptor Signaling: Methods and Protocols*, Methods in Molecular Biology, vol. 2111, https://doi.org/10.1007/978-1-0716-0266-9_7,

T-bet (*Tbx21*) and the cytokine interferon-γ (IFNγ) and are specialized for defense against intracellular viral and bacterial pathogens [4, 5]. Th2 cells express GATA-3 and the cytokines interleukin (IL)-4 and IL-13; these cells provide protective immunity against helminths but are also strongly implicated in allergic immunity and asthma [6, 7]. All Th17 cells express RORγt and the cytokines IL-17A and IL-17F and mediate immunity to extracellular bacteria and fungi [8]; however, an additional subclass of "pathogenic" Th17 cells (pTh17) has unique differentiation requirements and is thought to be responsible for many autoimmune conditions [9, 10]. Unlike the above subsets, regulatory T (Treg) cells serve to dampen immune responses and are characterized by the lineage-defining transcription factor Foxp3 [11, 12]. Lastly, an additional subset of Th0 cells can be used as a control for activated, non-polarized $CD4^+$ T cells in vitro.

Here, we provide a method for robust differentiation and culture of the six $CD4^+$ T cell subsets described above from mouse naïve T cells. Cells cultured using this protocol expand rapidly, exhibit high activation rates, and efficiently express both signature cytokines and transcription factors. Our methods have been successfully used to generate cells for both functional and genomic assays, resulting in insight that is well-correlated with in vivo function [13, 14]. Thus, these methods are suitable for a wide range of potential assays investigating the biology of $CD4^+$ T cells.

2 Materials

2.1 Coating Plates for T Cell Activation

1. Sterile Dulbecco's phosphate-buffered saline (DPBS) without calcium or magnesium.
2. Tissue-culture-treated plates (*see* **Note 1**).
3. Affinity-purified polyclonal anti-hamster IgG (*see* **Note 2**).

2.2 Lymphocyte Harvest

1. Complete culture medium (*see* **Note 3**):
   (a) Complete RPMI: RPMI 1640 medium with L-glutamine supplemented with 10 mM HEPES, 2 mM glutamine, 1 mM sodium pyruvate, 100 U/mL penicillin-streptomycin, 50 μg/mL gentamicin, and 10% heat-inactivated fetal bovine serum (FBS) (*see* **Note 4**).
   (b) Complete IMDM: Iscove's Modified Dulbecco's Medium supplemented with 2 mM glutamine, 100 U/mL penicillin-streptomycin, 50 μg/mL gentamicin, and 10% heat-inactivated FBS.
2. Sterile 40 μm nylon cell strainer (Falcon no. 352340 or similar).

3. Sterile 6 well plate.
4. 3 mL Sterile syringe; one per condition.
5. Stainless steel forceps. Clean instruments may be autoclaved or treated with 70% ethanol prior to use.
6. Stainless steel surgical scissors.
7. Ammonium-chloride-potassium (ACK) RBC Lysis Buffer, sterile, tissue culture grade.
8. 15 mL Sterile polypropylene conical tubes; one per condition.
9. Ice.
10. 5 mL Sterile serological pipettes.

2.3 Magnetic Selection

1. MACS buffer (DPBS without calcium or magnesium (pH 7.2), 0.5% bovine serum albumin (BSA), and 2 mM EDTA).
2. Mouse CD4 T cell magnetic negative selection kit (Invitrogen MagniSort or similar).
3. Compatible magnet for magnetic selection kit; one per condition.
4. 48 μm Nylon mesh squares, 2.5 cm × 2.5 cm (sterile); one per condition (*see* **Note 5**).
5. 5 mL Sterile capped polystyrene FACS tube; two per condition.
6. 15 mL Sterile polypropylene conical tubes; one per condition.

2.4 Naïve T Cell FACS Sort

1. Sterile DPBS, without calcium or magnesium.
2. Fluorophore-conjugated antibodies: anti-CD25 (clone PC61), anti-CD4 (clone RM4-5), anti-CD44 (clone IM7), anti-CD62L (clone MEL-14).
3. Viability dye.
4. MACS buffer (*see* Subheading 2.3, **item 1**).
5. 15 mL Sterile polypropylene conical tubes; one per condition.

2.5 T Cell Plating and Activation/Polarization

1. Room-temperature DPBS without calcium or magnesium.
2. Complete RPMI or IMDM (*see* Subheading 2.2, **item 1** and **Note 6**).
3. Activating antibodies: functional-grade anti-CD3 (clone 17A2) and anti-CD28 (clone 37.51).
4. Cytokine neutralizing antibodies: functional-grade anti-IFNγ (clone XMG1.2) and/or anti-IL-4 (clone 11B11) that limits polarization to alternative effector fates.
5. Polarizing cytokines: *see* Table 2 (*see* **Note 7**).

Table 1
Volumes and seeding numbers for T cell culture

Plate size	Anti-hamster IgG coating volume (μL)	Final culture volume (mL)	No. of cells to seed (for a 72 h culture)
12 well	450	2	3×10^5
24 well	225	0.8–1	1.2×10^5
48 well	105	0.4	5×10^4
96 well	50	0.2	2×10^4

3 Methods

Carry out all procedures involving sterile solutions and cell suspensions using aseptic technique in a class II biological safety cabinet:

3.1 Coating Plates for T Cell Activation

1. Prepare the anti-hamster IgG coating solution at 25 μg/mL in DPBS (*see* **Note 8**). For recommended volumes, refer to Table 1.
2. Add the appropriate volume of diluted anti-hamster IgG/DPBS to the wells of the plate. Swirl to ensure even coating.
3. Place plates in a 37 °C tissue culture incubator for at least 4 h. Alternatively, plates may be coated by overnight incubation in a 4 °C refrigerator.

3.2 Lymphocyte Harvest

1. Prepare for secondary lymphoid tissue harvest by placing a 40 μm cell strainer in one well of a 6 well plate. Add 5 mL complete media to the strainer, and place the plate on ice.
2. Sacrifice a mouse via carbon dioxide inhalation or another IACUC-approved method.
3. Flay the mouse, keeping the peritoneum intact, and isolate the cervical, brachial, axial, and inguinal lymph nodes (LN). Combine these into the media within the cell strainer.
4. Open the peritoneum, and isolate the mesenteric and lumbar LNs, as well as the spleen. Combine these into the cell strainer (*see* **Note 9**).
5. Use the plunger from a 3 mL syringe (rubber side) to first grind the LNs, followed by the spleen, through the cell strainer into the media (*see* **Note 10**).
6. Using a 5 mL serological pipette, resuspend and collect the flow-through from the strainer, and run it through a fresh 48 μm nylon mesh square into a 15 mL conical tube. Rinse the strainer with 1 mL of MACS buffer, collect the flow-

through, and run it through the same filter into the 15 mL conical tube.

7. Spin to pellet the cells at 400 × *g* for 5 min at 4 °C.
8. Aspirate the supernatant, and resuspend the cell pellet in 1 mL ACK lysis buffer. Incubate for 1–2 min at room temperature.
9. Add 10 mL cold MACS buffer to the 15 mL conical, and invert to dilute the lysis buffer.
10. Spin to pellet cells again as above.
11. Proceed to magnetic selection.

3.3 Magnetic Selection (See Note 11)

1. Resuspend the cells in 225 μL of MACS buffer *per mouse* (i.e., 450 μL for 2 mice), and filter through a 48 μm nylon mesh square into a fresh, sterile 5 mL FACS tube.
2. Rinse the conical tube with another 225 μL of MACS buffer *per mouse*, and combine this through the 48 μm nylon mesh filter into the FACS tube. The total volume of cells plus buffer should now be approximately 500 μL. Remove and discard the mesh filter.
3. Add 100 μL of antibody cocktail *per mouse* from the Invitrogen CD4 T cell selection kit. This mixture contains antibodies that are selective for non-CD4+ T cell lineages that are targeted for depletion.
4. Pulse vortex five times at a level that allows complete mixing without allowing the solution into the cap of the tube.
5. Incubate for 10 min at room temperature for antibody binding.
6. Add MACS buffer to the cells in the FACS tube to increase the total volume to 4 mL. Cap and invert several times to wash.
7. Spin to pellet the cells at 400 × *g* for 5 min at 4 °C.
8. Aspirate the supernatant, and resuspend the cell pellet in 450 μL of MACS buffer *per mouse*. If there are clumps, either remove them carefully with a pipette or refilter (as above) into a fresh 5 mL FACS tube.
9. Thoroughly resuspend the magnetic selection beads from the Invitrogen selection kit using a 1 mL pipette.
10. Add 100 μL *per mouse* of the resuspended selection beads to the cell suspension. Pulse vortex five times at a level that allows complete mixing without allowing the solution into the cap of the tube.
11. Incubate for 5 min at room temperature to allow beads to capture antibody-bound cells.
12. Increase the volume to 2.5 mL by adding MACS buffer. Mix well by pipetting. Do not vortex.

13. Insert the FACS tube containing the bead and cell suspension into the selection magnet.
14. Incubate for 5 min at room temperature. During this time, bead-bound cells will be drawn to the sides of the tube, leaving the $CD4^+$ T cell-enriched population in solution.
15. Without removing the tube from the magnet, in one motion, carefully pour off the cell suspension into a clean 15 mL conical tube. Place the cells on ice. This is the $CD4^+$ T cell-enriched fraction.
16. Remove the tube from the magnet and repeat **steps 12–15**, combining both decanted $CD4^+$ T cell fractions into the same 15 mL tube.
17. Proceed to FACS sort procedure.

3.4 Naïve T Cell FACS Sort

1. Spin down the enriched $CD4^+$ T cells to pellet at 400 × *g* for 5 min at 4 °C.
2. During the spin, prepare the staining solution:
   (a) Allow 500 μL of MACS buffer or DPBS per mouse, with antibodies and viability dye added using predetermined optimal dilutions (*see* **Note 12**).
3. Aspirate the supernatant, and resuspend each sample in 500 μL of the antibody staining solution.
4. Incubate on ice for 20 min.
5. Wash the cells with 10 mL of MACS buffer, and spin to pellet at 400 × *g* for 5 min at 4 °C.
6. Resuspend the cells in 350 μL MACS buffer, and filter the cells through a 48 μm nylon mesh square into a fresh, sterile 5 mL FACS tube for sorting.
7. To prepare the collection tubes, add 350 μL of heat-inactivated FBS to each fresh, sterile 5 mL FACS tube. Cap and vortex on high first with the tube right side up and then with the tube upside down to coat the sides with FBS.
8. Sort live, $CD4^+CD25^-CD44^-CD62L^+$ cells to obtain naïve $CD4^+$ T cells for culture and polarization. *See* Fig. 1 for a typical gating strategy.

3.5 T Cell Plating for Activation and Polarization

1. Prepare cells in complete medium at an appropriate concentration for the plate and well size being used. Adjust the concentration so that the full number of cells can be added in half the final culture volume (e.g., 5×10^4 cells in 200 μL for a 48 well plate).
2. Prepare a 2× concentrated antibody and cytokine mixture in complete medium according to Table 2 (*see* **Note 13**).

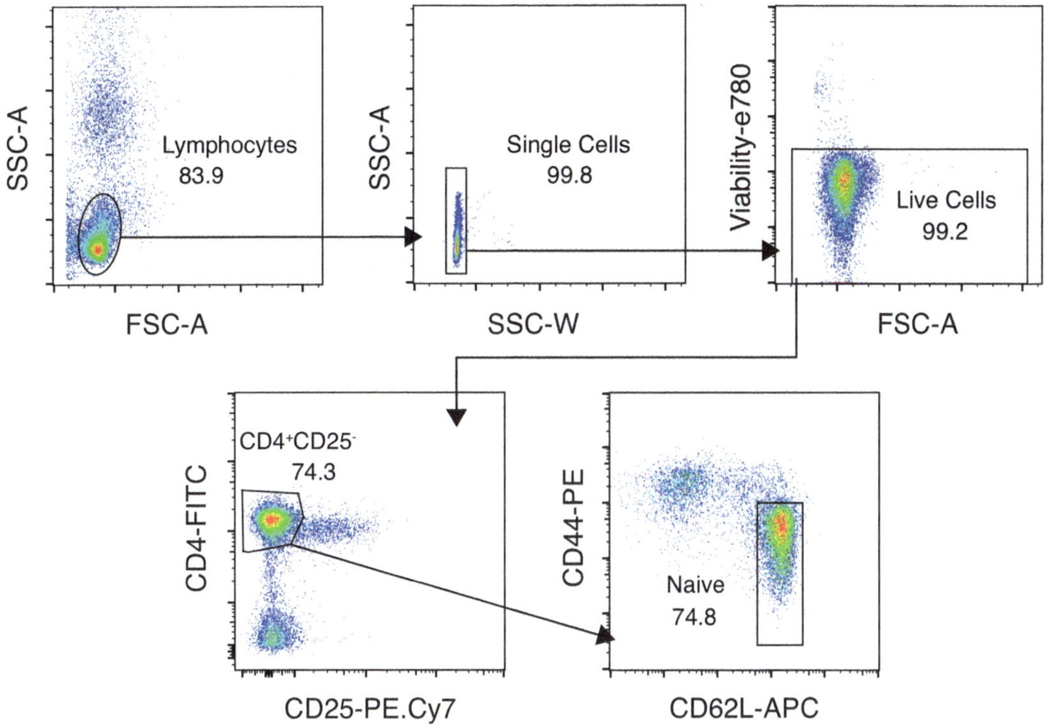

Fig. 1 Gating strategy for sorting naïve CD4$^+$ T cells. Lymphocytes are gated based on forward and side scatter; then single cells are gated based on side scatter width and forward scatter area. Cells should then be gated as CD4 positive and CD25 negative to exclude Tregs. Finally, CD44 negative cells and CD62L positive cells are gated on to select for naïve CD4$^+$ T cells

Table 2
Final concentrations of antibodies and cytokines for T-helper polarization media (*See* Note 7)

Antibody/ cytokine	Th0	Th1	Th2	Th17	pTh17	iTreg
Anti-CD3	0.25 μg/mL	0.25 μg/mL	0.25 μg/mL	0.25 μg/mL	0.25 μg/mL	0.25 μg/mL
Anti-CD28	1 μg/mL	1 μg/mL	1 μg/mL	1 μg/mL	1 μg/mL	1 μg/mL
Anti-IFNγ	2 μg/mL		2 μg/mL	2 μg/mL	2 μg/mL	2 μg/mL
Anti-IL-4	2 μg/mL	2 μg/mL		2 μg/mL	2 μg/mL	2 μg/mL
IL-2	50 U/mL	50 U/mL	50 U/mL	–	–	50 U/mL
IL-1β	–	–	–	–	20 ng/mL	–
IL-4	–	–	2 ng/mL	–	–	–
IL-6	–	–	–	10 ng/mL	–	–
IL-12	–	10 ng/mL	–	–	–	–
IL-23	–	–	–	–	25 ng/mL	–
TGFβ1	–	–	–	0.3 ng/mL	–	5 ng/mL

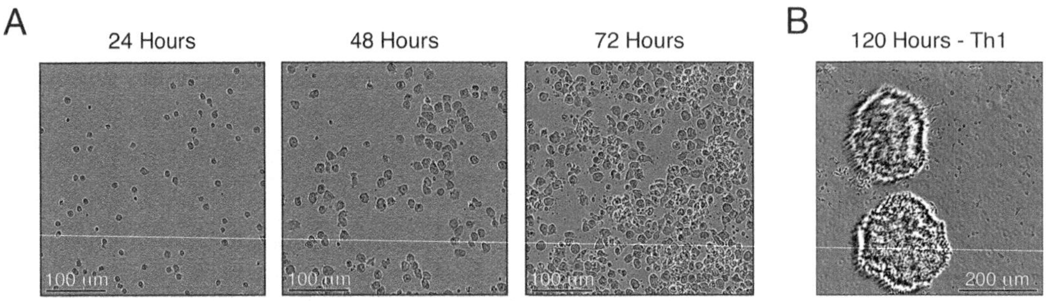

Fig. 2 Changes in cellular morphology during activation. (**a**) Cells uniformly begin blasting by 24 h post activation and continue to enlarge until 48 h post activation. Between 48 and 72 h, blasting cells then undergo substantial proliferation that will continue for several days. Th0 conditions are shown but similarly apply to other subsets. Failure to observe these morphological changes is an early indicator that cells have not been activated appropriately. (**b**) Th1 cells at 120 h post activation; 72 h of TCR stimulation followed by an additional 48 h off of TCR stimulation. Th1 cells uniquely form large clusters following removal from TCR, as depicted here

3. Aspirate the anti-hamster coating solution, and wash plates twice with a volume of room-temperature DPBS or MACS buffer that is greater than the original coating volume. Swirl to wash, and ensure that plates do not dry between washes.
4. Add cells to the appropriate wells.
5. Add the 2× concentrated antibody and cytokine mixture to wells. Tilt the plate back and forth gently to mix.
6. Place the culture plate into a humidified 37 °C incubator under 5% carbon dioxide.
7. Monitor the cells daily by visualizing cultures under a light microscope to confirm activation and differentiation. *See* Fig. 2 for a reference.
8. For cultures propagated greater than 3 days, split the cells at 72 h in the following split ratios for optimal performance (*see* **Note 14**):
   (a) Th0—replate all.
   (b) Th1—1:3 to 1:4.
   (c) Th2—1:4 to 1:8.
   (d) Th17—1:6 to 1:8.
   (e) pTh17—1:6 to 1:8.
   (f) iTreg—1:3 to 1:4.
9. Th0, Th17, pTh17, and iTreg cells should be replated in the presence of anti-CD3, anti-CD28, and polarizing cytokines, without blocking antibodies. Th1 and Th2 cells should be replated in the presence of polarizing cytokines, without anti-CD3, anti-CD28, or blocking antibodies (*see* **Notes 15–17**).

4 Notes

1. Different manufacturers should be tested as certain plastics have shown to be ill-suited for T cell activation under these conditions described in this protocol.
2. The quality of anti-hamster antibody is essential due to lot-to-lot variability. We recommend testing multiple lots and buying in bulk. The optimum coating dilution should be determined empirically for that which supports T cell activation (survival and proliferation) and differentiation.
3. Appropriate selection of FBS is critical to replicate these results. Several FBS lots should be evaluated for performance (using the metrics of activation and differentiation) prior to selection of FBS. For example, some lots of FBS induce low-level Treg differentiation under Th0 conditions, likely due to above-average levels of TGFβ (unpublished observations).
4. Heat inactivation of FBS (heating to 56 °C for 30 min) is necessary to inactivate complement proteins that may be present in the serum, thus preventing complement-mediated lysis of cells in the presence of culture antibodies.
5. 48 μm Nylon mesh filters are cut into 2.5 cm × 2.5 cm^2 from 12 in. × 12 in. sheets and sterilized by autoclaving.
6. We recommend RPMI for Th1, Th2, and iTreg cultures and IMDM for Th17 and pTh17 cultures. Th0 cultures should be activated in the corresponding media as the subsets for which they are serving as a control. All subsets can differentiate in either medium, and optimal conditions should be tested and verified.
7. Either human or mouse IL-2 maybe used in this protocol, whereas human TGFβ1 was found to be more effective in polarization compared to mouse. For all other cytokines, only the mouse homologs have been validated.
8. The optimal coating concentration of anti-hamster IgG varies from lot to lot and should be determined empirically.
9. You can combine LNs and the spleen from up to two mice per strainer, if needed.
10. Lifting the lip of the strainer onto the side of the well to provide a flat surface above the liquid level prevents one from having to "chase" the submerged lymph nodes in the media when processing cells.
11. Although the Invitrogen MagniSort kit is described here, there are several other similar selection kits that are compatible with this protocol, including column-based selection protocols.

Please refer to the manufacturer's protocol for magnetic selection if using other kits.

12. Antibodies should be titrated to achieve optimal resolution. This is specific to the fluorophores selected and cell sorting instrument used. The antibody master mix should be prepared using MACS if protein-insensitive viability dyes (i.e., DAPI, 7AAD, etc.) are used. DPBS is only used if the viability dye is not compatible with protein-containing buffers.
13. Addition of IL-2 is optional in Th0 polarization conditions; however, it supports the proliferation and longevity of the cultures.
14. Th0, Th17, pTh17, and iTreg cells should be split into freshly coated anti-hamster wells as described in Subheading 3.1. Th1 and Th2 cells should be split into uncoated wells in the absence of TCR stimulation.
15. Remove Th1 and Th2 cells from TCR stimulation at 72 h to facilitate further differentiation and cytokine production [15, 16]. At this point, Th1 culture plates may be slightly raised on one side to culture wells at an angle. This will facilitate the formation of cells into clusters that maximize contacts and full differentiation. Conversely, Th2 cells should be closely monitored to avoid overcrowding. This is essential for maximal cytokine production.
16. Following differentiation, you should observe the following cytokine production profiles (Fig. 3):
    (a) Th0—no cytokine expression.
    (b) Th1—50–80% IFNγ expression at 120 h post activation.
    (c) Th2—20–40% IL-4 expression at 120 h post activation.
    (d) Th17—60–80% IL-17A expression at 72 h post activation.

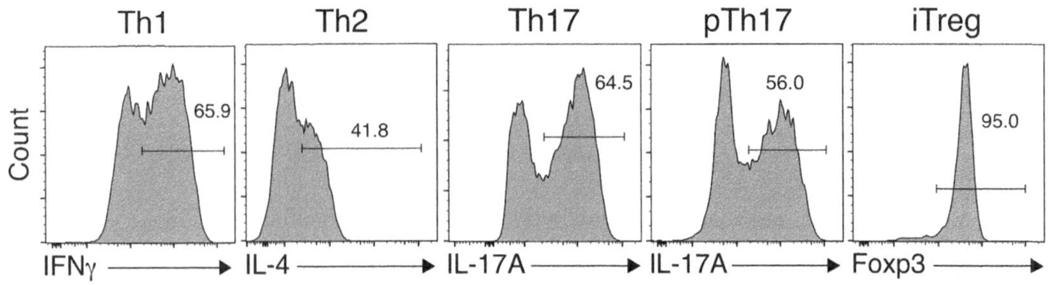

Fig. 3 Typical cytokine profiles of differentiated cells. Representative histograms showing expected cytokine production profiles at 120 h post activation from Th1 and Th2 conditions, 96 h post activation from pTh17 conditions, and 72 h post activation from Th17 and iTreg conditions. Cells were stimulated for 4 h with phorbol ester and ionomycin in the presence of monensin, followed by intracellular staining and flow cytometry

(e) pTh17—50–70% IL-17A expression at 96 h post activation,

(f) iTreg—80–100% Foxp3 expression at 72 h post activation,

17. Th2 cells should be fixed with a final concentration of 2% PFA and permeabilized with a final concentration of 0.5% saponin to achieve optimal intracellular cytokine staining for IL-4. It should be noted that transcription factor staining cannot be done with this method of fixation and permeabilization.

Acknowledgments

This work was supported by NIH grant R01 GM115474 to M.C.

References

1. Jelley-Gibbs DM et al (2000) Two distinct stages in the transition from naive CD4 T cells to effectors, early antigen-dependent and late cytokine-driven expansion and differentiation. J Immunol 165(9):5017–5026
2. Murphy KM, Reiner SL (2002) The lineage decisions of helper T cells. Nat Rev Immunol 2(12):933–944
3. Wilson CB, Rowell E, Sekimata M (2009) Epigenetic control of T-helper-cell differentiation. Nat Rev Immunol 9(2):91–105
4. Hsieh CS et al (1993) Development of TH1 CD4+ T cells through IL-12 produced by listeria-induced macrophages. Science 260 (5107):547–549
5. Lighvani AA et al (2001) T-bet is rapidly induced by interferon-gamma in lymphoid and myeloid cells. Proc Natl Acad Sci U S A 98(26):15137–15142
6. Mosmann TR et al (1986) Two types of murine helper T cell clone. I. Definition according to profiles of lymphokine activities and secreted proteins. J Immunol 136 (7):2348–2357
7. Kanhere A et al (2012) T-bet and GATA3 orchestrate Th1 and Th2 differentiation through lineage-specific targeting of distal regulatory elements. Nat Commun 3:1268
8. Bettelli E et al (2008) Induction and effector functions of T(H)17 cells. Nature 453 (7198):1051–1057
9. Lee Y et al (2012) Induction and molecular signature of pathogenic TH17 cells. Nat Immunol 13(10):991–999
10. Ghoreschi K et al (2010) Generation of pathogenic T(H)17 cells in the absence of TGF-beta signalling. Nature 467(7318):967–971
11. Ohkura N, Kitagawa Y, Sakaguchi S (2013) Development and maintenance of regulatory T cells. Immunity 38(3):414–423
12. Schmitt EG, Williams CB (2013) Generation and function of induced regulatory T cells. Front Immunol 4:152
13. Ciofani M et al (2012) A validated regulatory network for Th17 cell specification. Cell 151 (2):289–303
14. Carr TM et al (2017) JunB promotes Th17 cell identity and restrains alternative CD4+ T-cell programs during inflammation. Nat Commun 8(1):301
15. Agarwal S, Rao A (1998) Modulation of chromatin structure regulates cytokine gene expression during T cell differentiation. Immunity 9 (6):765–775
16. Bandukwala HS et al (2012) Selective inhibition of CD4+ T-cell cytokine production and autoimmunity by BET protein and c-Myc inhibitors. Proc Natl Acad Sci U S A 109 (36):14532–14537

Chapter 8

CD4$^+$ T-Cell Differentiation In Vitro

Wenyong Yang, Xueying Chen, and Hongbo Hu

Abstract

CD4$^+$ T helper cells play crucial roles in adaptive immune response against pathogens, as well as in host immune homeostasis. Upon TCR activation, naïve CD4$^+$ T cells differentiate into one of several lineages of Th cells, with hallmark transcription factors, cytokine production, and functions in vivo, according to the particular cytokine milieu. To study the regulating mechanism and function of Th cells, in vitro CD4$^+$ T-cell differentiation is crucial. The following protocols describe the methods to induce naïve CD4+ T-cell differentiate into Th1, Th2, Th17 and Treg by activating TCR, together with the different cytokines and blocking antibodies in vitro. The efficiency of T helper cell differentiation is examined by detecting the expression of hallmark cytokines and transcription factors.

Key words CD4$^+$ T cells, T-cell differentiation, Cytokines, TCR activation, Flow cytometry

1 Introduction

CD4$^+$ T cells play crucial roles in adaptive immunity, including facilitating B cells for antibody production, enhancing CD8$^+$ T-cell-mediated cytotoxic T lymphocyte (CTL) response, and maintaining immune homeostasis. Upon antigen stimulation, naïve CD4$^+$ T cells proliferate and differentiate into several subsets. The cytokine milieu is the major factor to determine the commitment of T-cell differentiation. Based on their immunological functions, signature transcription factors, and cytokine production, T helper cells can be divided into different lineages, including Th1, Th2, Th9, Th17, Th22, Tfh and Treg [1].

Th1 cells are the major resource of IFN-γ in the immune response against intracellular pathogens, such as viruses and intracellular bacteria. Th1 cells are also involved in autoimmune diseases, causing tissue damage in inflammatory bowel disorders (IBD) and rheumatology arthritis (RA), etc. [2, 3]. IL-12 is the key cytokine for Th1 differentiation, and autocrine IFN-γ also plays an important role [4, 5]. IL-12 and IFN-γ promote expression of the master transcription factor of Th1 cells, T-bet, through STAT4

Chaohong Liu (ed.), *T-Cell Receptor Signaling: Methods and Protocols*, Methods in Molecular Biology, vol. 2111,
https://doi.org/10.1007/978-1-0716-0266-9_8,

and STAT1, respectively. Then T-bet further induces more IFN-γ and T-bet expression to fully promote Th1 cell development [6]. Th2 cells secret IL-4, IL-5 and IL-13 that are important for eosinophil recruitment, B-cell response, immunoglobulin E (IgE) production, as well as mucosal expulsion [7]. Th2 polarization from naïve $CD4^+$ T cells requires IL-4 that triggers JAK1-STAT6 activation and finally leads to expression of GATA3, a master regulator of Th2 cells [8, 9]. The IFN-γ produced by Th1 represses IL-4-mediated Th2 polarization strongly, whereas IL-4 produced by Th2 suppresses IL-12/IFN-γ-driven Th1 differentiation [10]. Th17 cells, which produce IL-17 family cytokines, are responsible for immune response against extracellular pathogen and mucosal homeostasis. Dysregulated Th17 function also causes autoimmunity, such as multiple sclerosis, IBD and psoriasis [11]. A unique transcription factor expressed by Th17 cells is retinoic acid-related orphan receptor γ t (RORγt) [12]. Transforming growth factor β (TGF-β) and IL-6 as well as IL-1β are important for Th17 differentiation from naïve $CD4^+$ T cells [13, 14]. In addition, IL-23 is another key cytokine in maturation and maintenance of Th17 lineage [15]. Regulatory T cell (Treg) is a unique T helper subset to suppress immune response, maintaining the self-tolerance and homeostasis [16]. Treg can naturally occur in the thymus, called nTreg, or be induced from naïve $CD4^+$ T cells in the periphery, called iTreg [17, 18]. The suppressive mechanisms of Treg include secreting inhibitory cytokines, such as IL-10, IL-35 and TGF-β, metabolic disruption (such as IL-2 consumption), killing effector T cells, and inhibition of dendritic cell maturation and function [19]. TGF-β induces Treg differentiation from naïve $CD4^+$ T cells by employing Smad-dependent signaling pathway. The key transcription factor of Treg is forkhead box P3 (Foxp3), which is required for Treg differentiation, maintenance and function [20, 21].

Despite the in vivo models have been developed to study the function of Th cells in immunity response, in vitro T-cell differentiation system is irredundant for investigating the molecular mechanism of Th differentiation. In this protocol, we will describe the methods of Th1, Th2, Th17and Treg differentiation from naïve $CD4^+$ T cells in vitro. The in vitro differentiation of Th9, Th22 and Tfh has not yet been established.

2 Materials

2.1 Animals

6–8 weeks old C57BL/6 mice are from Model Animal Research Centre of Nanjing University, China.

2.2 Materials for Naive $CD4^+$ T-Cell Sorting

1. Naive $CD4^+$ T-Cell Isolation Kit (Cat: 130-104-453, Miltenyi), MACS LS column (Cat: 130-042-401, Miltenyi), and MACS Separator (Cat: 33743, Miltenyi).

2. Isolation buffer: PBS, pH 7.2, containing 0.5% bovine serum albumin (BSA, Cat: B2064, SIGMA) and 2 mM EDTA (Cat: E6758, SIGMA).
3. Antibodies for naïve $CD4^+$ T-cell purity test: anti-mCD4-FITC (Cat:553729, clone GK1.5, BD), anti-mCD62L-APC (Cat: 104412, clone MEL-14, Biolegend), and anti-m/hCD44-PE (Cat: 103008, clone IM7, Biolegend).

2.3 Materials for Plate Coating and Cell Culture

1. 48-well plate (Cat: 150687, Thermo Nunc).
2. Anti-mouse CD3ε (Cat: 16-0031, clone 145-2C11, eBioscience).
3. DPBS/MODIFIED (Cat: SH30028.02, Hyclone).
4. Culture medium: RPMI medium (Cat:SH30809.01, Hyclone) with 10% FBS and 55 μM β-mercaptoethanol (Cat: 21985-023, Gibco).

2.4 Materials for T-Cell Differentiation and Examination

1. Antibodies

   Anti-mouse CD28 (Cat: 16-0281, clone 37.51, eBioscience), anti-mouse IFN-γ (Cat: 14-7311-85, clone XMG1.2, eBioscience), and anti-mouse IL-4 (Cat: 14-7041-85, clone 11B11, eBioscience). Anti-mouse CD4-APC (Cat:100412, clone GK1.5, eBioscience), anti-mouse IFN-γ FITC (Cat: 11-7311-82, clone XMG1.2, eBioscience), anti-mouse IL-4 PE (Cat: E02067-1633, clone 11B11, eBioscience), anti-mouse/rat IL-17A PE (Cat: 4276914, clone eBio17B7, eBioscience), and anti-mouse/rat Foxp3 PE-Cy7 (Cat: 25-5773-82, clone FJK-16S, eBioscience).
2. Cytokines

   Recombinant murine IL-2 (Cat: AF-212-12, Peprotech), recombinant murine IL-4 (Cat: AF-214-14, Peprotech), and recombinant murine IL-6 (Cat: AF-216-16, Peprotech). Recombinant murine IL-12 (Cat: CM39, Novoprotein) and recombinant mouse TGF-β1 (Cat: CK33, Novoprotein). Recombinant murine IL-1β (Cat: 211-11B, Peprotech) and recombinant murine IL-23 (Cat: 1887-ML, R&D Systems).
3. Reagents

   Phorbol 12-myristate 13-acetate ((PMA, Cat: P1585, SIGMA) stock concentration is 50 μg/mL, and final concentration is 50 ng/mL), inomycine ((Cat: 10634, SIGMA) stock concentration is 1 mg/mL, and final concentration is 1 μg/mL), and Monensin ((Cat: 00-4505-51, 1000×, eBioscience) final concentration is 2 μM).
4. Intracellular staining kits

   Intracellular cytokine staining kit (fixation/permeabilization kit, Cat: 554714, BD) and Foxp3-staining buffer kit (Cat: 00-5523-00, Invitrogen).

3 Methods

3.1 Isolation of Naïve $CD4^+$ T Cells

Miltenyi Naïve $CD4^+$ T-Cell Isolation Kit is used to isolate naïve $CD4^+$ T cells that are $CD4^+CD25^-CD44^{low}CD62L^{high}$.

1. Mouse is sacrificed by cervical dislocation, and whole body is rinsed by 75% ethanol.
2. Fix the limbs of mouse in supine position by sterile dissection pins or needles. Cut the skin longitudinally from inferior belly to chin, avoiding puncture of the peritoneal cavity. Separate skin by forceps, and pin the skin. Take lymph nodes without fat tissue by forceps. Then cut the peritoneum, and the spleen is on the left side. Take the spleen by forceps.
3. Put lymph nodes and spleen into 35 mm tissue culture dish, with 2 mL PBS with 1% FBS. Grind lymph nodes and spleen into single-cell suspension by pressing with a plunger of syringe on 40 μm filter.
4. Collect and centrifuge cells at 400 × *g* for 5 min, then discard supernatant.
5. Resuspend the cell pellet with 2 mL red blood cell lysis buffer, and incubate at room temperature for 5 min (*see* **Note 1**).
6. Stop cell lysis by adding 8 mL PBS with 1% FBS, then filter cells with 40 μm filter, and count cell number.
7. Centrifuge cells at 400 × *g* for 5 min, then discard supernatant.
8. Resuspend pellet in 400 μL isolation buffer per 10^8 total cells.
9. Add 100 μL of biotin-antibody cocktail (include anti-CD8a, anti-CD11b, anti-CD11c, anti-CD19, anti-CD25, anti-CD45R (B220), anti-CD49b (DX5), anti-CD105, anti-MHC Class II, anti-Ter-119, and anti-TCRγ/δ antibodies) per 10^8 total cells. Mix cells and antibody cocktail, then incubate for 5 min on ice.
10. Add additional 200 μL isolation buffer per 10^8 total cells.
11. Add 200 μL Anti-Biotin MicroBeads and 100 μL CD44 MicroBeads per 10^8 total cells. Mix cells and microbeads, then incubate for 10 min on ice.
12. Wash the cells with 5 mL isolation buffer and centrifuge cells at 400 × *g* for 5 min, then discard supernatant.
13. Resuspend 10^8 cells in 500 μL isolation buffer.
14. Place LS column in MACS separator, and rinse column with 3 mL isolation buffer.
15. Apply cells onto the column, and collect flow-through.

Table 1
Recommended cell concentration and culture time for Th cell differentiation

Cell subsets	Recommended cell concentration (10^6/mL)/per well (48-well plate)	Days
Th1	0.2–0.3	3
Th2	0.1–0.15	5
Th17 (TGF-β1)	0.3	5
Th17 (IL-23)	0.2–0.3	5
Treg	0.3	4

16. Wash column with 3 mL isolation buffer and collect flow-through (*see* **Note 2**). Save some cells to test the purity of isolated naive CD4+ T cells.
17. Centrifuge cells at 400 × *g* for 5 min, discard supernatant.
18. Resuspend pellet in RPMI medium with FBS and β-mercaptoethanol for further experiment.

3.2 Differentiation of Th and Treg Cells

1. Coat a well of 48-well plate with 200 μL anti-mouse CD3ε antibody (1 μg/mL) in DPBS overnight at 4 °C.
2. Aspirate DPBS from well completely, and add 1 mL resuspended naïve CD4+ T cell from **step 18**; the optimal cell concentration for Th differentiation is listed in Table 1. Add 1 μL anti-CD28 (1 μg/mL) antibody to each well.
3. Add cytokines and blocking antibodies for different Th cell differentiation accordingly (summarized in Table 2). For Th0, no cytokine or blocking antibody is added.
4. Put the plate into tissue culture incubator for 3–5 days (*see* **Note 3**).

3.3 Analysis of Differentiation of Th Cells In Vitro

Intracellular cytokine staining (ICS) and real-time quantitative PCR are used to test the differentiation of CD4+ cells (*see* **Note 4**).

3.3.1 Intracellular Staining and Flow Cytometry Analysis

Staining strategy: anti-mCD4-APC, anti-mIFN-γ FITC, and anti-mIL-4 PE are used for Th1 and Th2cells; anti-mCD4-APC, anti-mIFN-γ FITC, and anti-m/R IL-17A PE are used for Th17 cells; anti-mCD4-APC and anti-m/R Foxp3 PE-Cy7 are used for Treg cells.

1. After 3–5 days, add 1 μL PMA (50 ng/mL), 1 μL inomycine (1 μg/mL), and 1 μL monensin (2 μM, intracellular protein transport inhibitor) to the wells for Th1, Th2, and Th17

Table 2
Recommended concentration of cytokines and blocking antibodies for Th cell differentiation

Cell subsets	Cytokines or antibodies	Final concentration (ng/mL)
Th0	Anti-CD28	1000
Th1	Anti-CD28 IL-12 Anti-IL-4 IL-2	1000 10 2500 10
Th2	Anti-CD28 IL-4 Anti-IFN-γ IL-2	1000 20 5000 10
Th17 (TGF-β)	Anti-CD28 TGF-β1 IL-6 Anti-IFN-γ Anti-IL-4	1000 1 10 5000 5000
Th17 (IL-23)	Anti-CD28 IL-1β IL-6 IL-23 Anti-IFN-γ Anti-IL-4	1000 10 10 20 5000 5000
Treg	Anti-CD28 TGF-β1 IL-2 Anti-IFN-γ Anti-IL-4	1000 1 10 5000 5000

differentiation. Note, there is no need to add PMA, inomycine, nor monensin for Foxp3 staining (Treg cell differentiation).

2. Continue to culture those cells at 37 °C for 4–6 h in 5% CO_2 tissue culture incubator.
3. Harvest cells into 1.5 mL tubes, centrifuge cells at 400 × *g* for 5 min, and then discard supernatant.
4. Resuspend the cell pellet in 100 μL PBS with 1% FBS, and add 1 μL anti-mCD4-APC antibody, mix and incubate at 4 °C for 30 min in dark.
5. Wash cells with 1 mL PBS with 1% FBS, centrifuge at 400 × *g* for 5 min at 4 °C, and then discard supernatant.
6. For cytokine intracellular staining, fixation/permeabilization kit is used. Add 250 μL fixation/permeabilization solution per tube, and fix cells for 20 min at 4 °C. Wash cells with

1 mL 1× BD Perm/Wash™ buffer (dilute 10× BD Perm/Wash™ buffer with distilled H_20) twice. Resuspend the cell pellet in 100 μL 1× BD Perm/Wash™ buffer, and add antibody mix for different Th cell described above. Incubate cells at 4 °C for 30 min in dark. Wash cells with 1 mL 1× BD Perm/Wash™ buffer per tube twice. Resuspend the cell pellet in PBS with 1% FBS.

7. For Foxp3 staining of Treg cells, Foxp3-staining kit is used. Resuspend cell pellet with 500 μL 1× Foxp3 fixation/permeabilization working buffer (Foxp3 fixation/permeabilization concentrate: Foxp3 fixation/permeabilization diluent = 1:3). Incubate the cells for 30 min at 4 °C in the dark. Add 1 mL 1× permeabilization buffer (dilute 10× permeabilization buffer with distilled H_20.) to wash. Centrifuge cells at 600 × *g* for 5 min at 4 °C, and discard the supernatant. Resuspend cell pellet with 100 μL 1× permeabilization buffer, and add 1 μL anti-m/R Foxp3 PE-Cy7 antibody. Incubate the cells at 4 °C for 30 min in the dark. Wash cells with 1 mL 1× permeabilization buffer. Centrifuging cells at 600 × *g* for 5 min at 4 °C and discard the supernatant. Resuspend stained cells in PBS with 1% FBS.
8. Analyze cells using flow cytometry.

3.3.2 mRNA Extraction and Real-Time Quantitative PCR Analysis

Real-time quantitative PCR (qPCR) is used to detect the signature transcription factors and cytokines of differentiated $CD4^+$ Th cells. The mRNA levels of *Tbx21* and *Ifng* are tested for Th1 cells, while *Gata3* and *Il4* are tested for Th2 cells. *Rorc* and *Il17a* mRNA levels are tested for Th17 cells, and *Foxp3 and Tgfb1* mRNA levels are tested for Treg cells. *Rn18s* servers as reference gene (Table 3).

4 Note

1. In the process of naïve $CD4^+$ T-cell isolation, keep cells on ice all the time which will significantly increase the cell viability.
2. The purity of naïve $CD4^+$ T cells is critical for T-cell differentiation. MACS-isolated naïve $CD4^+$ T cell can be further purified as $CD4^+CD44^LCD62^HCD25^-$ by using flow cytometry-associated sorting.
3. Check T-cell proliferation every day, and add fresh culture medium with cytokines/antibodies when medium turns to yellow.
4. Th0 wells should be included as control, which only have anti-CD3/CD28 antibodies, without any cytokines nor blocking antibodies.

Table 3
Sequences of qPCR primers to detect mRNA expression of cytokines and transcription factors

Cell subsets	Genes		qPCR primers
Th1	*Tbx21*	F	5′-GCCAGGGAACCGCTTATATG-3′
		R	5′-GACGATCATCTGGGTCACATTGT-3′
	Ifng	F	5′-CAGCAACAGCAAGGCGAAA-3′
		R	5′-CTGGACCTGTGGGTTGTTGAC-3′
Th2	*Il4*	F	5′-CGCCATGCACGGAGATG-3′
		R	5′-CGAGCTCACTCTCTGTGGTGTT-3′
	Gata3	F	5′-GCCTGCGGACTCTACCATAA-3′
		R	5′-AGGATGTCCCTGCTCTCCTT-3′
Th17	*Il17a*	F	5′-AGCTGGACCACCACATGAAT-3′
		R	5′-AGCATCTTCTCGACCCTGAA-3′
	Rorc	F	5′-CAAGTCATCTGGGATCCACTAC-3′
		R	5′-TGCAGGAGTAGGCCACATTACA-3′
Treg	*Foxp3*	F	5′-TTCGAGGAGCCAGAAGAGTTTC-3′
		R	5′-GGGCCTTGCCTTTCTCATC-3′
	Tgfb1	F	5′-GCTCTTGTGACAGCAAAGATAACAA-3′
		R	5′-CGCCCCGACGTTTGG-3′
	Rn18s	F	5′-CTTAGAGGGACAAGTGGCG-3′
		R	5′-ACGCTGAGCCAGTCAGTGTA-3′

Acknowledgement

This study was supported by grant from the Ministry of Science and Technology (the National Key Research and Development Program 2016YFA0502203) and National Natural Science Foundation of China (no. 91740111 and 81871232).

References

1. Zhu J, Yamane H, Paul WE (2010) Differentiation of effector CD4 T cell populations (∗). Annu Rev Immunol 28:445–489
2. Wu W, He C, Liu C et al (2015) miR-10a inhibits dendritic cell activation and Th1/Th17 cell immune responses in IBD. Gut 64:1755–1764
3. Wang D, Liu Y, Li Y et al (2017) Galphaq regulates the development of rheumatoid arthritis by modulating Th1 differentiation. Mediat Inflamm 2017:4639081
4. Bradley LM, Dalton DK, Croft M (1996) A direct role for IFN-gamma in regulation of Th1 cell development. J Immunol 157:1350–1358
5. Hu H, Wang H, Xiao Y et al (2016) Otud7b facilitates T cell activation and inflammatory responses by regulating Zap70 ubiquitination. J Exp Med 213:399–414
6. Yamane H, Paul WE (2012) Memory CD4$^+$ T cells: fate determination, positive feedback and plasticity. Cell Mol Life Sci 69:1577–1583
7. Mowen KA, Glimcher LH (2004) Signaling pathways in Th2 development. Immunol Rev 202:203–222
8. Pulendran B, Artis D (2012) New paradigms in type 2 immunity. Science 337:431–435
9. Hirose K, Iwata A, Tamachi T et al (2017) Allergic airway inflammation: key players beyond the Th2 cell pathway. Immunol Rev 278:145–161
10. Hwang ES, Szabo SJ, Schwartzberg PL et al (2005) T helper cell fate specified by kinase-

mediated interaction of T-bet with GATA-3. Science 307:430–433

11. Yang J, Sundrud MS, Skepner J et al (2014) Targeting Th17 cells in autoimmune diseases. Trends Pharmacol Sci 35:493–500
12. Ivanov II, Mckenzie BS, Zhou L et al (2006) The orphan nuclear receptor RORgammat directs the differentiation program of proinflammatory IL-17+ T helper cells. Cell 126:1121–1133
13. Veldhoen M, Hocking RJ, Atkins CJ et al (2006) TGFbeta in the context of an inflammatory cytokine milieu supports de novo differentiation of IL-17-producing T cells. Immunity 24:179–189
14. Mangan PR, Harrington LE, O'quinn DB et al (2006) Transforming growth factor-beta induces development of the T(H)17 lineage. Nature 441:231–234
15. Mcgeachy MJ, Chen Y, Tato CM et al (2009) The interleukin 23 receptor is essential for the terminal differentiation of interleukin 17-producing effector T helper cells in vivo. Nat Immunol 10:314–324
16. Hu H, Sun SC (2016) Ubiquitin signaling in immune responses. Cell Res 26:457–483
17. Sakaguchi S, Yamaguchi T, Nomura T et al (2008) Regulatory T cells and immune tolerance. Cell 133:775–787
18. Rudensky AY (2011) Regulatory T cells and Foxp3. Immunol Rev 241:260–268
19. Vignali DA, Collison LW, Workman CJ (2008) How regulatory T cells work. Nat Rev Immunol 8:523–532
20. Fontenot JD, Gavin MA, Rudensky AY (2003) Foxp3 programs the development and function of CD4+CD25+ regulatory T cells. Nat Immunol 4:330–336
21. Hori S, Nomura T, Sakaguchi S (2003) Control of regulatory T cell development by the transcription factor Foxp3. Science 299:1057–1061

Chapter 9

Characterization of Immune Cell Subset Expansion in Response to Therapeutic Treatment in Mice

Jakub Tomala and Jamie B. Spangler

Abstract

Flow cytometry has revolutionized the field of molecular immunology, enabling the monitoring and characterization of immune events at the single-cell level. Here, we describe a flow cytometry-based workflow to quantify the activation of specific immune cell subsets in mice in response to a molecular intervention. Compared to laborious long-term disease models, this technique allows for relatively rapid evaluation of candidate therapeutics designed to elicit a targeted immune response. This approach has the range to address both disease applications in which an immunostimulatory effect would be desired (e.g., cancer, infectious disease) or those in which an immunosuppressive effect would be desired (e.g., autoimmune disorders, transplantation medicine). Overall, our technique presents a powerful and accessible strategy for preliminary in vivo assessment of potential immunotherapeutics.

Key words Immunology, T cell, Natural killer cell, Flow cytometry, Immunophenotyping, Immunotherapy

1 Introduction

Flow cytometry is a widely used and extremely versatile method for analyzing the expression of cell surface and intracellular molecules [1, 2]. This technology is used for a wide range of basic research, preclinical, and clinical applications, including studying protein interactions, measuring signal activation, and characterizing and quantifying various cell types within mixed populations [3]. In particular, the ability to discriminate between cell subsets within a heterogeneous milieu has led to transformative advances in the discipline of molecular immunology, as it affords researchers the single-cell resolution requisite to globally and locally interrogate the immune response.

The role of the immune system in infectious diseases and autoimmune disorders has long been recognized, but it is only within the past 20 years that the critical role of the immune response in cancer, cardiovascular disease, and neurodegenerative

Chaohong Liu (ed.), *T-Cell Receptor Signaling: Methods and Protocols*, Methods in Molecular Biology, vol. 2111,
https://doi.org/10.1007/978-1-0716-0266-9_9,

diseases has truly been appreciated. As our awareness and understanding of the immune system expands, so does our ability to manipulate it through targeted interventions, such as natural or engineered antibodies, cytokines, or growth factors [4, 5]. However, a major challenge in exploring the immune response and developing new immunotherapeutics is the low throughput of preclinical characterization. Although cell-based assays are useful for triaging promising compounds, in vitro studies are not necessarily representative of the complex immune environment observed in vivo. Co-culture or mixed immune population assays provide more context, but still fail to fully recapitulate important spatial and temporal features of the immune niche. Thus, in vivo models (typically murine) [6] serve as the gold standard for characterizing the performance of immunomodulatory agents; however, the requirement to employ lengthy and expensive animal disease studies for each compound would be prohibitive for the drug development process. Consequently, there is a high demand for a robust, reliable, and reasonably rapid platform to assess the effects of potential therapeutics on immune activation in vivo.

We present a relatively quick (<1 week) in vivo platform to identify the relative activation of different immune cell subsets in response to a therapeutic perturbation. This protocol is based on (1) treatment of mice with the compound of interest, (2) harvesting mouse immune cells and preparation of single-cell suspensions, (3) staining of cell surface antigens with antibodies, (4) fixation, permeabilization, and intracellular staining, (5) processing of samples via flow cytometry, and (6) analysis of the acquired data (Fig. 1). We primarily focus on analyzing cells from spleens or pooled lymph nodes, the most common tissues suitable for immunophenotyping in mice. These organs are easy to harvest and ideally suited to prepare single-cell suspensions required for flow cytometry. Other organs (e.g., liver, lungs, bone marrow) require different methods of harvesting and/or processing of the organs and are not covered by this protocol, although the staining and analysis are identical.

This protocol also provides the option to incorporate adoptive transfer of $CD4^+$ or $CD8^+$ cells into congenic mice prior to treatment to assess the phenotype of the activated transferred T cells compared to endogenous T cells, with only a slight modification to the surface staining procedure. Adoptive cell transfer is an exciting new approach for immunotherapy that shows particular promise in cancer [7], and its seamless incorporation into this protocol allows for simultaneous evaluation of the compound of interest, as well as insight into its potential incorporation into a relevant therapeutic strategy for oncology. Adoptive cell transfer is built on the OT-I ($CD8^+$)/OT-II ($CD4^+$) transgenic OVA-specific mouse model in conjunction with B6.PL congenic mice ($Thy1.1^+$) [8]. There are several other similar adoptive transfer systems which may be easily

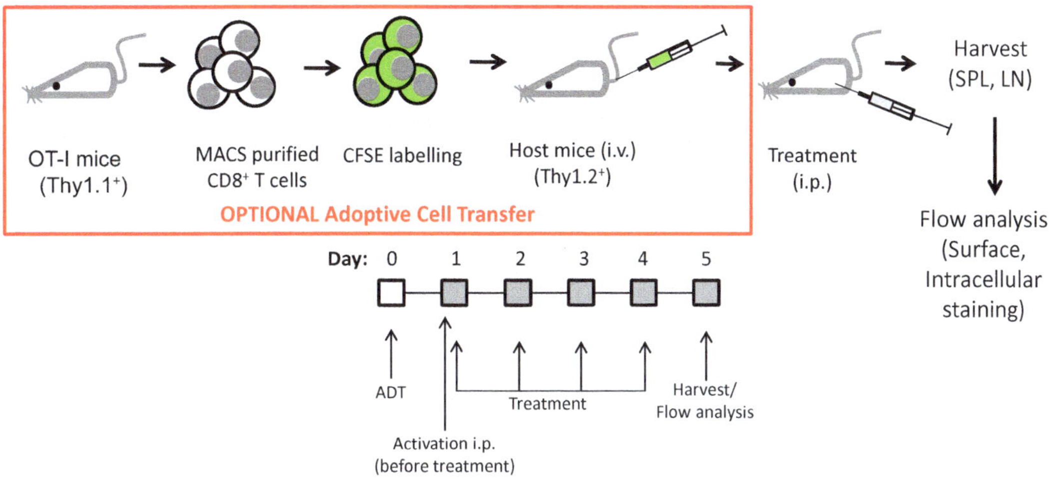

Fig. 1 Experimental setup, including optional adoptive transfer of purified T cells. Titrated amounts of selected agent(s) (treatment) are injected *i.p.* for 4 consecutive days. On day 5, spleen (SPL), pooled lymph nodes (LN), and/or selected LN are harvested, single-cell suspensions are prepared, and cells are stained according to the protocol. Samples will be then processed by flow cytometry and analyzed to identify and quantify immune cell subsets. This protocol also has an optional workflow, outlined in red, for adoptive cell transfer: 1 day prior to treatment, purified OT-I $CD8^+$ and/or OT-II $CD4^+$ T cells ($Thy1.1^+$) are labeled with CFSE and injected *i.v.* or *r.o.* in PBS into C57BL/6 recipients ($Thy1.2^+$). The following day, mice are injected *i.p.* with 2 nmol SIINFEKL peptide (OT-I) and/or OVA323–339 peptide (OT-II) 2 h before application of treatment. Alternatively, mice are injected *i.p.* with 100 μg/mouse of whole ovalbumin in PBS 8 h before application of treatment to activate transferred cells. PBS-treated mice (negative control) and mice administered activating peptide/protein plus poly(I:C) (75 μg/mouse) in PBS (positive control) should be used in each experiment

substituted into this experimental framework [9, 10] . The adoptive transfer additionally requires implementation of magnetic activated cell-sorting (MACS) procedures.

Minimal requirements for flow cytometry instrumentation are three-laser configuration to allow use of both violet laser- and red laser-excited fluorochromes (e.g., 405–407 nm or 633 nm, respectively). Four-laser flow cytometers are strongly recommended, as the addition of a yellow-green or similar laser (e.g., 561 nm) allows not only the use of a wider spectrum of fluorochromes but also improves the quality of data by decreasing the compensation level. Utilization of plate loader/high-throughput sampler (HTS) is highly recommended to facilitate sample handling during the staining and processing steps of the protocol. Conducting the experiment using individual tubes is also an option for experiments with low number of samples. Tubes have the advantages of reduced cross-contamination between samples (due to spillover in a plate) and superior control over sample processing, but require significantly more time and handling, which can limit experimental throughput and consistency.

This protocol has been successfully implemented to evaluate the immune response to several engineered immunotherapeutics, including agents that stimulate and agents that suppress immune activity [11–14]. As the immunotherapy space expands, we anticipate this protocol and other related ones will become valuable research tools to facilitate the development of novel drugs and their translation to the clinic.

2 Materials

2.1 Plastics

This protocol requires the use of 15/50 mL conical tubes, centrifuge tubes, 96-well V-bottom and/or 96-well U-bottom microtiter plates, 1.2 mL flow cytometry tubes (AlphaLaboratories), 30 μm and 70 μm CellTrics cell strainers (Sysmex), and 70 μm Falcon cell strainers (Corning).

2.2 Antibodies

All antibodies required for staining of surface and intracellular antigens are listed in Table 1.

2.3 Blocking Agents

1. Fc block (FG CD16/32, eBioscience).
2. Normal mouse serum (ThermoFisher Scientific).

2.4 Viability Dye

Fixable viability dye (FVD, eBioscience).

2.5 Buffers

Prepare all solutions using Milli-Q ultrapure water (Millipore Sigma). Prepare and store all buffers at 4 °C unless otherwise indicated. Diligently follow all waste disposal regulations when discarding waste materials.

1. Flow buffer: 1× PBS, 1% FBS, 2 mM ethylenediaminetetraacetic acid (EDTA) pH 8.0. Prepare a 1 M EDTA solution first and adjust the pH to 8.0 with NaOH.
2. ACK lysing buffer (Gibco).
3. Foxp3/Transcription Factor Staining Buffer Set (eBioscience): Contains Foxp3/Transcription Factor Fixation/Permeabilization Concentrate and Diluent, Permeabilization Buffer (10×).
4. Adoptive transfer reagents (*optional*).
   (a) MACS buffer (1× PBS, 0.5% BSA, 2 mmol EDTA).
   (b) $CD4^+$ mouse T-Cell Isolation Kit (Miltenyi Biotec).
   (c) $CD8a^+$ mouse T-Cell Isolation Kit (Miltenyi Biotec).
   (d) MidiMACS Separator (Miltenyi Biotec).
   (e) MACS LS columns (Miltenyi Biotec).
   (f) (*Optional*) Carboxyfluorescein succinimidyl ester (CFSE) labeling reagents (*optional*): Vybrant™ CFDA SE Cell Tracer Kit (ThermoFisher Scientific, #V12883).

Table 1
Panels of mAbs for staining of surface and intracellular markers

Panel	Epitope	Fluorochrome	Clone	Dilution	Vendor	Cat. #	Staining Volume
Effector T cell/memory phenotype T cell panel							
Surface staining	CD3	APC	17A2	200	Biolegend	100236	20
	CD8	BV 786	53-6.7	400	Biolegend	100750	
	FVD	Efluor 450	N/A	600	eBioscience	65-0863-14	
	CD44	R-A700	IM7	500	BD	565480	
	CD122	PE	TMb1	200	Biolegend	123210	
	CD62L	APC-Cy7	MEL14	200	Biolegend	104428	
	CD4/ CD90.1*	PerCP-Cy5.5	OX-7	300*	Biolegend	100434/ 202516	
Intracellular staining	Ki67	Alexa Fluor 488	16A8	200	Biolegend	652418	100
Regulatory T cell/natural killer cell/natural killer T cell panel							
Surface staining	CD3	PE	17A2	200	Biolegend	100206	20
	CD4	BV 786	GK1.5	600	Biolegend	100453	
	FVD	Efluor 450	N/A	600	eBioscience	65-0863-14	
	CD25	BV 650	PC61.5	500	Biolegend	102038	
	NK1.1/ CD90.1*	PerCP-Cy5.5	PK136/OX-7	300*	Biolegend	108728/ 202516	
	CD49b	PE-Cy7	DX5	200	Biolegend	108922	
Intracellular staining	Foxp3	APC	FJK-16s	600	eBioscience	17-5773-82	100
	Ki67	Alexa Fluor 488	16A8	200	Biolegend	652418	

Suggested panel compositions for characterization of effector and memory-phenotype (MP) T cells (*top*) and for regulatory T (T_{reg}), natural killer (NK) and natural killer T (NK-T) cells (*bottom*). *Congenic marker antibodies (e.g. CD90.1/Thy1.1) are required for optional adoptive cell transfer

2.6 Instruments

1. Centrifuges with refrigeration capable of spinning plates, 15/50 mL conical tubes, and 1.5 mL centrifuge tubes are required.
2. A flow cytometer (ideally four lasers and equipped with HTS plate loader) is needed.
3. Multichannel pipettes (ideally 12 channels) are useful for sample handling.
4. GentleMACS Dissociator (Miltenyi Biotec) and/or AutoMACS Pro Separator (Miltenyi Biotec) instruments are *optional*, but extremely helpful to expedite workflow.

3 Methods

3.1 Animal Treatments

Note that all animal treatments must be conducted under an approved protocol and animals must be housed in an approved vivarium, in compliance with institutional requirements. We recommend labeling animals for ready identification (*see* **Note 11**). Note that other animals may be substituted for mice in this protocol (*see* **Note 12**).

(*Optional*) *Adoptive transfer of CD4⁺ or CD8⁺ activated cells into congenic mice.*

If you are using the AutoMACS Pro Separator instrument, determine the cell count of your sample, and place it in one of the chill racks. Program the instrument to carry out the appropriate cell isolation protocol, and start the automatic labeling and separation process (*see* the autoMACS Pro Separator manual from Miltenyi Biotec for details).

If you are conducting manual labeling and separation, implement the following steps:

1. Prepare a single-cell suspension and determine the cell number. Work fast, keep cells at 4 °C, and use precooled solutions for all manipulations (2–8 °C). *Keep small sample of unlabeled cells for subsequent flow analysis of sorted cell purity.*
2. Resuspend cell pellet in 40 μL MACS buffer per 10^7 total cells. When working with fewer cells, use the same volumes as indicated. When working with higher cell numbers, scale up all reagent volumes and total volumes accordingly.
3. Add 10 μL of the appropriate Biotin-Antibody Cocktail (from the $CD4^+$/$CD8^+$ isolation kit) per 10^7 total cells.
4. Mix well and incubate for 5 min in the refrigerator (2–8 °C).
5. Add 30 μL of flow buffer per 10^7 total cells.
6. Add 20 μL of Anti-Biotin MicroBeads per 10^7 total cells.
7. Mix well and incubate for 10 min in the refrigerator (2–8 °C).

8. Proceed to subsequent magnetic cell separation. A minimum of 500 μL is required for magnetic separation. If necessary, add buffer to the cell suspension. Also, addition of no more than 10^8 total labeled cells per MACS LS column is recommended.
9. Place a sterile MACS LS column in the magnetic field of a suitable MACS Separator. For details, refer to the MACS LS column data sheet from Miltenyi.
10. Equilibrate the column by rinsing with 3 mL of buffer.
11. Apply the cell suspension from **step** 7 onto the column. Collect the flow-through containing unlabeled cells, representing the desired enriched $CD4^+/CD8^+$ T cells.
12. Wash column with 3 mL of buffer. Collect unlabeled cells that pass through, representing the desired enriched $CD4^+/CD8^+$ cells, and combine with the effluent from **step 11**.
13. Collect sample and evaluate the purity of enriched cells by flow cytometry.
14. (*Optional*) Transferred cells may be tracked by CFSE labeling. To implement this, label the enriched cells with CFSE using the ThermoFisher Scientific Vybrant™ CFDA SE Cell Tracer Kit, following the manufacturer's protocol.
15. Inject *i.v.* or *r.o.* 0.5–1 × 10^6 purified OT-I ($CD8^+$) or OT-II ($CD4^+$) cells from donor mice (from **step 13** [or **14** if implementing CFSE labeling]) into acceptor congenic mice 24 h prior to the first therapeutic treatment. Co-transfer of both $CD4^+$ and $CD8^+$ transgenic cells in one experiment is feasible (*see* **Note 1**), as is the use of donor OT-I/$CD90.1^+$ and/or OT-II/$CD90.1^+$ crossbred mice (*see* **Note 2**).
16. On the next day, before initiating therapeutic treatment, apply either selected activating peptide in PBS (OT-I: OVA257–264 [SIINFEKL], 2 nmol/mouse; or OT-II: OVA323–339, 25 nmol/mouse) or complete ovalbumin in PBS (100 μg/mouse) *i.p.* If peptide activation is used, treatment can be applied 2 h later. If complete ovalbumin is used, treatment has to be applied at least 8 h later due to the requirement for antigen presentation.
17. Separate experimental mice into cohorts according to the protocol. Apply treatment using dosing frequency and strategy appropriate for your compound of interest. For antibody/cytokine complexes, four daily injections are typical (*see* **Note 3**) [11, 12].
18. 18–24 h after the last treatment, euthanize mice by cervical dislocation (or another approved form of euthanasia; *see* **Note 4**), and harvest spleen or pooled lymph nodes (LN) into 50 mL conical tube with 5 mL of flow buffer, and incubate at room temperature (*see* **Note 5**). Alternatively, individual LN can be

harvested in identical fashion but instead using a reduced volume of PBS (1 mL) and centrifuge tubes in place of 50 mL conicals. As another alternative, GentleMACS C-tubes can be used instead of 50 mL conicals for spleens or pool of LN. The use of GentleMACS for single LN is not recommended.

3.2 Preparation of Single-Cell Suspensions

1. Pour out the liquid contents together with each organ (from Subheading 3.1, **step 2**) into a separate 70 μm Falcon cell strainer in a small petri dish.
2. Disrupt each organ against the nylon mesh of the strainer by plastic plunger of 1 mL insulin syringe. Wash the strainer by adding 1 mL of flow buffer, and transfer the cell suspension into a new 50 mL conical. Alternatively, run GentleMACS C-tubes on the instrument. Use the correct program for each respective organ to achieve ideal dissociation. Pour the suspension through 70 μm Falcon cell strainer in a 50 mL conical. *See* **Note 6** for other tissue disruption options.
3. Centrifuge the suspensions for 5 min in a centrifuge at 300 × *g* at room temperature (e.g., 1200 rpm on a Hettich Rotanta 460 R with swinging-bucket rotor). Pour out the supernatant.
4. Add 3 mL per sample of ACK lysing buffer to remove red blood cells. Incubate cells for 10 min at room temperature.
5. Neutralize the reaction by adding 30 mL (10× volume of ACK lysing buffer) of flow buffer per sample.
6. Centrifuge as described in **step 3**.
7. Resuspend the cell pellet in either 1 mL of flow buffer (for spleen) or 0.5 mL of flow buffer (pooled LN). Individual LN should be resuspended in 250 μL per LN. Strain each cell suspension with a 30 μm Cell Trics cell strainer into a 1.5 mL microcentrifuge tube. Keep on ice or at 4 °C.
8. Add 50 μL of spleen suspension or 100 μL of LN suspension to each tube or well of a 96-well V-bottomed plate (*see* **Note 7** regarding 96-well plate options). Do not forget to create sample wells/tubes for unstained controls and single-color controls (at least one for each fluorochrome used). Also, be sure to save cells from each sample to measure cell density, either by manual counting on a hemocytometer or with addition of counting beads via flow.
9. Centrifuge tubes/plate for 4 min at 350 × *g* (e.g., 1300 rpm in Hettich Rotanta 460R with swinging-bucket rotor) at 4 °C. Dump the supernatant (*see* **Note 8**), and blot the tube or plate with a delicate task wiper to remove excess liquid and to minimize cross-contamination between the samples.

10. Resuspend each cell suspension in Fc block (functional grade CD16/CD32 mAb, 100× dilution, 20 μL/sample). Alternatively, suspensions can be blocked in 10% normal mouse serum or both Fc block and normal mouse serum. Incubate the cells for 30 min on ice.
11. Centrifuge tubes/plate for 4 min at 350 × *g* at 4 °C. Dump the supernatant and blot the tube or plate with a delicate task wiper.

3.3 Staining of Surface Antigens with Fluorescent Antibodies

1. Add 20 μL appropriately diluted fluorochrome-conjugated monoclonal antibody master mix in flow buffer to each tube or well (*see* Table 1, surface, and **Note 9**), and mix thoroughly but carefully by pipetting up and down. Incubate samples in dark on ice for 30 min.
2. Wash by pipetting 200 μL of flow buffer into each well/tube. Centrifuge tubes/plate for 4 min at 350 × *g* at 4 °C. Dump the supernatant and blot the tube or plate with a delicate task wiper.
3. Repeat **step 2** twice.

3.4 Fixation, Permeabilization, and Staining of Intracellular Markers with Fluorescent Antibodies

1. Add 100 μL of 1× Foxp3/transcription factor fixation/permeabilization buffer (concentrate/diluent—1:3) to each tube or well (*see* **Notes 10** and **13**). Mix thoroughly but carefully by pipetting up and down. Incubate in the dark and on ice for 30 min.
2. To wash, pipette 100 μL of 1× permeabilization buffer into each well/tube. Centrifuge tubes/plate for 4 min at 350 × *g* at 4 °C. Dump the supernatant and blot the tube or plate with a delicate task wiper.
3. Repeat **step 2** instead using 200 μL of 1× permeabilization buffer.
4. Add 100 μL of appropriately diluted fluorochrome-conjugated monoclonal antibody master mix in permeabilization buffer to each tube or well (Table 1, intracellular staining), and thoroughly but carefully pipet up and down. Incubate in the dark and on ice for 30 min.
5. To wash, add 200 μL of 1× permeabilization buffer to tubes/wells. Centrifuge tubes/plate for 4 min at 350 × *g* at 4 °C. Dump the supernatant and blot the tube or plate with a delicate task wiper.
6. Repeat **step 5** twice.
7. Resuspend the samples in 100 μL of flow buffer. Keep cells on ice or at 4 °C in the dark until analysis by flow cytometry.

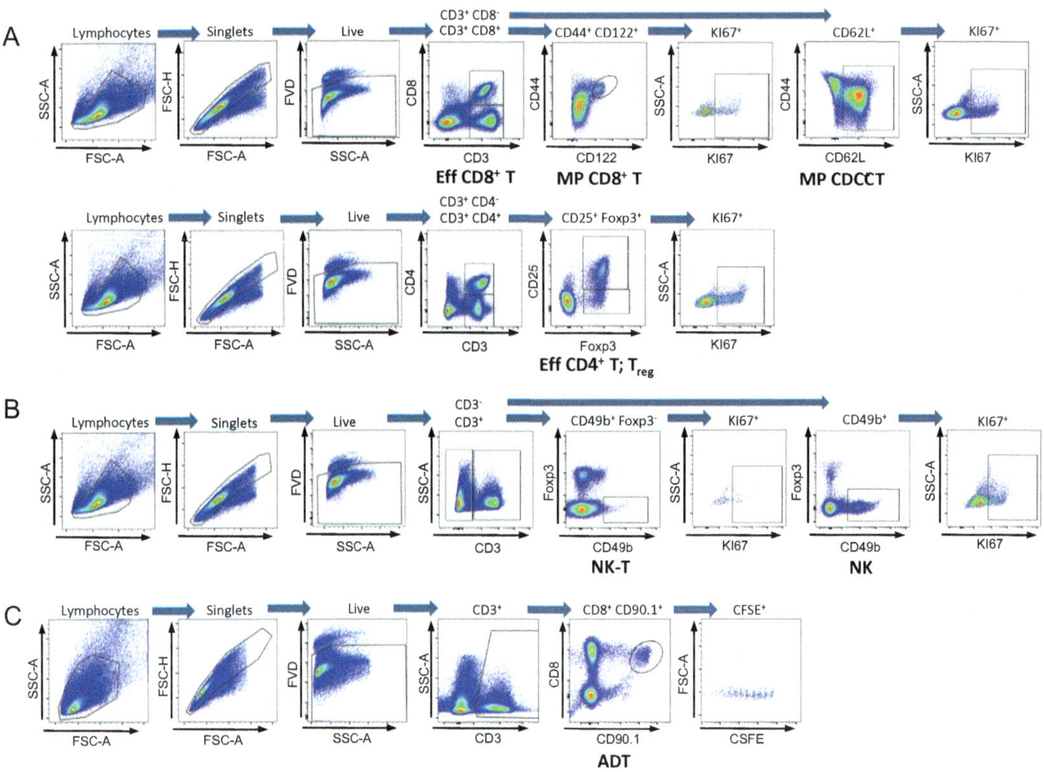

Fig. 2 Recommended gating strategy for immune cell subset analysis. Spleen samples were processed on a CytoFLEX LX instrument (Beckman Coulter) and analyzed in FlowJo X Software (Treestar). (**a**) Endogenous effector (Eff) T cell/memory phenotype (MP) T cell panel, (**b**) endogenous T_{reg} and natural killer (NK)/NK T-cell panel, and (**c**) adoptively transferred T cells (ADT)

3.5 Process Immune Cells via Flow Cytometry

1. Set the compensation matrix on the flow cytometry instrument according to the single-color controls.
2. Collect data for each tube/well on the flow cytometer. Collection of at minimum 50,000 events per sample is recommended. For the adoptive transfer option, collection of at least 100 events per sample in the adoptively transferred cell gate is advised.

3.6 Analyze Flow Cytometry Data

1. Process the acquired data using a flow cytometry analysis software, and analyze immune cell subset abundance and molecular expression profiles. FlowJo (Treestar) is recommended. A typical gating scheme is provided in Fig. 2.
2. Arrange the acquired values in charts using graphical analysis software. Microsoft Excel or GraphPad Prism is recommended. Typical data presentation formats are provided in Fig. 3. For statistical analysis, unpaired Student's *t*-test is recommended for comparison between two cohorts, and one-way ANOVA is recommended for comparison between multiple groups.

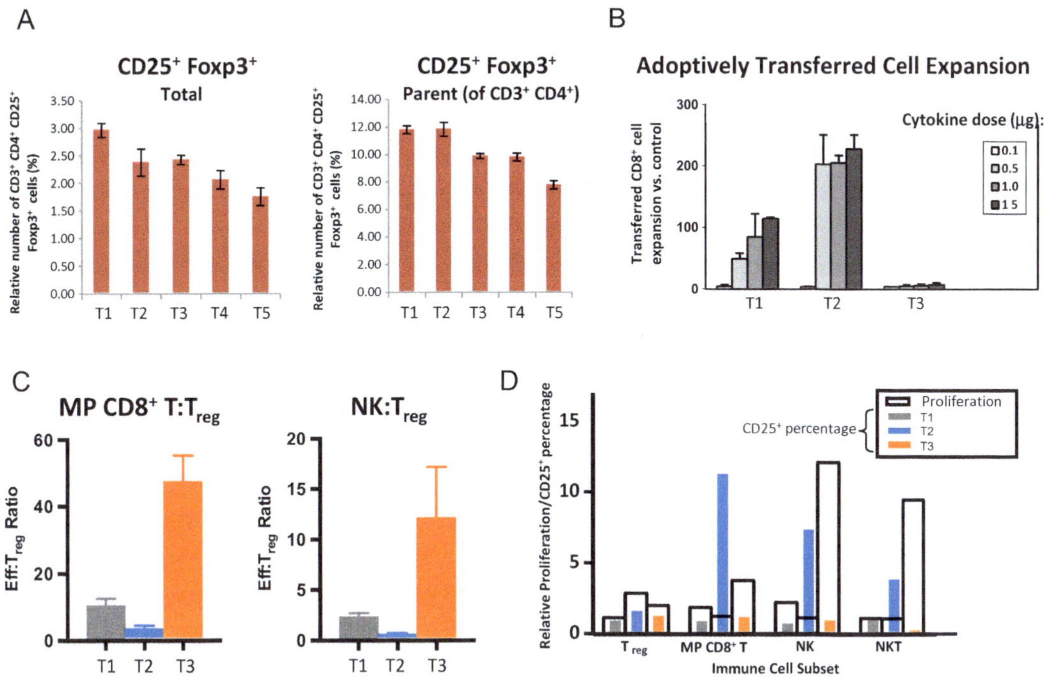

Fig. 3 Recommended data display formats. Flow cytometry data were gated and analyzed via FlowJo software and plotted in Microsoft Excel or Graphpad Prism software. (**a**) Relative cell counts drawn from parent gate or as a fraction of total of selected populations are analyzed using Excel. (**b**) Relative expansion of adoptively transferred cells relative to untreated control (fixed at 1) is analyzed in Excel. (**c**) Ratios of effector (Eff, specifically MP or NK cells) to T_{reg} cells are analyzed using GraphPad Prism. (**d**) Two-parameter relative display of proliferation and $CD25^+$ percentage (Untreated Control set to 1) is analyzed using GraphPad Prism. All panels display mean ± standard deviation. T1–T5, Treatments 1–5

4 Notes

1. You can co-transfer both $CD4^+$ and $CD8^+$ transgenic cells in one experiment if required. In that case, keep the ratio of $CD8^+$ to $CD4^+$ cells approximately 1:3 (e.g., 1×10^6 OT-I: 3×10^6 OT-II) to follow the natural distribution in the T cell compartment.
2. Donor OT-I/$CD90.1^+$ and/or OT-II/$CD90.1^+$ crossbred mice could be used as an alternative option in order to utilize naive C57BL/6 mice as acceptors for adoptive transfer of T cells.
3. Treatment schedules suggested in this protocol are suitable for larger proteins with half-life in circulation on the order of several hours, such as cytokine/antibody complexes, immunocytokines, antibody-drug conjugates, or cytostatic drugs. Dosing

amounts and frequency of treatment have to be optimized for smaller or larger compounds, in accordance with their respective circulation half-lives.

4. Suggested euthanasia method is cervical dislocation, but other means (e.g., exsanguination, exposure to CO_2, or overdose of anesthetic drugs) will work as alternatives, based on the layout of your experiment and institutional animal welfare policies.
5. Keep samples at room temperature during the harvest of organs, preparation of single-cell suspension, and red blood cell (RBC) lysis to reduce clumping due to temperature shifts. After RBC lysis and resuspension of clean single-cell suspension, keep samples on ice for preservation.
6. Tissue can also be disrupted by other means, e.g., glass homogenizator disruption, mashing between two frosted microscope slides, or digestion by collagenase D.
7. 96-well U-bottomed microtiter plates may be also used, especially if cell densities are high. For lower cell densities, 96-well V-bottomed plates are recommended to minimize cell attrition. Check your flow cytometer compatibility and settings for each type of plate.
8. Proper dumping of the supernatant during the wash steps is very critical to avoid spillover in between samples and/or losing cells in sample. Turn the plate up-side down with a wrist-rotating motion, keeping it parallel to the countertop surface, and swipe the plate immediately into the drain. Do the dumping just once with a brisk swipe, but do not be too harsh. Dry the plate by blotting on a prepared layer of delicate task wipers. Avoid shallow drains so that the splashing liquid does not contaminate your plate. An aspirator may be used as an alternative to dumping.
9. Choice of fluorochrome-antibody conjugates could be altered to minimize fluorochrome compensations. In this protocol, one should keep this in mind, especially in the case of co-transfer of $CD4^+$ and $CD8^+$ T cells. Nevertheless, PerCP-Cy5.5 channel is versatile (i.e., you can select a congenic marker in case of adoptive transfer, a direct stain of $CD4^+$/$CD8^+$ T cells, or NK1.1 for more precise stain of natural killer [NK] and NK T cells).
10. Fixation of cells for Foxp3 staining with solutions other than the eBioscience fixation/permeabilization solution is not recommended. For routine fixation without Foxp3 analysis, other fixation/permeabilization solutions may be used (e.g., BD Fix/Perm or solutions from An Der Grub or other manufacturers). The use of 2–4% paraformaldehyde is also an option, but this will not preserve the morphology of the cells unless carefully optimized.

11. Identification of experimental animals is optional, but highly suggested. Labeling methods range from simple ones like an ethanol-resistant marker to create stripes or other patterns on the tail of the mouse (works on other rodents as well, but be careful regarding skin allergy) to more complicated methods such as ear punctures or snips, ear tags, finger snips, or tattooed signs. Choose a numbering system corresponding to the method you employ, or create your own beforehand to avoid confusion.
12. This protocol can be adapted to other experimental animals wherein a sufficient number of antibodies for flow cytometry are available (e.g., rats). Executing adoptive transfer of $CD4^+$ or $CD8^+$ T cells would also be possible in this case, depending on the availability of T cell transgenic/congenic strains of the selected type of animal.
13. Optimize your experiment beforehand if you want to use alternate procedures. Staining, fixing volumes, and time intervals can be reduced to save antibody, other reagents, or time. For example, surface staining in 10 μL instead of 20 μL is possible (with carefully titrated mAbs and mixing of sample after addition). Reducing the volume of fixation/permeabilization reagent to 50 μL instead of 100 μL is also an option. 20 min rather than 30 min incubation steps are also acceptable if carefully synchronized with other parameters.

References

1. Muirhead KA, Horan PK, Poste G (1985) Flow cytometry: present and future. Nat Biotechnol 3(4):337–356
2. Adan A, Alizada G, Kiraz Y, Baran Y, Nalbant A (2017) Flow cytometry: basic principles and applications. Crit Rev Biotechnol 37 (2):163–176
3. Macey MG (ed) (2007) Flow cytometry: principles and applications [Internet]. Humana: Totowa, NJ. http://www.myilibrary.com?id=97216. Accessed 3 Apr 2019
4. Spangler JB, Moraga I, Mendoza JL, Garcia KC (2015) Insights into cytokine-receptor interactions from cytokine engineering. Annu Rev Immunol 33(1):139–167
5. Kureshi R, Bahri M, Spangler JB (2018) Reprogramming immune proteins as therapeutics using molecular engineering. Curr Opin Chem Eng 19:27–34
6. Taebel DW (1990) The importance of animals in biomedical research. Wis Med J 89(4):155. 158
7. Rosenberg SA, Restifo NP, Yang JC, Morgan RA, Dudley ME (2008) Adoptive cell transfer: a clinical path to effective cancer immunotherapy. Nat Rev Cancer 8(4):299–308
8. Ford ML, Koehn BH, Wagener ME, Jiang W, Gangappa S, Pearson TC and Larsen CP (2007) Antigen-specific precursor frequency impacts T cell proliferation, differentiation, and requirement for costimulation. J Exp Med 204(2):299–309.
9. Svedova M, Masin J, Fiser R, Cerny O, Tomala J, Freudenberg M, Tuckova L, Kovar M, Dadaglio G, Adkins I, Sebo P (2016) Pore-formation by adenylate cyclase toxoid activates dendritic cells to prime CD8+ and CD4+ T cells. Immunol Cell Biol 94 (4):322–333
10. Skopova K, Tomalova B, Kanchev I, Rossmann P, Svedova M, Adkins I, Bibova I, Tomala J, Masin J, Guiso N, Osicka R, Sedlacek R, Kovar M, Sebo P (2017) Cyclic AMP-elevating capacity of adenylate cyclase toxin-hemolysin is sufficient for lung infection but not for full virulence of Bordetella pertussis. Infect Immun 85(6):e00937

11. Spangler JB, Tomala J, Luca VC, Jude KM, Dong S, Ring AM, Votavova P, Pepper M, Kovar M, Garcia KC (2015) Antibodies to interleukin-2 elicit selective T cell subset potentiation through distinct conformational mechanisms. Immunity 42(5):815–825
12. Spangler JB, Trotta E, Tomala J, Peck A, Young TA, Savvides CS, Silveria S, Votavova P, Salafsky J, Pande VS, Kovar M, Bluestone JA, Garcia KC (2018) Engineering a single-agent cytokine/antibody fusion that selectively expands regulatory T cells for autoimmune disease therapy. J Immunol 201 (7):2094–2106
13. Trotta E, Bessette PH, Silveria SL, Ely LK, Jude KM, Le DT, Holst CR, Coyle A, Potempa M, Lanier LL, Garcia KC, Crellin NK, Rondon IJ, Bluestone JA (2018) A human anti-IL-2 antibody that potentiates regulatory T cells by a structure-based mechanism. Nat Med 24(7):1005–1014
14. Sockolosky JT, Trotta E, Parisi G, Picton L, Su LL, Le AC, Chhabra A, Silveria SL, George BM, King IC, Tiffany MR, Jude K, Sibener LV, Baker D, Shizuru JA, Ribas A, Bluestone JA, Garcia KC (2018) Selective targeting of engineered T cells using orthogonal IL-2 cytokine-receptor complexes. Science 359 (6379):1037–1042

Chapter 10

Primary T-Cell Transduction to Study Follicular Helper T-Cell Differentiation

Yang Zhang, Xuehui Long, and Xiaoming Wang

Abstract

Follicular helper T (Tfh) cells constitute a specialized $CD4^+$ T-cell subset that localizes in close proximity to B cells and are essential in the production of high-affinity, class-switched antibodies, and their dysregulations are also involved in autoimmune diseases. Modulating gene expression patterns in primary T cells is an important approach to understanding Tfh cell differentiation and function. In this chapter, we describe a protocol to evaluate Tfh cell differentiation with OT-II TCR transgenic T cells by retrovirally transducing gene of interest. This protocol adopts the recombinant retrovirus-based transduction of primary $CD4^+$ T cells, and it also includes procedures for adoptive transfer, immunization, and flow cytometry analysis.

Key words Retroviral vector, Spinfection, Follicular helper T cells, Adoptive transfer, OVA-Alum, Flow cytometry

1 Introduction

B cells activated by foreign antigen can enter germinal centers (GCs) for generating high-affinity antibodies via positive selection [1]. Efficient GC selection requires a unique subset of CD4 helper T cells called follicular helper (Tfh) cells, which provide activating signals for B cells in GC [1]. The Tfh cells are characterized by the expression of CXC chemokine receptor 5 (CXCR5), B-cell lymphoma 6 (Bcl6), inducible costimulatory (ICOS), and programed death 1 (PD-1) [2].

Tfh cell differentiation is a multistep and multifactorial process that starts at initial dendritic cell (DC) priming of a naïve $CD4^+$ T cell [2]. The $CD4^+$ T cells differentiate into different effector cells under the control of specific transcription factors. Transcription repressor Bcl6 was identified as the master regulator of Tfh cell differentiation. Tfh cell differentiation does not occur in the absence of Bcl6 in vivo, whereas constitutive Bcl6 expression induces $CD4^+$ T-cell differentiation into Tfh cells [3–5]. CD28 costimulatory signaling from DC is required for Bcl6 induction [6]. ICOS is

Chaohong Liu (ed.), *T-Cell Receptor Signaling: Methods and Protocols*, Methods in Molecular Biology, vol. 2111, https://doi.org/10.1007/978-1-0716-0266-9_10,

another costimulatory molecule expressed on all effector CD4$^+$ T cells and is required for Tfh cell differentiation and maintenance [7]. CXCR5 is a specific Tfh cell surface marker, which enables Tfh cells migrating into B-cell zone [8]. ASCL2 promotes early migration and differentiation of Tfh cells by upregulating CXCR5 and downregulating CCR7 [9]. TCF-1 and LEF-1 coordinate Tfh cell differentiation and maintenance by directly targeting several vital signaling factors, such as Bcl6 and ICOS [10, 11].

In the meanwhile, Tfh cell differentiation is also negatively regulated by other regulators, including Foxp1, Blimp-1, and KLF2. Foxp1 negatively regulates Tfh cell differentiation by directly binding to the promoter of *Il21* to inhibit its expression and indirectly suppressing the expression of ICOS [12]. Bcl6 expressed in Tfh cells and Blimp-1 expressed in non-Tfh compartments are mutual antagonists, as Blimp-1 negatively regulates Tfh cell differentiation by opposing the action of Bcl6 [13]. Krüppel-like factor 2 (KLF2) negatively regulates Tfh cell differentiation by repressing CXCR5 and activating CCR7, Selplg, and S1PR1. KLF2 could also induce Blimp-1 expression to antagonize Bcl6, thus inhibiting Tfh cell differentiation [14, 15]. Cytokine IL-2 is a potent inhibitor of Tfh cell differentiation [16]. The signaling of IL-2/CD25 activates Blimp-1 expression via phosphorylation of STAT5, negatively regulating Tfh cell differentiation [7, 16, 17].

The in vivo activation of primary CD4$^+$ T cells in thymus-dependent responses can be mimicked by stimulating naïve CD4$^+$ T cells with anti-CD3/CD28. Once an activated T cell is infected with retrovirus, the exogenous gene will integrate into the genome of the T cells, leading to the stable expression of the exogenous gene in the T cells. In this protocol, we constructed a retroviral vector with green fluorescent protein (GFP) as a reporter and transduced the retrovirus into OT-II TCR transgenic CD4$^+$ T cells, the TCR of which is MHC class II I-A^b restricted and specific for chicken ovalbumin 323–339. Transduced T cells were adoptively transferred into congenially distinguishable host mice, which were then immunized with OVA antigen to induce Tfh differentiation.

2 Materials

2.1 Construction of Retroviral Vector

1. Retroviral expression vector: MSCV2.2-IRES-GFP (Addgene).
2. Retroviral packaging vector: gag-pol-env plasmid (G/P/E).
3. Molecular biology reagents, such as DNA polymerase, restriction enzymes, DNA ligase, gel extraction kit, and plasmid DNA preparation kit.
4. *E. coli* DH5α competent cells.
5. LB media.

2.2 Mice

1. Donor mice (CD45.1): ovalbumin (OVA)-specific T-cell antigen receptor (TCR) transgenic (OTII) mice, 6–8 weeks old.
2. Recipient mice (CD45.2): C57BL/6 mice, 6–8 weeks old.

2.3 Recombinant Retrovirus Preparation and Cell Culture

1. Complete DMEM medium (cDMEM): DMEM high-glucose medium with 10% (vol/vol) fetal bovine serum (FBS), 1 × penicillin/streptomycin.
2. Opti-MEM reduced serum medium (Invitrogen).
3. Retrovirus packaging cell line: Plat-E cells [18].
4. Lipofectamine 3000 Transfection Reagent (Thermo Fisher Scientific).
5. 20 mg/ml DNase I: dissolve in phosphate buffered saline (PBS).
6. 100 mg/ml collagenase IV: dissolve in PBS.
7. Anti-CD3 (145-2C11; BioXCell).
8. Anti-CD28 (37.51; BioXCell).
9. MojoSort Mouse CD4 T-Cell Isolation Kit (Biolegend).
10. MojoSort Magnet (Biolegend).
11. Complete RPMI growth medium (cRPMI): RPMI 1640 medium with 10% (vol/vol) FBS, 10 mM HEPES, 2 mM L-glutamine, 1 × penicillin/streptomycin, 50 μM β-mercaptoethanol.
12. 1 M HEPES.
13. 8 mg/ml polybrene: dissolve polybrene in PBS.
14. Recombinant human interleukin-2 (R&D Systems).
15. Routine benchtop centrifuges (SL16R, Thermo Fisher Scientific).
16. 100 μm cell strainers.
17. 15/50 ml falcon tubes.
18. 6-well and 24-well cell culture plates.
19. Cell culture incubator.

2.4 Adoptive Transfer, Immunization, and Flow Cytometry Analysis

1. 1 mg/ml ovalbumin: dissolve in PBS.
2. Aluminum Hydroxide Gel Adjuvant (Aluminum content 6.5 mg/ml; Accurate Chemical and Scientific Corporation).
3. Syringe and needle for intraperitoneal injections.
4. BD 1 ml insulin syringe for intravenous injections.
5. FACS buffer: PBS with 2% fetal calf serum (FCS), 0.01% NaN_3, and 50 mM EDTA.

Table 1
The staining panel for transduced Tfh cells

Antibodies	Clone	Dilution
APC-Cy7 anti-mouse CD4	GK1.5	1:400
Pacific blue anti-mouse CD45.1	A20	1:400
PE-Cy7 anti-mouse PD-1	RMP1–30	1:200
Purified rat anti-mouse CXCR5	2G8	1:100
Biotin-conjugated anti-rat IgG	–	1:400
PE streptavidin	–	1:400

6. RPMI 2% (R2): RPMI medium, 2% FCS.
7. 96-well round-bottomed plates.
8. Flow cytometer.
9. Normal rat serum.
10. Antibodies for flow cytometry (Table 1).

3 Methods

3.1 Construction of Retroviral Expression Vector

MCSV2.2-IRES-GFP is an optimized vector for expressing target genes in multiple cell lines and primary cells; it drives the expression of exogenous genes by the 5′LTR promoter of murine stem cell virus. Besides the gene of interest, MCSV2.2-IRES-GFP also expresses green fluorescence protein (GFP) downstream IRES sequences. Cotransfection of MSCV2.2-IRES-GFP with gag-pol-env expression vectors (G/P/E) into retroviral packaging cell lines (here we use Plat-E cells) can generate retrovirus particles in the culture medium of transfected cells (Fig. 1).

1. Amplify defined DNA fragment from cDNA.
2. Gel purify the DNA fragment and MSCV2.2-IRES-GFP that has digested with restriction enzymes. Ligate purified DNA fragment with the linearized vector using DNA ligase.
3. Transform *E. coli* DH5α competent cells with ligation products.
4. Plate the transformation mixture on LB plates containing ampicillin. Extract candidate clones of the recombinant retroviral vector.
5. Confirm the identity of the clones by restricting enzyme digestion and sequencing.

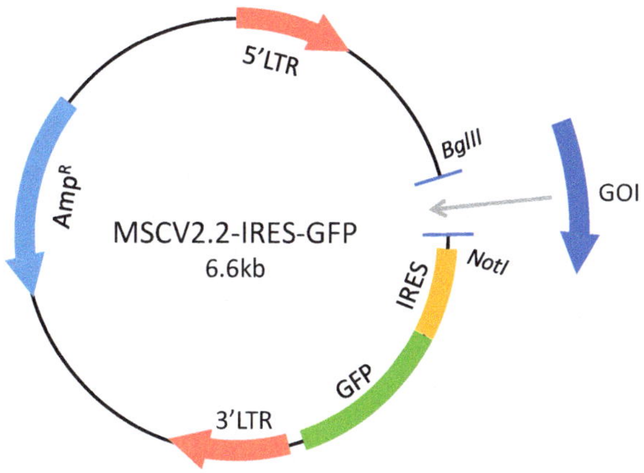

Fig. 1 Diagram of the retroviral vector and the strategy for cloning exogenous genes into the retroviral vector. Amplify the exogenous gene with cDNA of homologous species as template. The purified exogenous gene is cloned into MSCV2.2-IRES-GFP retroviral vector. GOI gene of interest

Table 2
Workflow for preparing recombinant retroviruses and infecting CD4⁺ T cells. Tasks involving handing of Plat-E cells are listed in the second row, whereas those involving handing of CD4⁺ T cells are listed in the last row

Day 0	Day 1	Day 2	Day 3	Day 4
Seed plat-E cells	Transfect plat-E cells	Replace the medium	Prepare retroviral suspension	
	Coat the plates	Stimulate naïve $CD4^+$ T cells with anti-CD3/CD28	Spinfect the cultured $CD4^+$ T cells	Add rIL2

3.2 Preparation of Recombinant Retrovirus, in Vitro Culture of Primary CD4⁺ T Cells, and Spinfection of the Cultured Primary CD4⁺ T Cells with Retrovirus

We performed retroviruses generation and T-cell stimulation simultaneously to obtain highly efficient transduction. We make a timeline flowchart for better implementing the experiment. This part of the experiment includes preparing recombinant retroviruses, setting up the T-cell culture, and infecting T cells with the retroviruses; it takes about 4 days (Table 2). The following protocol is given for culturing Plat-E cells in 6-well plates and T cells in 24-well plates.

3.2.1 Preparation of Recombinant Retrovirus

1. (Day 0) Seed 0.85 million Plat-E cells in 2 ml complete DMEM medium per well, so that the cell density will be at about 70–90% confluency at the day of transfection (*see* **Note 1**).
2. (Day 1) Transfect the Plat-E cells. For each well, replace 2 ml cDMEM freshly 30–60 min before transfection (*see* **Note 2**).

3. Dilute 8 μl Lipofectamine 3000 Reagent in 200 μl Opti-MEM reduced serum medium in a 1.5 ml Eppendorf tube; mix well. Dilute 3 μg retroviral plasmid DNA and 1 μg G/P/E plasmid DNA in 200 μl Opti-MEM reduced serum medium in another 1.5 ml Eppendorf tube, then add 8 μl P3000 Reagent, and mix well (*see* **Note 3**). Add diluted DNA to diluted Lipofectamine 3000 Reagent. Mix and incubate at room temperature for 10–15 min (not more than 15 min) to allow DNA-lipid complexes to form.
4. Add the DNA-lipid complex mixture dropwise onto the medium in each well, and homogenize the mixture by gently swirling the plate.
5. Return the transfected Plat-E cells to culture incubator at 37 °C with 5% CO_2.
6. (Day 2) Replace the medium of the transfected Plat-E cells with fresh medium 12–18 h post transfection.

3.2.2 Purification of Mouse $CD4^+$ T Cells (See **Note 4**)

1. (Day 2) Extract whole spleen from the OVA-specific OTII TCR transgenic mouse, and place in a 6 cm Petri dish with 2 ml complete RPMI 1640 medium. Add 10 μl DNase I (20 mg/ml) and 10 μl collagenase IV (100 mg/ml), and mix by gently shaking the dish. Then inject the medium into the spleen with a 1 ml syringe for better breaking the peptide bonds in collagen. Place the dish in a cell culture incubator at 37 °C with 5% CO_2 for 15 min.
2. Add 20 μl 5 mM EDTA and mix well at room temperature for 5 min (*see* **Note 5**). Mash spleen through a 100 μm cell strainer to get single cell suspension by the bulb of a 1 ml sterile syringe. Pipette the cell suspension 8–9 times to break tissue particles, and collect the cell suspension in 4 ml cRPMI. Then pass the cell suspension through the cell strainer again to remove unbroken cell aggregates. In the final wash of the splenocyte sample, resuspend the cells in 1× MojoSort buffer by adding up to 7–8 ml in a 15 ml conical tube. Centrifuge the cell suspension at 300 × *g* at 4 °C for 5 min.
3. Remove the supernatant and resuspend the cell pellet in an appropriate volume of 1 × MojoSort buffer. Count and adjust the cell concentration to 1×10^8 cells/ml. Add biotin-conjugated anti-CD4 antibody to the splenocyte sample. Use 10 μl of antibody per 1×10^7 cells. For example, add 100 μl biotin-antibody cocktail (biotin anti-CD8a, CD11b, CD11c, CD19 CD24, CD45R/B220, CD49b CD105, MHC II, TER-119, TCR-γδ) for separating 1×10^8 cells in 1 ml of 1 × MojoSort buffer. Tap the tube to mix, and incubate for 15 min at 4 °C.

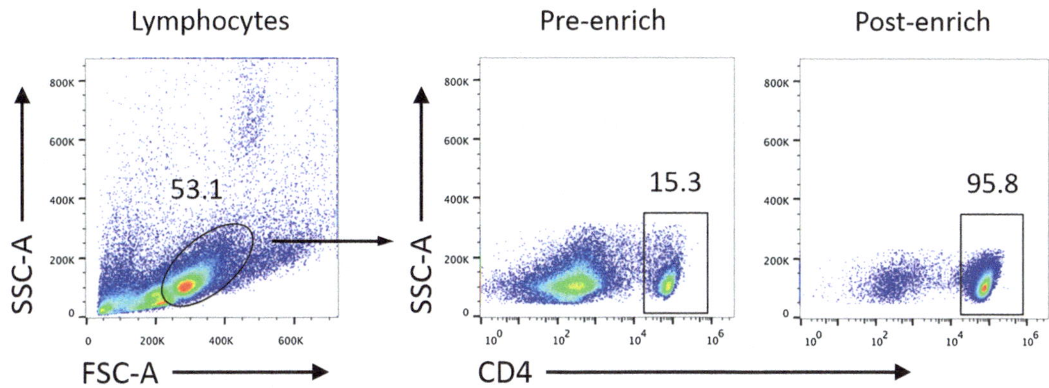

Fig. 2 Checking the purity of CD4$^+$ T cells post enrichment. Flow cytometry analysis of CD4$^+$ T cells from pre-enrich and post-enrich groups

4. Wash the cells once with 1× MojoSort buffer. Centrifuge the cell suspension at 300 × *g* at 4 °C for 5 min.
5. Resuspend the cell pellet in 1 ml 1× MojoSort buffer. Resuspend Streptavidin Nanobeads by pipetting up and down and gentle vortexing. Add 100 μl of Streptavidin Nanobeads for 1 × 10^8 cells. Tap the tube to mix, and incubate for 15 min at 4 °C.
6. Wash the cells once with 1 × MojoSort buffer. Centrifuge the cell suspension at 300 × *g* at 4 °C for 5 min.
7. Remove the supernatant and resuspend the cell pellet in 3 ml of 1 × MojoSort buffer. Transfer the cells to a 5 ml (12 × 75 mm) polypropylene tube.
8. Place the tube in the precooled magnet on ice for 5 min. The beads will bind to the sides of the tube.
9. Pour out and collect the liquid.
10. Pool the unlabeled fractions and centrifuge at 300 × *g* at 4 °C for 5 min. Remove the supernatant; resuspend the cells to a concentration of 1.3–1.6 × 10^6 cells/ml in cRPMI.
11. The purity of the cell sample can be assessed by flow cytometry after staining with fluorescently conjugated anti-CD4 antibody (Fig. 2). Refer to Subheading 3.4.2 for the staining methods. The purity of CD4$^+$ T cells using this protocol is usually above 95%.

3.2.3 Stimulation of Primary CD4$^+$ T Cells (See ***Note 6****)*

1. (Day 1) The 24-well plate was coated with anti-CD3 and anti-CD28 antibodies (5 μg/ml for both, diluted in PBS) 500 μl/well overnight at 4 °C.
2. (Day 2) Add 1 ml cRPMI with 1.3–1.6 × 10^6 purified cells to each well of the coated plate.
3. Place the plate in a cell culture incubator at 37 °C with 5% CO_2.

3.2.4 Spinfection and Rest of CD4+ T Cells

1. (Day 3) Harvest virus 24 h after T-cell stimulation. Briefly, transfer the medium of the transfected Plat-E cells into new tubes; centrifuge 1000 × *g* at room temperature for 5 min to remove Plat-E cell debris (*see* **Note 7**). Transfer the supernatant containing retroviral particles into new tubes, add 1/100 volume 1 M HEPES, and polybrene to a final concentration of 8 μg/ml.
2. Centrifuge the plate with cultured T cells at 750 × *g* at room temperature for 5 min. Carefully transfer the growth media to wells in a new plate. Save the media in cell culture incubator.
3. Add 1 ml/well viral supernatant to the wells with T-cell pellet. Centrifuge at 1000 × *g* at room temperature for 2 h.
4. Carefully aspirate the supernatant after spinfection, and immediately add back the growth media which was saved in **step 2** to the cell pellets.
5. Return the plate back to 37 °C incubator for further culture.
6. (Day 4) 24 h after spinfection, discard the medium and replace the fresh cRPMI with 100 U/ml human recombinant IL2.
7. Return the plate back and incubate for 2 more days.

3.3 Adoptive Transfer and OVA-Alum Immunization

3.3.1 Adoptive Transfer of CD4+ T Cells

1. Collect and wash the cells once with RPMI supplemented with 2% FCS (R2). Centrifuge the cell suspension at 300 × *g* at room temperature for 5 min. Resuspend 2.5 × 10^6 cells/ml with R2.
2. Use BD 1 ml insulin syringe to transfer 0.2 ml of the cell suspension (5 × 10^5 cells/mouse) into sex-matched recipient mice (CD45.2) by intravenous injection into the inner canthus vein plex.

3.3.2 OVA-Alum Immunization of Recipient Mice

1. 24 h after cell transfer, freshly prepare a suspension containing 1 mg/ml of ovalbumin in PBS, mix equal volume with aluminum hydroxide gel adjuvant (aluminum content 6.5 mg/ml).
2. Strongly vortex the mixture for 20 min at room temperature so that ovalbumin was efficiently precipitated onto alum.
3. Use 1 ml syringes to intraperitoneally inject 200 μl OVA-Alum mixture per mouse.

3.4 Analysis of Transduced Cells

3.4.1 Cell Preparation from Spleen

1. Harvest spleen from mice at days 3–5 post immunization, and place spleen in 6 cm Petri dishes containing 4 ml RPMI supplemented with 2% FCS (R2).
2. Gently mash spleen through a 100 μm cell strainer to get a single-cell suspension by the bulb of a 1 ml sterile syringe.
3. Count the cell number. Dilute single-cell suspensions to 1 × 10^7 cells/ml in R2.

3.4.2 Surface Staining

1. Load 100 μl of cell suspension to a 96-well round-bottomed plate (1×10^6/well).
2. Pellet the cells by centrifuging the plate at 1000 × *g* at 4 °C for 2 min. Decant the medium by quickly flicking the plate.
3. Dilute purified rat anti-mouse CXCR5 1:100 in FACS buffer, 50 μl per well (*see* **Note 8**). Vortex well and centrifuge at 15,000 × *g* at 4 °C for 5 min to remove particulates.
4. Resuspend the cells with 50 μl purified rat anti-mouse CXCR5 per well, mix well by pipetting up and down, and incubate for 25 min on ice in the dark.
5. Wash the cells twice with 200 μl of FACS buffer per well. Centrifuge the plate at 1000 × *g* at 4 °C for 2 min.
6. Dilute biotin-conjugated anti-rat IgG 1:400 in FACS buffer. Vortex well and centrifuge at 15,000 × *g* at 4 °C for 5 min.
7. Resuspend the cells with 50 μl biotin-conjugated anti-rat IgG per well, mix well by pipetting up and down, and incubate for 25 min on ice in the dark.
8. Wash the cells twice with 200 μl of FACS buffer per well. Centrifuge the plate at 1000 × *g* at 4 °C for 2 min.
9. Block excessive anti-rat IgG with 5% normal rat serum (NRS) for 20 min on ice, centrifuge, and discard supernatant.
10. Prepare fluorescent antibody cocktail with FACS buffer containing 2% NRS at an appropriate dilution: APC-Cy7 anti-mouse CD4, 1:400; Pacific Blue anti-mouse CD45.1, 1:400; PE-Cy7 anti-mouse PD-1, 1:200; PE Streptavidin, 1:400 (*see* Table 1).
11. Centrifuge the prepared antibody cocktail at 15,000 × *g* at 4 °C for 5 min (*see* **Note 9**).
12. Resuspend the cells with 50 μl fluorescent antibody cocktail, mix well by pipetting up and down, and incubate for 25 min on ice in the dark.
13. Wash the cells twice with 200 μl of FACS buffer per well. Centrifuge the plate at 1000 × *g* at 4 °C for 2 min.
14. Resuspend the cells in 200 μl of FACS buffer.

3.4.3 Flow Cytometer Analysis

To verify how well this transduction protocol works, we transduced OT-II $CD4^+$ T cells with retroviral vector MSCV2.2-IRES-GFP expressing Bcl6 which is the master transcription factor for Tfh cell differentiation.

1. Run the samples on a flow cytometer and analyze the data using FlowJo software. The gating strategy is shown in Fig. 3.
2. Gate on lymphocytes using FSC-A and SSC-A.
3. Gate on single cells using FSC-H and FSC-A.

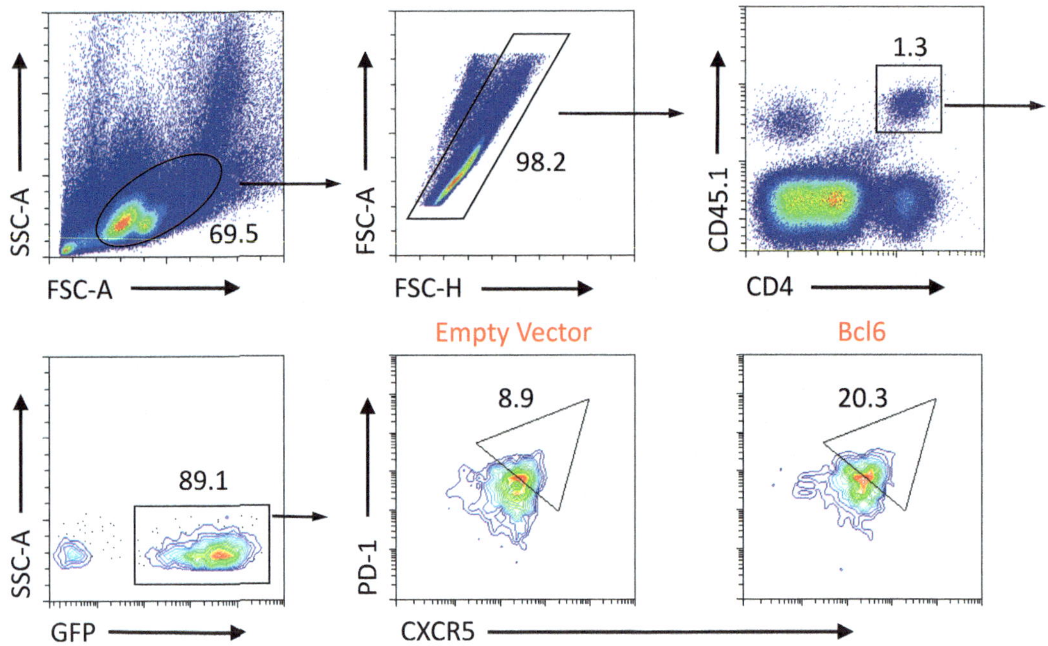

Fig. 3 Gating strategy for Tfh cells in adoptive transfer assay. Forward scatter (FSC) and side scatter (SSC) were used to gate lymphocytes. FSC-H and FSC-A were used to gate for single cells. The empty vector or gene of interest overexpression was determined by GFP expression. The frequency of splenic Tfh cells ($CD4^+CXCR5^+PD\text{-}1^+$) in $CD45.1^+GFP^+$-transduced T cells were assessed at day 3 post OVA-Alum immunization. As an example, Bcl6 transduction significantly promoted Tfh differentiation compared with the empty vector

4. Gate on $CD4^+$ $CD45.1^+$ T cells within single cells.
5. Determine the cells transduced with retroviral vector as GFP^+. Gate on GFP^+ cells within $CD4^+$ $CD45.1^+$ T cells.
6. Determine Tfh cells as $CXCR5^+PD\text{-}1^+$. Gate on Tfh cells within GFP^+ cell.

4 Notes

1. Plat-E cells should be in good shape.
2. Lipofectamine3000 Transfection Reagent leverages lipid nanoparticle technology; it has low toxicity to cells and improved cell viability.
3. The protein products of retroviral genes gag, pol, and env (G/P/E) are necessary for the replication and packaging of recombinant retroviruses. Gag is a polyprotein and processed to matrix and other core proteins that determine retroviral core, pol is the reverse transcriptase, and env is the envelope protein, resides in lipid layer, and determines viral tropism.

Although Plat-E cells harbor the stably expressed G/P/E expression vectors, we further cotransfect more G/P/E plasmid DNA with MSCV2.2-IRES-GFP at a mass ratio of 1:3 to ensure good virus packaging.

4. The T-cell purification is performed in sterile environment, which is beneficial for the subsequent culture. Therefore, this part is carried out in clean hood.
5. DNase I and collagenase IV are dissolved in PBS. EDTA terminates collagenase digestion by chelating calcium and magnesium which are required for collagenase activity. Enzyme termination is important, and any residual collagenase would be detrimental for the cell culture.
6. We found that resting the cells for a few days in the IL2-containing media after stimulation and spinfection helps to reduce the experimental variability, although it would also work if we adoptively transfer OT-II cells immediately after spinfection.
7. We can obtain 50–80% transduction efficiency by using unconcentrated fresh retroviruses. For higher efficiency, we would recommend virus concentration with polyethylene glycol (PEG) 8000.
8. We used a three-step process to stain CXCR5: purified rat anti-mouse CXCR5 antibody followed by a biotinylated anti-rat IgG antibody and fluorophore-conjugated streptavidin antibody. This method enables a clearer distinction between $CXCR5^{+}$ and $CXCR5^{-}$ populations than directly using fluorophore-conjugated anti-CXCR5 antibody.
9. Store prepared fluorescein antibody cocktails in dark tubes or wrap in tin foil on ice. Fluorescein quenches easily when exposed to light.

References

1. Crotty S (2014) T follicular helper cell differentiation, function, and roles in disease. Immunity 41:529–542
2. Vinuesa CG, Linterman MA, Yu D, MacLennan IC (2016) Follicular helper T cells. Annu Rev Immunol 34:335–368
3. Johnston RJ, Poholek AC, DiToro D, Yusuf I, Eto D, Barnet B, Dent AL, Craft J, Crotty S (2009) Bcl6 and Blimp-1 are reciprocal and antagonistic regulators of T follicular helper cell differentiation. Science 325:1006–1010
4. Nurieva RI, Chung Y, Martinez GJ, Yang XO, Tanaka S, Matskevitch TD, wang YH, Dong C (2009) Bcl6 mediates the development of T follicular helper cells. Science 325:1001–1005
5. Yu D, Rao S, Tsai LM, Lee SK, He Y, Sutcliffe EL, Srivastava M, Linterman M, Zheng L, Simpson N, Ellyard JI, Parish IA, Ma CS, Li QJ, Parish CR, Mackay CR, Vinuesa CG (2009) The transcriptional repressor Bcl-6 directs T follicular helper cell lineage commitment. Immunity 31:457–468
6. Weber JP, Fuhrmann F, Feist RK, Lahmann A, Al Baz MS, Gentz LJ, Vu Van D, Mages HW, Haftmann C, Riedel R, Grün JR, Schuh W, Kroczek RA, Radbruch A, Mashreghi MF, Hutloff A (2015) ICOS maintains the T follicular helper cell phenotype by down-regulating Krüppel-like factor 2. J Exp Med 212:217–233
7. Choi YS, Kageyama R, Eto D, Escobar TC, Johnston RJ, Monticelli L, Lao C, Crotty S

(2011) ICOS receptor instructs T follicular helper cell versus effector cell differentiation via induction of the transcriptional repressor Bcl6. Immunity 34:932–946

8. Ansel KM, McHeyzer-Williams LJ, Ngo VN, McHeyzer-Williams MG, Cyster JG (1999) In vivo-activated CD4 T cells upregulate CXC chemokine receptor 5 and reprogram their response to lymphoid chemokines. J Exp Med 190:1123–1134
9. Liu X, Chen X, Zhong B, Wang A, Wang X, Chu F, Nurieva RI, Yan X, Chen P, van der Flier LG, Nakatsukasa H, Neelapu SS, Chen W, Clevers H, Tian Q, Qi H, Wei L, Dong C (2014) Transcription factor achaete-scute homologue 2 initiates follicular T-helper-cell development. Nature 507:513–518
10. Xu L, Cao Y, Xie Z, Huang Q, Bai Q, Yang X, He R, Hao Y, Wang H, Zhao T, Fan Z, Qin A, Ye J, Zhou X, Ye L, Wu Y (2015) The transcription factor TCF-1 initiates the differentiation of T(FH) cells during acute viral infection. Nat Immunol 16:991–999
11. Choi YS, Gullicksrud JA, Xing S, Zeng Z, Shan Q, Li F, Love PE, Peng W, Xue HH, Crotty S (2015) LEF-1 and TCF-1 orchestrate TFH differentiation by regulating differentiation circuits upstream of the transcriptional repressor Bcl6. Nat Immunol 16:980–990
12. Wang H, Geng J, Wen X, Bi E, Kossenkov AV, Wolf AI, Tas J, Choi YS, Takata H, Day TJ, Chang LY, Sprout SL, Becker EK, Willen J, Tian L, Wang X, Xiao C, Jiang P, Crotty S, Victora GD, Showe LC, Tucker HO, Erikson J, Hu H (2014) The transcription factor Foxp1 is a critical negative regulator of the differentiation of follicular helper T cells. Nat Immunol 15:667–675
13. Johnston RJ, Poholek AC, DiToro D, Yusuf I, Eto D, Barnett B, Dent AL, Craft J, Crotty S (2009) Bcl6 and Blimp-1 are reciprocal and antagonistic regulators of T follicular helper cell differentiation. Science 325:1006–1010
14. Weber JP, Fuhrmann F, Feist RK, Lahmann A, Al Baz MS, Gentz LJ, Mages HW, Haftmann C, Riedel R, Grun JR, Schuh W, Kroczek RA, Radbrunch A, Mashreghi MF, Hutloff A (2015) ICOS maintains the T follicular helper cell phenotype by down-regulating Krüppel-like factor 2. J Exp Med 212:217–233
15. Lee JY, Skon CN, Lee YJ, Oh S, Taylor JJ, Malhotra D, Jenkins MK, Rosenfeld MG, Hogquist KA, Jameson SC (2015) The transcription factor KLF2 restrains $CD4^+$ T follicular helper cell differentiation. Immunity 42:252–264
16. Johnston RJ, Choi YS, Diamond JA, Yang JA, Crotty S (2012) STAT5 is a potent negative regulator of TFH cell differentiation. J Exp Med 209:243–250
17. Nurieva RI, Podd A, Chen Y, Alekseev AM, Yu M, Qi X, Huang H, Wen R, Wang J, Li HS, Watowich SS, Qi H, Dong C, Wang D (2012) STAT5 protein negatively regulates T follicular helper (Tfh) cell generation and function. J Biol Chem 287:11234–11239
18. Morita S, Kojima T, Kitamura T (2000) Plat-E: an efficient and stable system for transient packaging of retroviruses. Gene Ther 7:1063–1066

Chapter 11

In Vitro Generation of Stem Cell Memory-Like T Cells from Activated T Cells

Makoto Ando, Mari Ikeda, Akihiko Yoshimura, and Taisuke Kondo

Abstract

Adoptive T-cell therapy is an attractive strategy for cancer immunotherapy. The transfer of in vitro expanded tumor-associated antigen (TAA)-specific T cells from patients may effectively fight against the original tumor cells. The chimeric antigen receptor-engineered T (CAR-T) cells are also shown to be a promising therapy for hematologic malignancies. However, one of the limitations of these T-cell-based therapies is a rapid acquisition of tolerant (anergy, deletion, dysfunctional and/or exhausted) phenotypes of T cells during activation in vitro and/or after transfer in vivo. We and others found that stem cell memory T (T_{SCM}) cells are strongly resistant against such tolerance, showing strong expansion and persistence in vivo, and provide long-lasting antitumor effects. Here we describe a protocol for the generation of phenotypically T_{SCM}-like cells (iT_{SCM} cells), which can be induced by simple co-culture of activated T cells with OP9 stroma cells expressing a Notch ligand. We also showed the methods of cancer immunotherapy by using NSG mice.

Key words Adoptive T-cell therapy, Stem cell memory T cells, Notch signaling, Co-culture with feeder cells, EB virus-specific T cells, NSG mice

1 Introduction

Adoptive T-cell-based therapy is a well-known strategy as a recent technological breakthrough for cancer therapy [1, 2]. The procedure includes (1) collection of potential tumor-associated antigen (TAA)-specific T cells from tumor tissues and peripheral blood of patients, (2) the expansion of them in vitro, and (3) the infusion of expanded T cells into the patient. In place of TAA-specific T cells, chimeric antigen receptor-engineered T (CAR-T) cells have been successfully applied for hematological cancers [3–5]. However, several technical limitations remain unsolved. In vitro expansion of T cells provides strong effector functions, whereas it often contributes terminally differentiated phenotypes and easily induces dysfunctionality upon transfer because of repeated antigen stimulation [6]. To manufacture TAA-specific T cells or CAR-T cells, T-cell

Chaohong Liu (ed.), *T-Cell Receptor Signaling: Methods and Protocols*, Methods in Molecular Biology, vol. 2111, https://doi.org/10.1007/978-1-0716-0266-9_11,

receptor (TCR) and cytokine signaling continuously stimulates T cells, resulting in T-cell exhaustion and dysfunction [7]. These dysfunctional or exhausted T cells are easily expired and elevate the risk of relapse [8].

On the other hand, less differentiated T cells showing naïve T cell or central memory phenotypes are known to govern to the outcomes of adoptive T-cell therapy [9, 10]. Especially, stem cell memory T (T_{SCM}) cells, which show overlapping phenotypes between naïve T cell and central memory (T_{CM}) cells, have been shown to provide superior antitumor ability [9, 11]. Therefore, a number of methods to generate TAA-specific T_{SCM} cells or CAR-T_{SCM} cells have been reported [12–16]. However, all of these methods are generation of T_{SCM} cells from naïve T cells. There are few naïve T cells but more memory T-cell pool in aged people [17]. Therefore, it would be beneficial if direct conversion of activated T cells but not naïve T cells into (T_{SCM})-like cell is possible. We previously reported the methods for generating T_{SCM}-like (iT_{SCM}) cells from endogenous memory phenotype-activated T cells [18]. In brief, activated T cells were converted into iT_{SCM} cells by co-culturing with a Notch ligand-expressing feeder cells.

In this report, we describe a step-by-step protocol for generating antigen (EB virus)-specific human iT_{SCM} cells, including reagents, experimental setting, and detailed procedures. First, whole CD8 T cells are isolated from peripheral blood mononuclear cells and stimulated by the co-culture with autologous lymphoblastoid cell line (LCL). After stimulation, proliferated LCL-specific T cells are transferred on a Notch ligand-expressing feeder (OP9-hDLL1) layer and co-cultured for 11 days (Fig. 1). The co-culture with OP9-hDLL1 cells induces iT_{SCM} cells. We previously reported that co-culture of EBV-specific T cells with OP9-hDLL1 cells induced EBV-specific iT_{SCM} cells, and these iT_{SCM} cells actually exhibited strong in vivo antitumor effect [19].

2 Materials

2.1 Cells

1. OP9-hDLL1 cell line: human delta-like 1 (DLL1) cDNA was subcloned into a lentiviral vector CS-II and transduced into the OP9 cell line (ATCCR CRL-2749™). OP9 cells stably expressing hDLL1 (OP9-hDLL1) were isolated by FACS using anti-hDLL1 antibody.
2. Autologous LCL, as referred in previous reports (*see* **Note 1**) [20, 21].
3. Peripheral blood mononuclear cells (PBMC) (*see* **Note 2**), *see* Subheading 3.

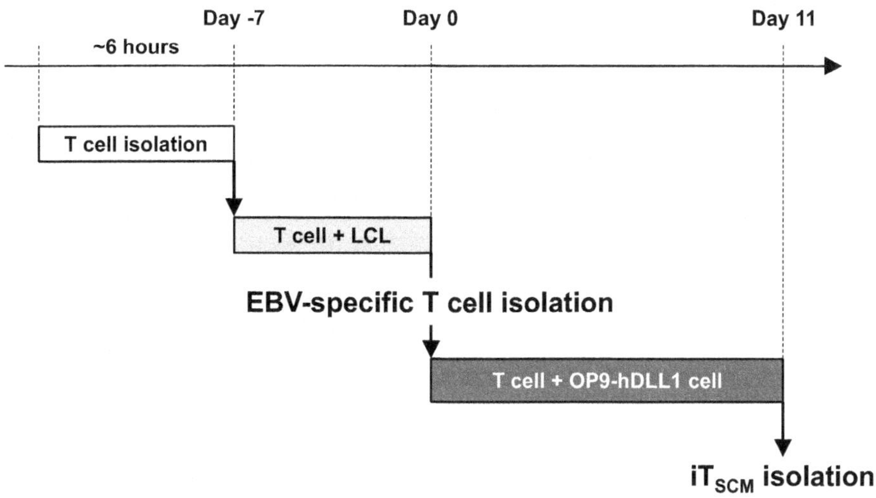

Fig. 1 Schematic for scheduling to induce iT_{SCM} cells. Peripheral $CD8^+$ T cells are isolated at first and continuously co-cultured with autologous lymphoblastoid cell line (LCL) for 7 days. Next, Epstein-Barr virus (EBV)-specific T cells are purified by cell sorting and transferred onto human Delta-like 1-expressing feeder cells, OP9-hDLL1 cells, for 11 days. Evaluation of iT_{SCM} is performed 11 days after OP9-hDLL1 co-culture. (Data and figures are from ref. 19)

2.2 Cell Culture

1. Density gradient media, i.e., Ficoll-Paque, Lymphoprep.
2. Dulbecco's phosphate buffer of saline (PBS).
3. Pooled normal human male AB serum, which is obtained from healthy donors with serotype AB (Innovative Research, Inc.) (*see* **Note 3**).
4. Fetal bovine serum (FBS) for OP9-hDLL1 cell culture (*see* **Note 4**).
5. Human IL-7 (Peprotech).
6. $CD8^+$ T cell isolation kit, human (Miltenyi Biotec).
7. LS column (Miltenyi Biotec).
8. MACS buffer: 0.5% [w/v] BSA and 2 mM EDTA in PBS.
9. OP9-hDLL1 cell medium: minimum essential medium alpha (αMEM) containing 20% FBS and antibiotics.
10. T-cell prime medium: RPMI1640 medium supplemented with 10% AB serum, antibiotics, 1 mM sodium pyruvate, 1 mM HEPES buffer solution, MEM nonessential amino acid solution, 2 mM L-glutamine, and 55 μM 2-mercaptoethanol.
11. iT_{SCM} induction medium: αMEM containing 20% FBS, antibiotics, 55 μM 2-mercaptoethanol, and 10 ng/ml human IL-7.

2.3 Antibodies (Ab) and Fluorophores

1. Anti-human CD8α PerCP-Cy5.5 (BioLegend).
2. Anti-human CD45RA PE-Cy7 (BioLegend).

3. Anti-human CCR7 Alexa Fluor 647 (BioLegend).
4. CellTrace CFSE cell proliferation kit (Thermo Fisher Scientific).

2.4 Equipment

1. QuadroMACS separator (Miltenyi Biotec).
2. Cell incubator (temperature 37 °C, CO_2 level 5%).
3. Water bath (temperature 37 °C).
4. Centrifuge.
5. X-ray irradiator: for example, CELLRAD (FAXITRON).
6. FACS Canto II cytometer (BD Biosciences).
7. FACS Aria II cell sorter (BD Biosciences).
8. FlowJo software v10 (Tree Star, Ashland).

2.5 Mice

1. 8-week-old NOD.Cg-*PrkDC* ^{cid}*Il2rg*^{tm1Wjl} (NSG) mice (Charles River Laboratories).

3 Methods

Appropriate institutional regulatory board permission must be obtained before starting experiments. Written informed consent was obtained from all individuals.

3.1 Human CD8$^+$ T-Cell Isolation

1. Collect 40 ml of whole blood from an EBV-seropositive healthy donor.
2. Separate PBMC from 40 ml samples of whole blood by centrifugation (900 × *g*, 35 min, 20 °C, slow acceleration, and slow deceleration) through Lymphoprep (Alere Technologies AS) or other density centrifugation media (e.g., Ficoll-Paque) according to the manufacturer's instructions (Fig. 2) (*see* **Note 5**).
3. Collect PBMC band between serum and Lymphoprep layers.
4. Wash banded PBMC twice in MACS buffer, and resuspend in T-cell prime medium in 15 ml centrifuge tubes until starting the next step (*see* **Note 6**).
5. Obtain a cell count for PBMC.
6. Enrich human CD8$^+$ T cells by negative selection (Fig. 2). We use a CD8$^+$ T cell isolation kit, human from Miltenyi Biotec.
7. Remove supernatant by centrifugation (400 × *g*, 5 min, 4 °C).
8. Resuspend PBMC pellet in 40 μl of MACS buffer per 10^7 total PBMC.
9. Add 10 μl of biotin-antibody cocktail per 10^7 total PBMC.

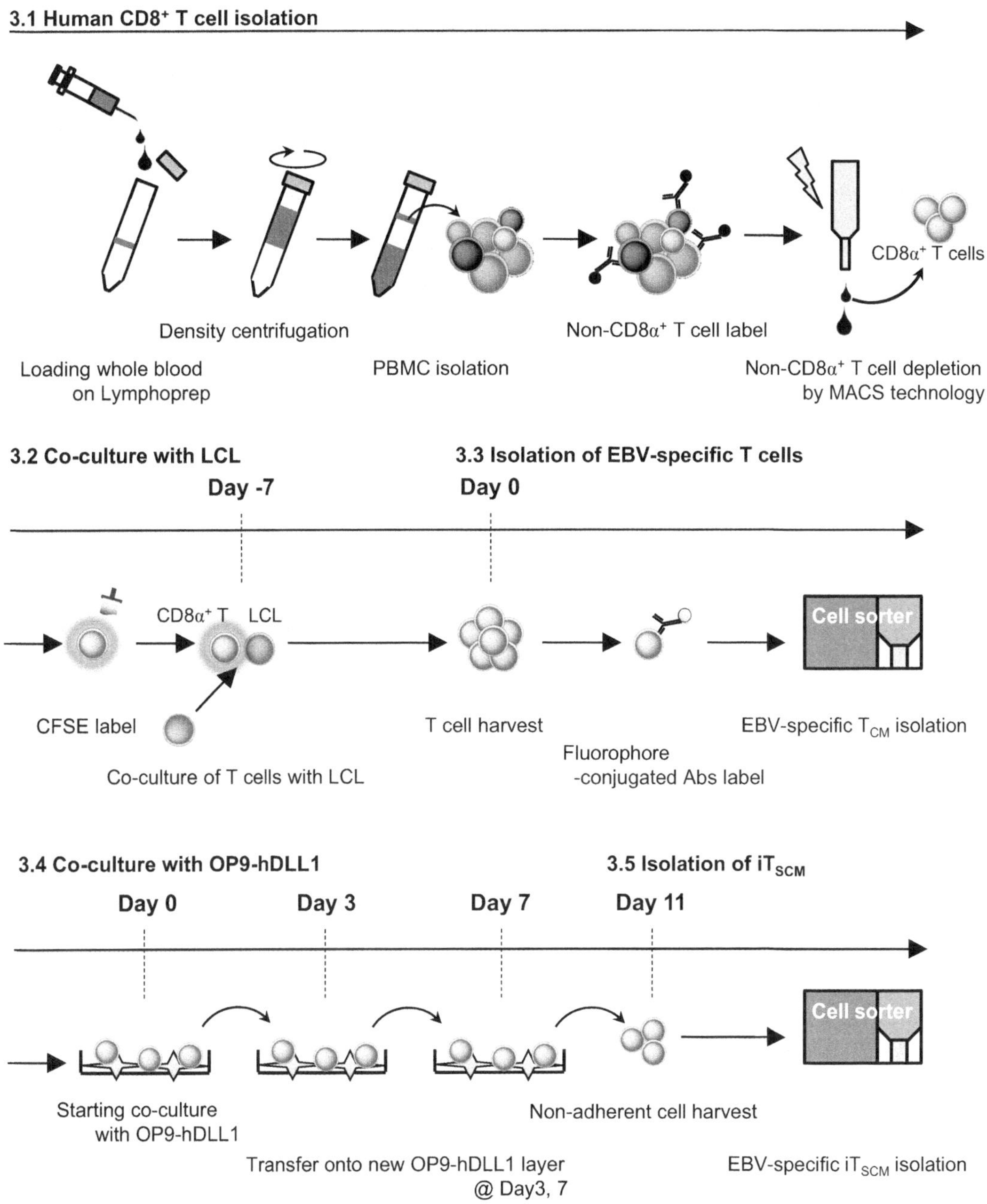

Fig. 2 Detailed protocols for iT_{SCM} generation. (*Top*) Start with $CD8^+$ T-cell isolation on Day 7 of the prime step. Whole blood cells of healthy donors are loaded on Lymphoprep reagent and density-based centrifuged. Next, peripheral blood mononuclear cell (PBMC) layer is carefully isolated from the centrifuge tube. And then, peripheral $CD8^+$ T cells are negatively isolated from PBMC by $CD8^+$ T-cell isolation kit. (*Middle*) Isolated $CD8^+$ cells are labeled by cell trace dye (e.g., carboxyfluorescein succinimidyl ester [CFSE]), followed by the addition of labeled $CD8^+$ T cells to irradiated autologous LCL (T-to-LCL ratio of 4:1) and the start of co-culture in a 96-well round-bottom plate (100 μl per well). EBV-specific T cells with memory phenotypes, which are defined as cell trace dye-diluted $CD45RA^-CCR7^+CD8\alpha^+$ cells, are purified by cell sorting (also *see* Fig. 3a). (*Bottom*) Purified EBV-specific T cells (1×10^5 cells/ml) are co-cultured with OP9-hDLL1 cells (6-, 12-, 24-, and

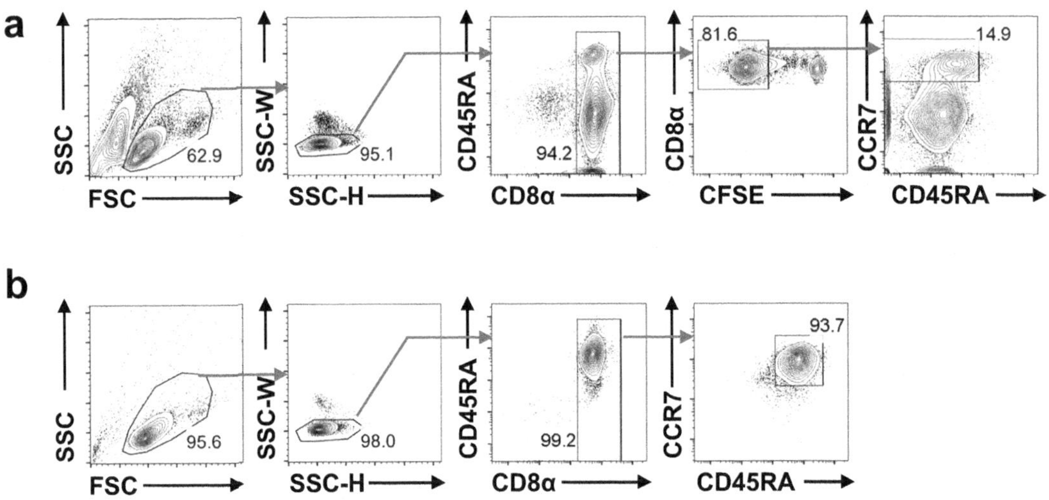

Fig. 3 Gating strategy in EBV-specific iT_{SCM} generation. (**a**) CFSE-labeled $CD8^+$ T cells were co-cultured with 40 Gy irradiated EBV-transformed autologous LCL for 7 days. EBV-specific activated T cells mainly showed T_{EM} ($CD8\alpha^+CFSE^{lo/-}CCR7^-$) cell phenotypes and T_{CM} ($CD8\alpha^+CFSE^{lo/-}CCR7^+$ cell phenotypes (Day 0). (**b**) T_{CM} cells were sorted and then co-cultured with OP9-hDLL1 cells for 11 days. Flow cytometry analysis of $CD8\alpha^+$ cells 11 days after OP9-hDLL1 cell co-culture

10. Mix the contents well and incubate for 5 min in the refrigerator (4 °C).
11. Add an additional 30 μl of MACS buffer per 10^7 total PBMC.
12. Add an additional 20 μl of $CD8^+$ T-Cell MicroBead Cocktail per 10^7 total PBMC.
13. Mix the cell suspension well and incubate for 10 min in the refrigerator (4 °C).
14. Add an additional 10 ml of MACS buffer and centrifuge (400 × *g*, 5 min, 4 °C).
15. Place a QuadroMACS separator on the clean bench.
16. Place an LS column in the magnetic field of the QuadroMACS Separator.
17. Equilibrate the LS column by rinsing with 3 ml of MACS buffer.
18. Resuspend the pellet in 1 ml of MACS buffer (*see* **Note** 7).
19. Apply cell suspension onto the LS column.

Fig. 2 (continued) 48-well flat-bottom plate or 10 cm dish) in the presence of human IL-7 (10 ng/ml). Harvesting $CD8\alpha^+$ T cells, adjusting cell density (1 × 10^5 cells/ml), and transferring onto a new OP9-hDLL1 layer are done on Days 3 and 7. Co-culture with OP9-hDLL1 cells for 11 days induced iT_{SCM} cells, defined as $CD8\alpha^+CD45RA^+CCR7^+$ cells (also *see* Fig. 3b)

20. Collect negatively isolated flow-through that contains enriched $CD8^+$ T cells (Fig. 2).
21. Rinse 15 ml centrifuge tube twice with 1 ml of MACS buffer, and also apply the buffer onto the column.
22. Wash the column twice with 3 ml of MACS buffer and completely collect unlabeled cells in the flow-throw.
23. Collect a total of 9 ml of cell suspension that contains enriched $CD8^+$ T cells.
24. Remove supernatant by centrifugation (400 × *g*, 5 min, 4 °C).
25. Resuspend the pellet in T-cell prime medium in 15 ml centrifuge tube until starting the next step.

3.2 Co-culturing of Human $CD8^+$ T Cells with Autologous LCL

1. Obtain a count of human $CD8^+$ T cells.
2. Stain T cells in a 37 °C water bath by cell trace dye (e.g., CFSE) according to the manufacturer's instructions (Fig. 2) (*see* **Note 8**).
3. Harvest autologous LCL.
4. Irradiate autologous LCL at a dose of 40 Gy.
5. Remove the supernatant of LCL by centrifugation (400 × *g*, 5 min, 4 °C).
6. Resuspend with T-cell prime medium and obtain a cell count for LCL.
7. Transfer T cells and LCL (T-cell-to-LCL ratio of 4:1) into fresh T-cell prime medium at 5×10^5/ml of T cells and 1.25×10^5/ml of LCL in 100 μl T-cell prime medium in a 96-well round-bottom plate (Fig. 2).
8. Spin down the T-cell culture plate that contains T cells and LCL (*see* **Note 9**).
9. Place the tissue culture plate and start with incubation in a humidified CO_2 incubator at 37 °C.
10. Remove the medium and resuspend T cells in T-cell prime medium and incubate at 37 °C every 4 days.
11. Scale up the culture medium volume and culture wells. T cells should be cultured at 5×10^5/ml.
12. Analyze and isolate EBV-specific T cells by flow cytometry analysis. EBV-specific T cell could be detected by cell trace dye-diluted T cells (Fig. 2).

3.3 Isolation of EBV-Specific T Cells with Central Memory Phenotypes

1. Harvest activated T cells from the culture plate, and transfer the cell suspension into 1.5 ml tubes at an appropriate date at around 7 days after activation (Fig. 2) (*see* **Note 10**).
2. Harvest the supernatant containing T cells.
3. Transfer cell suspension to 15 ml or 50 ml centrifuge tubes.

4. Remove the supernatant by centrifugation (400 × *g*, 5 min, 4 °C).
5. Prepare fluorophore-conjugated Ab staining solution.
   In brief, Ab staining solution; fluorophore-conjugated anti-CD8α, anti-CD45RA, anti-CCR7 antibodies (dilution factor; 1:100) in MACS buffer.
6. Resuspend pellet in 100 μl Ab staining solution and stain cells by fluorophore-conjugated Ab for 15 min at 4 °C (Fig. 2).
7. Add 10 ml of MACS buffer and centrifuge (400 × *g*, 5 min, 4 °C).
8. Remove supernatant and resuspend cells in 1 ml of MACS buffer.
9. Isolate phenotypically T_{CM} cells by cell sorter (we normally use a FACS Aria sorter) gated as $CD8\alpha^+CD45RA^-CCR7^+$ cells (Figs. 2 and 3a) (*see* **Note 11**).

3.4 Co-culturing of EBV-Specific T Cells with OP9-hDLL1 Cells

1. Prepare iT_{SCM} induction medium containing 10 ng/ml recombinant human IL-7 during cell sorting.
2. Centrifuge collected cell suspension immediately (400 × *g*, 5 min, 4 °C).
3. Resuspend pellet in iT_{SCM} induction medium.
4. Obtain cell count for isolated T_{CM} cells.
5. Resuspend T cells into fresh iT_{SCM} induction medium at 1×10^5/ml (*see* **Note 12**).
6. Remove the supernatant of OP9-hDLL1 culture plate or dish and immediately transfer T-cell suspension onto OP9-hDLL1 cells (Fig. 2) (*see* **Note 13**).
7. Place culture plate or dish and start with incubation in a humidified CO_2 incubator at 37 °C.
8. Change medium and resuspend T cells in iT_{SCM} induction medium, replate T cells onto new OP9-hDLL1 layer, and incubate at 37 °C to pass T cells on new feeder cells for every 3–4 days (Fig. 2) (*see* **Notes 14** and **15**).

3.5 Isolation and Analysis of iT_{SCM} Cells

Anticipated results by this method were described in the Note section (*see* **Note 16**).

1. Gently harvest the supernatant of the co-culture with OP9-hDLL1 cells, and place it in a 50 ml centrifuge tube. It will contain enriched iT_{SCM} cells (*see* **Note 17**).
2. Remove the supernatant by centrifugation (400 × *g*, 5 min, 4 °C).
3. Prepare fluorophore-conjugated Ab staining solution.

4. Resuspend pellet in 100 μl Ab staining solution and stain cells by fluorophore-conjugated Ab for 15 min at 4 °C.
5. Add 10 ml of MACS buffer and centrifuge (400 × *g*, 5 min, 4 °C).
6. Remove supernatant and resuspend cells in 1 ml of MACS buffer.
7. Analyze by flow cytometry (e.g., FACS Canto II) or isolate phenotypically iT_{SCM} cells by cell sorter (we normally use a FACS Aria sorter) gated as $CD8\alpha^+CD45RA^+CCR7^+$ cells (Figs. 2 and 3b).
8. Centrifuge collected cell suspension immediately (400 × *g*, 5 min, 4 °C).
9. Obtain cell count for isolated iT_{SCM} cells and continuously use iT_{SCM} cells for appropriate experiments.
10. All data obtained by flow cytometer are analyzed by FlowJo software.

3.6 Adoptive iT_{SCM} Cell Transfer to Tumor-Bearing NSG Mice

1. LCL (5×10^6) cells were subcutaneously inoculated into NSG mice 12 days before T-cell transfer.
2. EBV-specific iT_{SCM} cells (5×10^5) were adoptively transferred through i.v. into tumor-bearing mice.
3. Mice were intraperitoneally injected with human IL-2 (500 ng per mouse) twice, 24 and 48 h after T-cell transplantation.
4. Tumor sizes were measured with a vernier caliper every 3 or 4 days, and tumor volume was calculated according to the following formula: volume $= 0.5 \times$ length $\times$ width^2. Figure 4 shows tumor volumes and survival rates of LCL-bearing mice [19].

4 Notes

1. LCL is a B-cell line transformed by Epstein-Barr (EB) virus. Established cell lines generally express CD19 molecule and present EB virus-specific antigens. Co-culturing of memory T cells with autologous LCLs induces endogenous EB virus antigen-specific T cells. In addition, the maintenance of a Biosafety Level 2 (BSL-2) facility is required for starting experiments involving the EB virus. More information and guidelines on biosafety and practices are provided by the CDC and NIH [22]. This study was performed using BCL-2 equipment in a BCL-2 facility.
2. Checking Epstein-Barr virus' seropositivity is needed before starting experiments. Endogenous EBV-specific T cells in

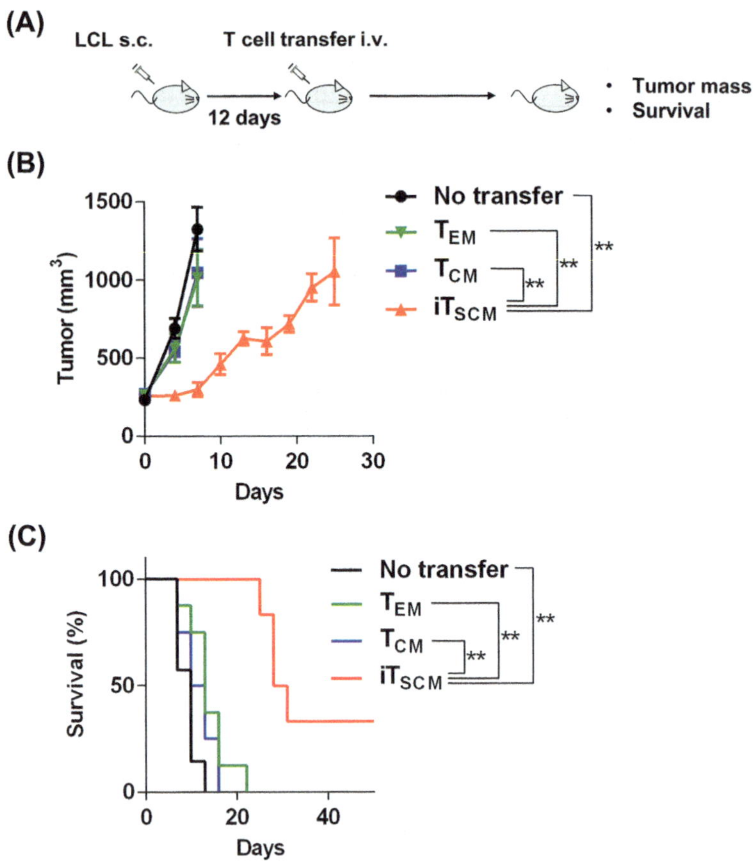

Fig. 4 Antitumor potential of human iT_{SCM} cells. (**a**) Schematic for generating humanized tumor model mice for adoptive T-cell therapy. Severe immunodeficient (NOD.Cg-*PrkDC* ^{cid}*Il2rg*^{tm1Wjl}*/Szj*, NSG) mice were subcutaneously inoculated with 5×10^6 EBV-transformed LCL. 5×10^5 T_{EM}, T_{CM}, and iT_{SCM} cells were adoptively transferred into LCL-bearing mice 12 days after LCL inoculation. (**b**) Tumor volumes of LCL-bearing mice. (**c**) Survival rates of LCL-bearing mice. ($n = 7$ for no transfer and T_{EM}; $n = 4$ for T_{CM}; $n = 6$ for iT_{SCM}) $**P < 0.01$ (one-way ANOVA [B], the Kaplan-Meier method [C]). Data are representative of at least two independent experiments. Error bars show s.e.m. (Data and figures are from ref. 19)

memory T-cell pool could react with EBV-specific antigens which are presented on cell surface of LCL.

3. As serum quality is critical for primary T-cell culture, testing of several lots of serum is necessary for culture optimization and experimental quality.
4. As serum quality is critical for OP9 feeder cell culture, testing of several lots of serum is necessary for culture optimization and experimental quality.

5. Centrifugation at the appropriate temperature is critical because low or high temperatures can change the density of the centrifugation media. Slow acceleration and deceleration are also critical for creating a clear PBMC band. We always use 15 ml centrifuging tube made of polyethylene terephthalate, PET. PBMC band is clearly visible in PET tubes.
6. It is important for centrifugation by more than $500 \times g$ during the first wash step because of contaminating the density gradient media. Centrifugation by low gravity is not sufficient to generate a pellet.
7. It is necessary for resuspending more than 1 ml of MACS buffer if dealing with more than 5×10^7 PBMC if dealing with a clogging column.
8. A small number of EBV-specific T cells are contained in whole $CD8^+$ T cells. To distinguish EBV-specific T cells and nonresponding T cells, a detection system of cell proliferation is necessary.
9. Accumulation of both T cells and LCL at the bottom of the 96-well round-bottom plate is necessary for strongly responding EBV-specific T cells.
10. It is necessary for preparation of OP9-hDLL1 cell. Harvest and replate $2–4 \times 10^4$/ml OP9-hDLL1 cells into a 6-well tissue culture plate or a 10 cm culture dish the day before harvesting activating T cells.
11. Maintain the collection tube at 4 °C. Chilling collected cells is critical for cell survival.
12. Concentration is clearly critical for iT_{SCM} induction. If overloading on OP9-hDLL1 layer, iT_{SCM} induction may not succeed.
13. Immediate removal of the supernatant of OP9-hDLL1 cells and transference of the T-cell suspension onto the OP9-hDLL1 layer are critical to avoid desiccation of the OP9-hDLL1 plate.
14. During the induction step, T cells are robustly expanded by threefold compared to their initial cell count. Count T-cell number at every passage and maintain the T-cell density at 1×10^5/ml. Maintaining the culture of T cells at low density is essential for iT_{SCM} induction because the contact of single T cell and OP9-hDLL1 cells is critical.
15. Do not allow OP9-hDLL1 cells to become over confluent. OP9-hDLL1 cells easily die after reaching over confluence, and this causes T-cell damage. Before OP9-hDLL1 reaches over confluence, transfer T cells onto a new OP9-hDLL1 layer.

16. If isolation of all $CD8^+$ T cells has been performed without any defects, the recovered cell number and purity should be greater than 0.5–3 × 10^6 cells (dependent on donors) and 90%, respectively, after the $CD8^+$ enrichment procedure using 10 ml of whole blood from a single healthy donor. If $CD8^+$ T-cell activation was also performed without any defects, more than 60% of T cells can be activated by TCR activation on Day 0 (Fig. 3a) and 15% of activated T cells will show T_{CM} phenotypes. Given that complete $CD8^+$ T-cell recovery from the $CD8^+$ T-cell enrichment procedure is 3 × 10^6 cells (Day 7) and the phenotypically T_{CM}-cell recovery from cell sorting is 15% (Day 7), the number of T_{CM} cells recovered from 10 ml of whole blood from a healthy donor should be 1–2 × 10^5 cells. After OP9-hDLL1 co-culture for 11 days without any defects, the number of iT_{SCM} cells recovered would be 0.5–2 × 10^6 cells, and the purity of iT_{SCM} cells would be greater than 80% (Fig. 3b).
17. Do not aggressively pipette the supernatant in the co-culture dish in order to prevent OP9-hDLL1 contamination into the T-cell suspension.

References

1. June CH (2007) Adoptive T cell therapy for cancer in the clinic. J Clin Invest 117 (6):1466–1476
2. Rosenberg SA, Dudley ME (2009) Adoptive cell therapy for the treatment of patients with metastatic melanoma. Curr Opin Immunol 21 (2):233–240
3. June CH et al (2018) CAR T cell immunotherapy for human cancer. Science 359 (6382):1361–1365
4. Dunbar CE et al (2018) Gene therapy comes of age. Science 359(6372):eaan4672
5. June CH, Sadelain M (2018) Chimeric antigen receptor therapy. N Engl J Med 379(1):64–73
6. Gattinoni L et al (2005) Acquisition of full effector function in vitro paradoxically impairs the in vivo antitumor efficacy of adoptively transferred CD8+ T cells. J Clin Invest 115 (6):1616–1626
7. Levine BL et al (2017) Global manufacturing of CAR T cell therapy. Mol Ther Methods Clin Dev 4:92–101
8. Louis CU et al (2011) Antitumor activity and long-term fate of chimeric antigen receptor-positive T cells in patients with neuroblastoma. Blood 118(23):6050–6056
9. Hinrichs CS et al (2009) Adoptively transferred effector cells derived from naive rather than central memory CD8+ T cells mediate superior antitumor immunity. Proc Natl Acad Sci U S A 106(41):17469–17474
10. Berger C et al (2008) Adoptive transfer of effector CD8+ T cells derived from central memory cells establishes persistent T cell memory in primates. J Clin Invest 118(1):294–305
11. Gattinoni L et al (2011) A human memory T cell subset with stem cell-like properties. Nat Med 17(10):1290–U325
12. Alvarez-Fernandez C et al (2016) A short CD3/CD28 costimulation combined with IL-21 enhance the generation of human memory stem T cells for adoptive immunotherapy. J Transl Med 14(1):214
13. Cieri N et al (2013) IL-7 and IL-15 instruct the generation of human memory stem T cells from naive precursors. Blood 121(4):573–584
14. Hurton LV et al (2016) Tethered IL-15 augments antitumor activity and promotes a stem-cell memory subset in tumor-specific T cells. Proc Natl Acad Sci U S A 113(48): E7788–E7797
15. Scholz G et al (2016) Modulation of mTOR signalling triggers the formation of stem cell-like memory T cells. EBioMedicine 4:50–61
16. Sabatino M et al (2016) Generation of clinical-grade CD19-specific CAR-modified CD8+ memory stem cells for the treatment of

human B-cell malignancies. Blood 128 (4):519–528

17. Nikolich-Zugich J (2014) Aging of the T cell compartment in mice and humans: from no naive expectations to foggy memories. J Immunol 193(6):2622–2629
18. Kondo T et al (2017) Notch-mediated conversion of activated T cells into stem cell memory-like T cells for adoptive immunotherapy. Nat Commun 8:15338
19. Kondo T et al (2018) Generation and application of human induced-stem cell memory T cells for adoptive immunotherapy. Cancer Sci 109(7):2130–2140
20. Kubuschok B et al (2002) Use of spontaneous Epstein-Barr virus-lymphoblastoid cell lines genetically modified to express tumor antigen as cancer vaccines: mutated p21 ras oncogene in pancreatic carcinoma as a model. Hum Gene Ther 13(7):815–827
21. Hui-Yuen J et al (2011) Establishment of Epstein-Barr virus growth-transformed lymphoblastoid cell lines. J Vis Exp 57:3321
22. Chosewood LC et al (2009) Biosafety in microbiological and biomedical laboratories, 5th edn. U.S. Dept. of Health and Human Services, Public Health Service, Centers for Disease Control and Prevention, National Institutes of Health, Washington, D.C., U.S.

Chapter 12

Artificial Antigen Presentosomes for T Cell Activation

Yi-Geng Pang and Chien-Chung Chang

Abstract

$CD8^+$ T cells constitute an essential component of the adaptive immune system. They are activated through T cell receptor (TCR) recognizing antigenic peptides presented by MHC class I molecules expressed by antigen-presenting cells, such as dendritic cells (DCs). Harvesting a large number of activated, antigen-specific human $CD8^+$ T cells for functional studies has been a laborious task for immunologists, largely because of the variables associated with DC preparations. Here we describe a robust, cost-effective DC-free antigen-presenting system capable of generating a large number of antigen-specific $CD8^+$ T cells in vitro.

Key words T cell activation, Major histocompatibility complex, Human leukocyte antigen, Antigen presentation

1 Introduction

In the mammalian immune system, T cells function to protect the host from viral infection or cancer cell expansion by recognizing viral or tumor antigen-derived peptides in the context of MHC molecules [1]. Activation of naïve T cells requires their physical interaction with antigen-presenting cells, such as dendritic cells (DCs), where T cell receptor (TCR) has direct contact with peptide-MHC complex on DCs [2], along with proper co-stimulation [3]. $CD8^+$ T cells, when activated, possess cytolytic activity and therefore play a major role in eliminating virus-infected or malignant tumor cells [4]. For research purposes, it is critical to obtain a large number of activated, antigen-specific $CD8^+$ T cells, especially human $CD8^+$ T cells, but this is hampered by the many variables that affect myeloid DC preparations in vitro, e.g., ineffective differentiation, poor antigen presentation, insufficient peptide loading, short life span, and costs.

To overcome these limitations, we extend a previous research [5] including our own [6] on the production of a 2-component (2-C) HLA class I/peptide-fused β_2m recombinant protein complex to generate an artificial antigen-presenting cell where the 2-C

Chaohong Liu (ed.), *T-Cell Receptor Signaling: Methods and Protocols*, Methods in Molecular Biology, vol. 2111, https://doi.org/10.1007/978-1-0716-0266-9_12, © Springer Science+Business Media, LLC, part of Springer Nature 2020

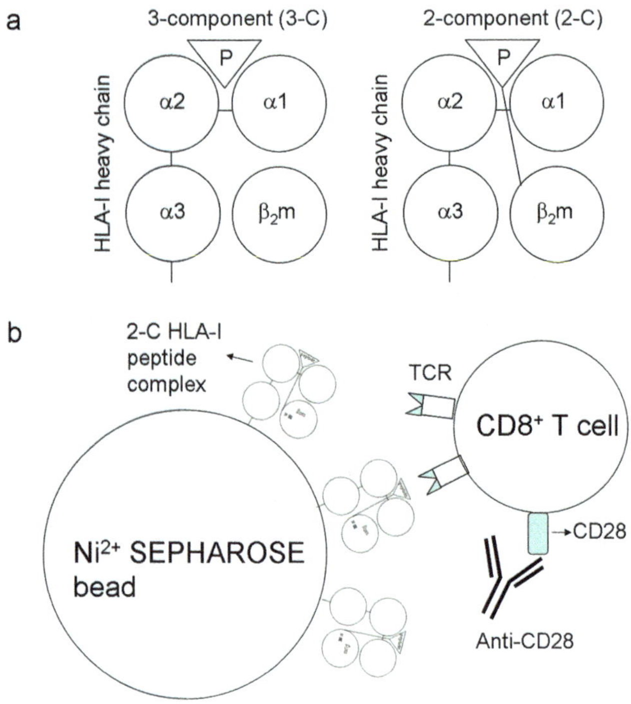

Fig. 1 The design of 2-component (2-C) HLA class I/peptide-fused β_2m-loaded artificial antigen presentosomes (AAPs). (**a**) HLA class I/peptide/β_2m complex in the native 3-C (left) vs. the 2-C (right) composition. A Gly/Ser-rich 15 a.a. linker connects the *C*-terminus of the peptide (P) with the *N*-terminus of β_2m in the 2-C composition. (**b**) Shown here is a representative model of interaction between antigen-specific CD8$^+$ T cells and a Ni^{2+} SEPHAROSE-based AAP loaded with recombinant 2-C HLA class I/peptide-fused β_2m complex in the presence of anti-CD28 co-stimulation

HLA class I/peptide-fused β_2m recombinant proteins are loaded onto Ni^{2+} SEPHAROSE beads, which we name it as artificial antigen presentosomes (AAPs) (Fig. 1). Co-culturing HLA class I-matched, human naïve CD8$^+$ T cells with 2-C AAPs results in HLA class I-restricted, antigen-specific CD8$^+$ T cells with both activity and numbers 5–25 times greater than those stimulated by plate-bound 2-C HLA class I/peptide-fused β_2m recombinant proteins. Our protocol is described as follows.

2 Materials

2.1 Plasmids and E. coli Strains

1. pET23a(+); pET28a(+) (Addgene, Watertown, MA, USA).
2. *E. coli* DH5α; *E. coli* BL21(DE3) (Yeastern Biotech, Taipei, Taiwan).

2.2 Antibodies

1. Mouse monoclonal antibody (mAb) CR11-351, IgG1, recognizing HLA-A2-, -A24, -A28 [7].
2. mAb L368, IgG1, recognizing human β_2m [8].
3. mAb LGIII 147.4.1, IgG1, recognizing all HLA-A allospecificities except -A23, -A24, -A25, -A32 [9].
4. mAb TP25.99, IgG1, recognizing HLA-A, -B, -C [10–12].
5. mAb OKT3, IgG2a, recognizing human CD3ε (BioLegend, San Diego, CA, USA).
6. FITC-conjugated anti-CD8α mAb, IgG1 (BioLegend).
7. anti-His_6 mAb, IgG1 (BioLegend).
8. Polyclonal rabbit anti-β_2m Ab FL-119 (Santa Cruz Biotechnology, Dallas, TX, USA).
9. R-PE-conjugated goat anti-mouse IgG polyclonal antibodies (Jackson ImmunoResearch Laboratories, West Grove, PA, USA).

2.3 Protein Purification

1. Ni^{2+} His Trap^{TM} HP column (GE Healthcare, Chicago, IL, USA).
2. ÄKTA™ fast performance liquid chromatography (FPLC) (GE Healthcare).

2.4 Flow Cytometry

1. FACSCalibur (BD Biosciences, San Jose, CA, USA).

2.5 ELISA

1. Coating buffer (44 mM $NaHCO_3$, 6 mM Na_2CO_3, pH 9.6).
2. Blocking buffer (1% bovine serum albumin (Sigma-Aldrich, St. Louis, MO, USA) in phosphate-buffered saline (PBS)).
3. Wash buffer (0.05% Tween 20 in PBS).
4. SureBlue™ 3,3′,5,5′-Tetramethylbenzidine (TMB) 1-Component Microwell Peroxidase Substrate (Kirkegaard & Perry Laboratories, Gaithersburg, MD, USA).
5. iMARK™ Microplate Absorbance Reader (Bio-Rad, Hercules, CA, USA).

3 Methods

3.1 Recombinant Protein Production

1. Five plasmid constructs were generated. They are pET28a$^+$-HLA-A*0201, pET28a$^+$-HLA-A*1101, pET28a$^+$-$HA_{540–548}$-linker-β_2m, pET28a$^+$-$M1_{58–66}$-linker-β_2m [6], and pET23a$^+$-$LMP2_{340–349}$-linker-β_2m [13], where the linker nucleotides encode 15 a.a., i.e., $G_3(G_2S)_4$: Gly Gly Gly Gly Gly Ser Gly Gly Ser Gly Gly Ser Gly Gly Ser. The a.a. sequence and other characteristics of the chosen peptides are listed in Table 1. All the encoded recombinant proteins contain a hexahistidine

Table 1
Antigenic peptides selected for construction of 2-C HLA class I/peptide-β_2m protein complexes

Target	Epitope	HLA allele	Peptide length	Amino acid sequence	IC_{50} (nM)
Influenza virus (H5N1)	Hemagglutinin$_{540-548}$ ($HA_{540-548}$)	HLA-A*0201	9	ALAIM VAGL	43.0
Influenza virus (H5N1)	Matrix protein$_{58-66}$ ($M1_{58-66}$)	HLA-A*0201	9	GILGFVFTL	18.8
Epstein-Barr virus	Latent membrane protein $2_{340-349}$ ($LMP2_{340-349}$)	HLA-A*1101	10	SSCSSCPL SK	8.2

(His_6) tag at their *C*-terminus. All these constructs were used to transform *E. coli* BL21 (DE3) and antibiotic-resistant single colonies were picked up and propagated. After $O.D._{600}$ of the selected bacterial growth reached 0.3, isopropyl β-D-1-thiogalactopyranoside (IPTG) was added with a final concentration of 1 mM to induce protein production. Different conditions were used to produce each protein, for instance, 8 h with 200 rpm shaking at 37 °C for HLA-A*0201, 12 h with 200 rpm shaking at 37 °C for HLA-A*1101, and 16 h with 200 rpm shaking at 37 °C for other recombinant β_2m protein (*see* **Note 1**).

2. The harvested *E. coli* pellets were resuspended in the lysis buffer (50 mM Tris–HCl, 100 mM NaCl, 1.4 mM β-mercaptoethanol (β-ME), 1× EDTA-free protease inhibitor cocktail set III (Merck Millipore, Burlington, MA, USA)), 30 mL/1 L culture, incubated on ice for 30 min, and then homogenized by high-pressure homogenizer Emulsiflex-C5 (Avestin Inc., Ottawa, ON, Canada). To harvest the inclusion bodies, we centrifuged the homogenized sample at 12,096×*g* for 20 min. The supernatant was transferred to another tube and the inclusion body pellets were solubilized with solubilization buffer (50 mM Tris–HCl, 100 mM NaCl, 1.4 mM β-ME, 6 M Urea), 50 mL/1 L culture at 4 °C for 2 h. The solubilized samples were centrifuged at 12,096×*g* for 20 min to remove the insoluble cell debris (*see* **Notes 2** and **3**).
3. To purify the recombinant proteins, we loaded samples to the His Trap™ HP affinity column (GE Healthcare) at a rate of 1 mL/min, washed columns with washing buffer (50 mM Tris–HCl, 100 mM NaCl, 1.4 mM β-ME, 6 M Urea, 20 mM Imidazole), and eluted the samples with the elution buffer (50 mM Tris–HCl, 100 mM NaCl, 1.4 mM β-ME, 6 M Urea, 300 mM Imidazole). The elution was collected fraction by fraction (1 mL/fraction) (*see* **Note 4**).

4. The purity of His_6-tagged purified recombinant proteins were checked by sodium dodecyl sulfate polyacrylamide gel electrophoresis (SDS-PAGE) (10% for recombinant heavy chain, 12% for recombinant β_2m). The loading volume of each lane is described as follows: supernatant, 24 μL; flow through, 30 μL; wash, 3 μL; and elution, 2 μL. The samples which were separated by electrophoresis on the SDS-PAGE were stained with the Coomassie Brilliant Blue R-250 staining buffer (3 mM Coomassie Brilliant Blue R-250, 45% methanol, 10% glacial acetic acid) for 30 min and distained with distaining buffer (20% methanol, 10% glacial acetic acid).

3.2 Refolding and FPLC Purification

1. We removed the urea from the solubilization buffer of purified recombinant proteins by dialysis against PBS. The final concentration of urea was less than 0.3 nM, which was $1/10^9$ of the original concentration. The recombinant protein (1.7 mg for heavy chain and 1.6 mg for β_2m) were injected forcefully to the 50 mL stirring refolding buffer (400 mM L-Arginine, 100 mM Tris, 2 mM EDTA, 5 mM reduced L-glutathione, 0.5 mM oxidized L-glutathione, pH 8.3) by a syringe as close to the stirring bar as possible at 4 °C. After that, we injected further 1.7 mg heavy chain recombinant protein two times to the stirring refolding buffer each for a 14-h mixing. Afterwards, we concentrated the refolding buffer from 50 mL to 3–4 mL and quantified the concentration by Bio-Rad Protein Assay (Bio-Rad) (*see* **Note 5**).
2. The refolded 2-C HLA class I/peptide-β_2m complexes were purified by FPLC. Each 2-C HLA class I/peptide-β_2m complex preparation was injected into the HiLoad 16/70 Superdex 75-gel filtration column (GE Healthcare) by ÄKTA™ purifier (GE Healthcare) and was separated with 120 mL PBS at a speed of 1 mL/min. The sample which had absorbance higher than 30 mAU at 280 nm was collected in 1.5 mL/fraction. The concentration of each sample was checked by Bio-Rad Protein Assay (Bio-Rad).
3. The purity of two refolded 2-C HLA class I/peptide-β_2m complexes, namely, HLA-A2/$M1_{58-66}$ and HLA-A11/$LMP2_{340-349}$, which had different FPLC profiles, was checked by 15% SDS-PAGE. Ten microliter of each fraction was used for checking. After electrophoresis, the gels were fixed by methanol and stained by silver staining. The sensitivity of silver staining is approximately 1–10 ng.
4. Non-reducing PAGE. Ten microliter of each fraction of peak 1 and peak 2 of FPLC-purified HLA-A2/$M1_{58-66}$ was used as samples for checking the composition. Each sample was mixed with 2.5 μL non-reducing sample buffer (100 mM Tris–HCl, pH 6.8, 0.2% bromophenol blue, and 20% glycerol), heated at

60 °C for 5 min, and separated by 10% non-reducing PAGE (8.4% acrylamide mixture, 390 mM Tris, pH 8.8, 0.1% ammonium persulfate) and non-reducing buffer (50 mM Tris, pH 8.9 and 380 mM) with 120 V for 1.5 h at room temperature. After electrophoresis, the non-reducing PAGE was stained by silver staining as previously described (*see* **Note 6**).

3.3 Direct and Competitive ELISA

1. For direct ELISA, we coated 100 μL mAb CR11-351 (1 μg/mL) or mAb LGIII 147.4.1 (1 μg/mL) for recombinant 2-C HLA-A*0201 or HLA-A*1101 complexes, respectively, onto 96-well plates with coating buffer (44 mM $NaHCO_3$, 6 mM Na_2CO_3, pH 9.6) for 14 h at 4 °C. Each well was then washed twice with 0.05% Tween 20/PBS, once with PBS, and blocked with 1% bovine serum albumin (BSA)/PBS for 1 h. After 1 h, wells were repeated with the wash steps and then incubated with each FPLC-purified recombinant 2-C HLA class I/peptide-β_2m complex at increasing concentrations for 1 h. Ten microliter of each fraction of peak 1 and peak 2 was added. After 1 h, wells were repeated with the wash steps and then incubated with rabbit anti-β_2m polyclonal antibodies, FL-119, at the dilution ratio of 1: 1, 000 for 1 h. After 1 h, wells were repeated with the wash steps and then incubated with horseradish peroxidase (HRP)-conjugated goat anti-rabbit antibodies at the dilution ratio of 1: 10,000 for 1 h. After 1 h, wells were repeated with the wash steps and then incubated with TMB substrate (Kirkegaard & Perry Laboratories) in the dark for color development. The O. D. values were then measured at 450 nm by the iMARK™ Microplate Absorbance Reader (Bio-Rad) (*see* **Note 7**).
2. For competitive ELISA, we coated the wells with recombinant 2-C HLA-A*0201 complexes (2 μg/mL) or BSA (2 μg/mL) for 14 h at 4 °C. Each well was washed as previously described and then added with mAb CR11-351 (1 μg) pre-incubated with the same 2-C HLA-A*0201 protein complex in solution at different concentrations (from 0 to 20 μ g) for 14 h at 4 °C. After 1-h incubation, each well was washed twice and then incubated with HRP-conjugated goat anti-mouse antibodies for 1 h. Afterwards, wells were again washed twice and then incubated with the TMB substrate in the dark until sufficient color developed. The O.D. values were measured at 450 nm. For calculating the inhibition percentage, the following formula was used:

$$\text{Percent inhibition}\,(\%) = \frac{\text{soluble HLAO.D.}_{450}\ (0\ \mu\text{g}) - \text{soluble HLAO.D.}_{450}\ (X\ \mu\text{g})}{\text{soluble HLAO.D.}_{450}\ (0\ \mu\text{g}) - \text{coated BSAO.D.}_{450}\ (2\ \mu\text{g})}$$

3.4 Immunoprecipitation

1. The conformation of refolded, FPLC-purified 2-C HLA class I/peptide-β_2m complexes was checked by immunoprecipitation. Each refolded complex (IP-input, 40 μg) was incubated with 40 μL 50% protein A/G agarose (Thermo Fisher Scientific, Waltham, MA, USA), which had been washed with 1 mL PBS three times, at 4 °C for 2 h, and then centrifuged at 9,391×*g* for 1 min. The supernatant, called pre-cleaned fraction, was collected and incubated with β_2m-specific mAb L368 for 14 h by inverting the tube back and forth smoothly. Then, the mixture was incubated with 40 μL pre-cleaned 50% protein A/G agarose for 2 h at 4 °C and then separated into IP-bound (precipitated pellet) and IP-unbound (supernatant) by centrifugation at 9,391×*g* for 2 min. The IP-bound sample was washed by 1 mL PBS three times.
2. Immunoblotting. Each IP-bound sample or IP-input (10 μg) was dissolved in 40 μL PBS and mixed with 16 μL sample buffer (100 mM Tris–HCl, pH 6.8, 5% β-ME, 4% SDS, 0.2% bromophenol blue, and 20% glycerol). These samples were heated at 95 °C for 15 min and separated by reducing SDS-PAGE. Then, we soaked the gels, polyvinylidene fluoride (PVDF) membranes (Merck Millipore), which had been activated by methanol for 2 min, and filter papers (Sartorius Stedim Biotech S.A., Goettingen, Germany) in transfer buffer (12 mM Tris–HCl, 96 mM glycine, and 20% methanol) for 10 min. The proteins were transferred from gel to PVDF membrane by Semi-Dry Electrophoretic Transfer Cell (Bio-Rad) for 45 min. The transferred PVDF membrane was blocked in 2% BSA/5% non-fat milk powder in PBS for 30 min. Then, we probed the membranes with mAb TP25.99 (10 μg/mL) and L368 (5 μg/mL) for 14 h at 4 °C with gentle shaking. Next, membrane was washed by PBS containing 0.05% Tween 20 twice and PBS once and then was incubated with HRP-conjugated goat anti-mouse antibodies for 30 min. Before adding the ECL substrate (PerkinElmer, Waltham, MA, USA), we repeated the wash step and exposed the membrane to the film (GE Healthcare).

3.5 AAP Construction

1. Different amounts (20 μg, 10 μg, 5 μg, 2.5 μg, 1.25 μg, and 0 μg) of FPLC-purified 2-C HLA class I/peptide-β_2m complexes were mixed with 5 μL 50% Ni^{2+} SEPHAROSE beads (34 μm, GE Healthcare) with shaking at 4 °C for 1 h. In Bio-Rad protein assay, we washed the complex-loaded Ni^{2+} SEPHAROSE beads, added protein assay dye, and detected the O. D. value at 600 nm. Results were expressed as O. D. values at 600 nm. In flow cytometry, the complex-loaded Ni^{2+} SEPHAROSE beads were washed with 0.05% Tween-20/PBS, stained with HLA-A2 conformation-specific mAb CR11-351

(0.5 μg/100 μL) for 1 h at 4 °C, and then stained with R-PE conjugated goat anti-mouse antibodies (1: 200) in the dark for 30 min, before being analyzed by FACSCalibur (BD Biosciences) with the CellQuest software (BD Biosciences). Results were expressed as mean fluorescence intensity (MFI) (*see* **Note 8**).

3.6 Antigen-Specific T Cell Activation and Proliferation

1. The peripheral blood lymphocytes (PBLs) were isolated from HLA-A2$^+$ (HLA-A11$^-$) and HLA-A11$^+$ (HLA-A2$^-$) healthy donors using Ficoll-Paque density gradient centrifugation (GE Healthcare). For stimulating antigen-specific CD8$^+$ T cells, we cultured PBLs (2 × 10^5/100 μL/test) with 2-C AAPs or with plate-bound 2-C HLA class I/peptide-β_2m complexes in the following conditions: for co-culturing with 2-C AAPs, 1 μL Ni^{2+} SEPHAROSE in 30 μL PBS was loaded with 1 μg 2-C HLA class I/peptide-β_2m complexes in 30 μL PBS in the wells for 14 h at 4 °C; for co-culturing with plate-bound 2-C HLA class I/peptide-β_2m complexes, 1 μg 2-C HLA class I/peptide-β_2m complexes was coated onto plates for 14 h at 4 °C. All the above groups were pre-coated with or without anti-CD28 mAb (1 μg/well). In addition, the test groups were also cross-controlled by HLA-unmatched 2-C AAPs or plate-bound 2-C HLA class I/peptide-β_2m complexes. The wells pre-coated with anti-CD3 and anti-CD28 mAb served as the positive control. After a 3-day co-culture at 37 °C, the PBLs and AAPs were spun down and the supernatants were collected for analyzing IFN-γ secretion with a human quantitative IFN-γ EIA kit. Briefly, we added 30 μL supernatant and 50 μL biotinylated IFN-γ-specific antibodies to each well which had been coated with a different IFN-γ-specific antibody for a 2-h incubation and washed the wells with washing buffer three times. Then, each well was incubated with streptavidin-HRP at room temperature for 30 min. After that, each well was repeated with the wash steps, incubated with the TMB substrate for 30 min, and measured for the O. D. value at 450 nm. The results were shown as IFN-γ concentration (mean ± SD) for different experimental groups (*see* Table 2).
2. PBLs were isolated from HLA-A2$^+$ (HLA-A11$^-$) and HLA-A11$^+$ (HLA-A2$^-$) healthy donors. Before adding PBLs for co-culturing, we had mixed the 5 μg HLA-A2/M1$_{58-66}$-fused β_2m complex or 5 μg HLA-A11/LMP2$_{340-349}$-fused β_2m complex with 5 μL pre-cleaned Ni^{2+} magnetic SEPHAROSE beads (GE Healthcare) for 14 h at 4 °C. The PBLs (8 × 10^5 cells/ 500 μL), which were isolated from the HLA-A2$^+$ and HLA-A11$^+$ (as a specificity control) healthy donors, were co-cultured with HLA-A2/M1$_{58-66}$-fused β_2m-loaded magnetic AAPs beads at a 24-well plate for 5 days at

Table 2
IFN-γ secretion and proliferation of antigen-specific CD8[+] T cells stimulated by 2-C AAPs vs. plate-bound 2-C HLA class I/peptide complexes[a]

	HLA-A*0201/ $M1_{58-66}$ 2-C AAP	HLA-A*1101/ $LMP2_{340-349}$ 2-C AAP	HLA-A*0201/ $M1_{58-66}$ plate-bound	HLA-A*1101/ $LMP2_{340-349}$ plate-bound
IFN-γ (pg/mL)[b]				
HLA-A*0201	957.9 ± 23.4	18.1 ± 5.6	57.2 ± 1.9	2.5 ± 1.1
HLA-A*1101	11.5 ± 3.4	1078.7 ± 16.3	4.0 ± 2.2	216.7 ± 2.1
Antigen-specific CD8[+] T cells (%)[c]				
HLA-A*0201	5.28	0.23	1.2	0.09
HLA-A*1101	0.21	6.29	0.1	1.7

[a]In the presence of anti-CD28 co-stimulation in both groups
[b]Determined by ELISA in triplicate
[c]Determined by flow cytometry with dual-color tetramer staining

37 °C in an incubator with 5% CO_2. The same procedure was performed for testing the HLA-A11/$LMP2_{340-349}$-fused β_2m magnetic AAPs. After incubation for 5 days, the AAPs were taken away by magnetic drill. The treated PBLs were stained with 2-C HLA-A2/$M1_{58-66}$-fused β_2m complex tetramerized by anti-His_6 mouse mAb for 1 h and washed three times with 1% BSA/PBS. Then, these PBLs were stained with R-PE goat anti-mouse antibodies (1: 200) in the dark for 30 min and washed three times with 1% BSA/PBS. Finally, these PBLs were stained with FITC-conjugated anti-CD8 mAb (BioLegend) (0.2 μL for each test) in the dark for 30 min, washed 3 times with 1% BSA/PBS, resuspended in PBS, fixed by PBS with 0.5% paraformaldehyde at room temperature for 20 min before being analyzed by FACSCalibur with the CellQuest software. Results were expressed as dot plot to compare the percentage of double positive population between the two groups (*see* Table 2).

4 Notes

1. It is highly advised the incubation temperature and time be optimized for each recombinant protein. The peak production interval could be shorter particularly for HLA-A2 and HLA-A11 heavy chain recombinant proteins. The incubation temperature could be lowered to 25–28 °C at the investigator's discretion to obtain a better yield.

2. To reduce the bacterial impurities, several wash steps could be performed for the inclusion body pellets using lysis buffer before solubilization.
3. The washed inclusion bodies could be stored at −80 °C up to 3 years without difficulty in performing subsequent purification procedures.
4. The best results could be obtained with an automated ÄKTA system set up with a gradient elution program at 4 °C.
5. The speed of refolding could be optimized and kept slow to reduce insoluble protein aggregations and to increase the yield of properly refolded 2-C HLA class I/peptide-fused β_2m complex. In our experience, the yield has been more than 30% based on FPLC profiles.
6. An alternative assay here is to perform dynamic light scattering (DLS) of the FPLC-defined refolded 2-C HLA class I/peptide-fused β_2m complexes to confirm their monomeric/polymeric status. DLS is a technique in physics for determining the size distribution profile of small particles in solution. When monochromatic and coherent laser hits the small particles, we can observe a time-dependent fluctuation in the scattering intensity. The fluctuation is caused by Brownian motion of each particle. We can calculate the diffusion coefficient by different intensity of scattering light at different time. Then, we can get the radius of small particles and suggest the polymerization condition by Stokes-Einstein equation.

Stokes–Einstein equation
$$D = \frac{KT}{6\pi\eta R}$$

D : diffusion coefficient ($m^2\ s^{-1}$)
K : Boltzmann constant ($J\ K^{-1}$)
T : temperature (K)
η : viscosity (kg/s m)
R : radius (m)

7. To reduce the background of ELISA, IgG-depleted BSA should be used.
8. The monomeric 2-C HLA class I/peptide-fused β_2m complex proteins could be stored at 4 °C in the presence of 5% glycerol up to 1 month without losing activity. Alternatively, the monomers could be stored in storage buffer (0.5 mM EDTA, 0.2% BSA, 10 mM Tris-HCl (pH 8.0), 150 mM NaCl, and 0.09% NaN_3) at 4 °C for 3 months. Loaded AAPs could be stored at 4 °C in sterile conditions in the presence of 0.5% BSA up to 3 weeks without losing activity. Adding preservatives to loaded AAPs is not recommended.

Acknowledgments

This work was supported in part by the National Science Council grant (NSC 98-2320-B-007-004-MY3) and the Ministry of Science and Technology grants (MOST 106-2320-B-007-001; MOST 107-2320-B-007-002) to C.-C. C.

References

1. Hennecke J, Wiley DC (2001) T cell receptor-MHC interactions up close. Cell 104:1–4
2. Pennock ND, White JT, Cross EW, Cheney EE, Tamburini BA, Kedl RM (2013) T cell responses: naive to memory and everything in between. Adv Physiol Educ 37:273–283
3. Chen L, Flies DB (2013) Molecular mechanisms of T cell co-stimulation and co-inhibition. Nat Rev Immunol 13:221–242
4. Zhang N, Bevan MJ (2011) $CD8^+$ T cells: foot soldiers of the immune system. Immunity 35:161–168
5. Tafuro S, Meier UC, Dunbar PR, Jones EY, Layton GT, Hunter MG et al (2001) Reconstitution of antigen presentation in HLA class I-negative cancer cells with peptide-beta2m fusion molecules. Eur J Immunol 31:440–449
6. Shen C, Chang CC, Zhang J, Guo W, Xia L, Meng F et al (2006) Structural and functional characterization of peptide-beta2m fused HLA-A2/MART1 (27-35) complexes. Biochem Biophys Res Commun 342:57–65
7. Tsujisaki M, Sakaguchi K, Igarashi M, Richiardi P, Perosa F, Ferrone S (1988) Fine specificity and idiotype diversity of the murine anti-HLA-A2, A28 monoclonal antibodies CR11-351 and KS1. Transplantation 45:632–639
8. Lampson LA, Fisher CA, Whelan JP (1983) Striking paucity of HLA-A, B, C and beta 2-microglobulin on human neuroblastoma cell lines. J Immunol 130:2471–2478
9. Wang X, Liang B, Rebmann V, Lu J, Celis E, Kageshita T et al (2003) Specificity and functional characteristics of anti-HLA-A mAbs LGIII-147.4.1 and LGIII-220.6.2. Tissue Antigens 62:139–148
10. Palmisano GL, Pistillo MP, Capanni P, Pera C, Nicolò G, Salvi S et al (2001) Investigation of HLA class I downregulation in breast cancer by RT-PCR. Hum Immunol 62:133–139
11. Rebmann V, Pfeiffer K, Pässler M, Ferrone S, Maier S, Weiss E et al (1999) Detection of soluble HLA-G molecules in plasma and amniotic fluid. Tissue Antigens 53:14–22
12. Desai SA, Wang X, Noronha EJ, Zhou Q, Rebmann V, Grosse-Wilde H et al (2000) Structural relatedness of distinct determinants recognized by monoclonal antibody TP25.99 on beta2-microglobulin-associated and beta2-microglobulin-free HLA class I heavy chains. J Immunol 165:3275–3283
13. Lee SP, Tierney RJ, Thomas WA, Brooks JM, Rickinson AB (1997) Conserved CTL epitopes within EBV latent membrane protein 2: a potential target for CTL-based tumor therapy. J Immunol 158:3325–3334

Chapter 13

Imaging Chimeric Antigen Receptor (CAR) Activation

Kendra A. Libby and Xiaolei Su

Abstract

The chimeric antigen receptor (CAR) has been extensively exploited in cancer immunotherapy. In spite of the success of CAR T cells in clinical applications, the molecular mechanism underlying CAR-T cell activation remains unclear. Key questions remain: how are CARs activated by tumor antigens? How do activated CARs transduce signaling to downstream pathways? Here we introduce a microscopy-based method for studying CAR signaling. We use an antigen-coated supported lipid bilayer to activate CARs and combine it with TIRF microscopy to visualize the initial activation process of CAR T cells. This enables monitoring CAR signaling at high spatial and temporal resolutions.

Key words Chimeric antigen receptor, T cell signaling, Microclusters, TIRF, CD19

1 Introduction

The development of chimeric antigen receptor (CAR)-armored T cells opened a new field in cancer immunotherapy [1]. T cells isolated from patients can be genetically engineered to express CARs for targeting specific cancer surface molecules. These engineered CAR-T cells are amplified in vitro before being sent back to patients to attack tumors. CAR-based therapy has been reported to successfully treat blood cancers including chronic lymphocytic leukemia (CLL), acute lymphoblastic leukemia (ALL), and non-Hodgkin lymphoma [2–4], and new CARs are being designed and tested for treating solid tumors [5].

Although CAR T cells can effectively kill cancer cells, the molecular mechanisms underlying CAR-T cell activation are not well understood. CAR is usually designed as a single-pass transmembrane receptor (Fig. 1a). Its extracellular domain contains a single-chain antibody that recognizes antigens on the cancer cell surface. CD19, a B cell surface marker, is a commonly used target for CAR. A transmembrane domain, usually from CD8α or CD28, connects the extracellular ligand-binding domain to the intracellular signaling domains. The intracellular region of CAR contains a

Chaohong Liu (ed.), *T-Cell Receptor Signaling: Methods and Protocols*, Methods in Molecular Biology, vol. 2111, https://doi.org/10.1007/978-1-0716-0266-9_13, © Springer Science+Business Media, LLC, part of Springer Nature 2020

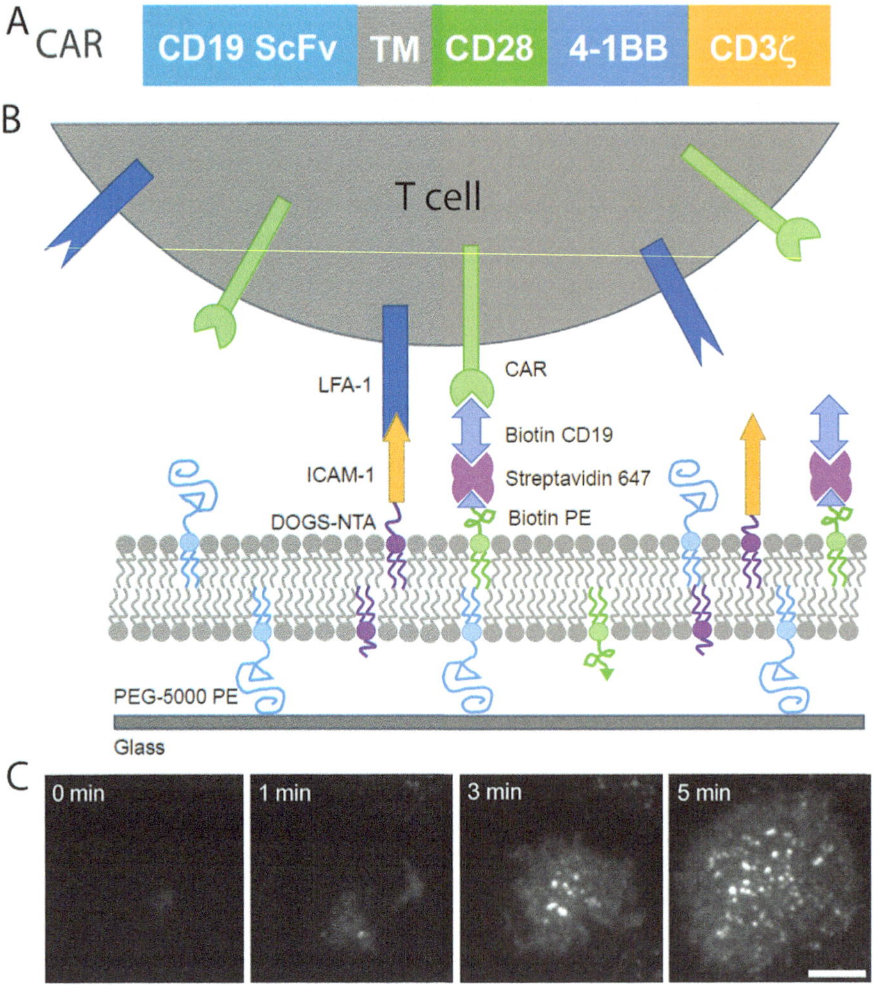

Fig. 1 A supported lipid bilayer system for CAR T cell activation. (**a**) The domain structure of CAR used in this study. (**b**) A functionalized lipid bilayer presents CD19 and ICAM-1 to CAR T cells. (**c**) TIRF microscopy reveals that CAR-GFP forms signaling microclusters as the T cell spreads on the CD19-coated bilayer. Scale bar: 5 μm

tandem fusion of the cytoplasmic domain of the T cell receptor (CD3ζ), and individual or combined co-receptors such as CD28 and 4-1BB. Intriguingly, the engagement of CAR with CD19 is sufficient to activate, at least qualitatively, all receptor signaling pathways [6], although the mechanism is not known.

To understand how CARs are activated by antigens and transduce signaling to downstream pathways, we developed a system for visualizing CAR signaling at high spatial and temporal resolutions. This system utilizes supported lipid bilayers to present antigens as well as other co-receptor ligands to CAR T cells. Supported lipid bilayers, mimicking the cell membranes of antigen-presenting cells, have been successfully used to present pMHC to activate the native

T cell receptor (TCR) [7]. It has also been utilized to study the mechanism of TCR signaling [8–10]. The supported lipid bilayer system has several features: (1) mobile antigens on membranes (in contrast to the immobile antigens attached to the glass surface) allow the reorganization of antigens and antigen-binding CARs to form higher order structures, e.g., signaling microclusters [11]; (2) a supported lipid bilayer that provides a planar surface, enabling high-quality imaging of CAR and other membrane-bound signaling protein dynamics by total internal reflection fluorescence (TIRF) microscopy; and (3) antigen density, co-receptor ligand density, and lipid composition that can be accurately controlled on the bilayer surface, enabling a quantitative understanding for CAR activation.

To prepare functionalized supported lipid bilayers, we attached the antigen CD19 to the supported lipid bilayer using a biotin-streptavidin method. To facilitate cell adhesion, we also attached ICAM-1, the ligand for integrin LFA-1, to the bilayer using a polyhistidine-NTA method (Fig. 1b). If desired, ligands for other (co)receptors (e.g., CD28, PD1) could be attached to bilayers as well. Once T cells expressing GFP-tagged CAR contact the CD19-functionalized membranes, they spread on the membranes. CARs, as visualized by TIRF microscopy, form microclusters to transduce signaling (Fig. 1c), reflecting an activation process which is similarly observed for the TCR [12]. The system could also be applied to study signaling molecules downstream CAR, including ZAP70, Gads, SLP76, Nck, or signaling reporters including calcium, MAPK, or NFAT. This would enable the monitoring of signaling kinetics at individual steps.

2 Materials

1. Lipid components: 1-palmitoyl-2-oleoyl-sn-glycero-3-phosphocholine (POPC) (e.g., Avanti, 850457C), 1,2-dioleoyl-sn-glycero-3-phosphoethanolamine-*N*-[methoxy(polyethylene glycol)-5000] (PEG-5000 PE) (e.g., Avanti, 880230C), 1,2-distearoyl-sn-glycero-3-phosphoethanolamine-*N*-[biotinyl(polyethylene glycol)-2000] (ammonium salt) (DSPE-PEG (2000) Biotin) (e.g., Avanti, 880,129) and 1,2-dioleoyl-sn-glycero-3-[(*N*-(5-amino-1-carboxypentyl)iminodiacetic acid) succinyl] (DOGS-NTA) (e.g., Avanti, 790404C) (*see* **Note 1**).
2. Proteins: His-tagged extracellular domain of ICAM-1 (e.g., Sino Biological, 10,346-H08H-50), Biotinylated extracellular domain of CD19 (e.g., Sino Biological, 11,880-H08H-B) (*see* **Note 2**) and Streptavidin Alexa Flour™ 647 (e.g., Thermo Fisher, S21374).

3. Tissue culture media: 10% heat-inactivated Fetal Bovine Serum (FBS) (e.g., Life Tech, 16140071) and 1% penicillin-streptomycin-glutamine (PSG) (e.g., Life Tech, MT30009CI) in RPMI 1640 (e.g., Thermo Fisher, 11875093).
4. 96-well glass bottom plate (requires flat glass, plastic wells with low fluorescence background, and resistance to treatment with acids, bases, and detergents, e.g., Matriplate MGB096-1-2-LG-L).
5. Hellmanex III (e.g., Sigma, Z805939).
6. Adhesive PCR Sealing Foil (e.g., Thermo Fisher, AB-0626).
7. Glass vials (e.g., National Scientific, B7800-2).
8. Gastight syringes 25 μL (e.g., Hamilton, 80,275) and 250 μL (e.g., Hamilton, 81,175).
9. Chloroform (e.g., Electron Microscopy Sciences, 12,550).
10. Argon.
11. NaOH.
12. PBS buffer: 155 mM NaCl, 3.0 mM Na_2HPO_4, 1.1 mM KH_2PO_4, pH 7.4.
13. High-speed centrifuge and polycarbonate centrifuge tubes.
14. Heat block.
15. Imaging media: 20 mM HEPES (e.g., Thermo Fisher, 15630-080) in RPMI 1640 no phenol red (e.g., Thermo Fisher, 11835-030).

3 Methods

3.1 Construction of Jurkat CAR T Cells

1. Prepare tissue culture media by adding 10% heat-inactivated FBS and 1% PSG to RPMI 1640; filter media if necessary. Store media at 4 °C and warm to 37 °C before usage.
2. Maintain cells in tissue culture media. Cells double roughly every 24 h. Keep the cell density below 0.5 million/mL. Split cells when necessary.
3. Produce lentivirus of CAR-GFP using HEK293 cells.
4. Infect T cells with the CAR virus.
5. Keep the culture for 10 days until the CAR expression is stabilized.
6. Select CAR-GFP$^+$ T cells by fluorescence-activated cell sorting (FACS).

3.2 Preparation of Small Unilamellar Vesicles (SUV) for Making Membranes (We follow our previous protocol for making SUV and membranes [8] with slight modifications)

1. Clean glass vials with 5% Hellmanex III, rinse them with milli Q water, and dry in oven.
2. Warm lipid stocks to room temperature.
3. Rinse the glass vial with chloroform. Pour ~1 mL chloroform into each vial.
4. Use glass syringes to prepare a lipid mix, about 4 μmol each vial. Lipid composition: 98% POPC, 2% DOGS-NTA, 0.1% DSPE-PEG (2000) Biotin, and 0.1% PEG-5000 PE (*see* **Note 3**).
5. Dry the lipid mix with a steady argon flow (*see* **Note 4**). Use a ~45 °C water bath while drying to maintain lipid solubility as the chloroform evaporates. Multiple white layers will form on the walls of the vial after the lipids dry.
6. Dry lipids further in a desiccator for at least 2 h.
7. Resuspend the dried lipids in 1.5 mL PBS. Vortex to mix.
8. Transfer resuspended lipids into two 1.5 mL microcentrifuge tubes, adding 750 μL to each tube.
9. Freeze the resuspended lipids in liquid nitrogen and thaw in a room temperature water bath. Repeat this freeze-thaw cycle 30 times until the cloudy solution becomes clear (*see* **Note 5**). If desired, freeze the resuspension at −80 °C to store for future use.
10. Centrifuge the resuspended lipids at 48,000 × *g* for 45 min at 4 °C.
11. The supernatant now contains SUVs; transfer this to a clean tube. Avoid disrupting the white pellet at the bottom of the tube. Cover the SUV solution with argon and store at 4 °C. The SUV should be used within 2 weeks (*see* **Note 6**).

3.3 Preparation of Antigen-Functionalized Supported Lipid Bilayers (See Note 7)

1. In a 1 L beaker, immerse glass imaging plate into 1 L of 5% Hellmanex III. Microwave to 50 °C. Stir and incubate overnight at room temperature.
2. The next morning, rinse each well 10 times with ultrapure water. Blow-dry all wells and seal plate with adhesive foil.
3. Use a blade to open wells to be immediately used. Add 250 μL freshly made 5 M NaOH to each well. Incubate 1 h at 50 °C on a heat block. Remove NaOH and repeat cleaning (*see* **Note 8**).
4. Add 250 μL 5 M NaOH to each well and incubate overnight at room temperature.
5. The next morning, remove NaOH from wells and rinse twice with 500 μL ultrapure water. Rinse twice with 500 μL PBS.
6. Add 200 μL PBS and 20 μL SUV to each well (*see* **Note 9**). Tap the tube containing SUV solution gently to mix before adding. Incubate for 1 h at 37 °C to allow bilayer formation (*see* **Note 10**).

7. Remove 100 μL from each well and wash three times with 500 μL PBS.
8. Prepare a solution of 10 nM streptavidin Ax647 in PBS and add 100 μL to each well. Incubate for 30 min at 37 °C.
9. Remove 100 μL from each well and wash three times with 500 μL PBS.
10. Prepare a solution of 10 nM CD19-biotin and 10 nM ICAM-1-his in PBS. Add 100 μL to each well and incubate for 1.5 h at 37 °C.

3.4 Imaging CAR T Cell Activation

1. Spin down CAR T cells to have ~0.1 million per well and resuspend in imaging media.
2. Remove 100 μL of solution from the well. Wash once with 500 μL imaging media.
3. Assess the bilayer quality in each well. A good bilayer will be fluid (can be tested by FRAP of streptavidin 647) and cover the surface completely (*see* **Note 11**).
4. Add ~0.1 million cells to the well and begin time lapse imaging immediately afterwards (*see* **Note 12**). Following landing, the cells will spread and microclusters will form in ~1–2 min, indicating an activation of the proximal signaling.

4 Notes

1. To prevent oxidation of lipids after opening, store the unused lipids in a glass vial, and fill the space above the lipids with argon. Store at −20 °C. Alternatively, we recommend using the bottling service provided by Avanti to order small aliquots of lipids.
2. ICAM-1 and CD19 used here are only the extracellular domain. They are purified from the Human cells, containing post-translational modifications.
3. POPC provides the structural component of the membrane. DOGS-NTA binds polyhistidine-tagged proteins. DSPE-PEG (2000) Biotin binds streptavidin, which can further recruit biotinylated molecules. PEG-5000 PE provides a cushion layer between the membrane and glass (Fig. 1b). A direct contact of membranes to the glass might reduce membrane fluidity. For any new composition of lipids, the membrane fluidity needs to be determined by FRAP (fluorescence recovery after photobleaching).
4. Dry the lipids using a gentle argon flow. Chloroform evaporation will take heat away and drop the temperature of the lipid mix. A water bath is necessary to maintain a stable temperature.

5. The exact cycle number may vary depending on lipid composition. Water bath-based sonication may help to reduce the freeze-thaw cycle number.
6. The SUV storage period may vary. To test how long one is able to store a specific lipid mixture, use FRAP to assess bilayer quality over time.
7. The key to making high quality bilayers is a sufficient cleaning of glass. Any dirt or dust on the glass will result in membrane defects (such as holes and tears) and reduce the mobility of lipids and membrane-associated proteins. We found a sequential cleaning using Hellmanex and NaOH is effective in removing dirt.
8. Ensure that the heat block remains around 50 °C, as high temperatures may melt the glue of the imaging plate, potentially leading to leaky wells. Similarly, avoid reheating the same plate too many times.
9. Depending upon how many lipids are lost during centrifugation, the lipid concentration after SUV preparation may vary. If the bilayer only covers part of the glass surface, increase the amount of SUV added.
10. Once a bilayer has been established, it must not dry out. Ensure that there is always at least 50 μL of solutions to cover the bottom of each well.
11. A fluid bilayer will look like television static as the individual streptavidin 647 molecules move around. There should be no dark holes or bright clumps. Any bilayer deficiencies are likely due to poor cleaning of the glass.
12. One can seed a few cells first to find the right TIRF angle and focal plane before adding the majority of cells for imaging. The number of cells one needs to add may vary depending upon how many actually interact with the bilayer; adjust the number of cells added as needed to get the desired density.

Acknowledgments

This work was supported by a Yale College First-Year Summer Research Fellowship in the Sciences and Engineering awarded to Kendra Libby. Xiaolei Su was supported by Grant#IRG 17-172-57 from the American Cancer Society, Hood Foundation Child Health Research Award, Gilead Sciences Research Scholars Award in Hematology/Oncology, The Andrew McDonough B+ Foundation, and The Rally Foundation.

References

1. June CH, Sadelain M (2018) Chimeric antigen receptor therapy. N Engl J Med 379(1):64–73
2. Ghobadi A (2018) Chimeric antigen receptor T cell therapy for non-Hodgkin lymphoma. Curr Res Transl Med 66(2):43–49
3. Lee DW et al (2015) T cells expressing CD19 chimeric antigen receptors for acute lymphoblastic leukaemia in children and young adults: a phase 1 dose-escalation trial. Lancet 385 (9967):517–528
4. Porter DL et al (2011) Chimeric antigen receptor-modified T cells in chronic lymphoid leukemia. N Engl J Med 365(8):725–733
5. Martinez M, Moon EK (2019) CAR T cells for solid tumors: new strategies for finding, infiltrating, and surviving in the tumor microenvironment. Front Immunol 10:128
6. Karlsson H et al (2015) Evaluation of intracellular signaling downstream chimeric antigen receptors. PLoS One 10(12):e0144787
7. Dustin ML et al (2007) Supported planar bilayers for study of the immunological synapse. Curr Protoc Immunol. Chapter 18: p. Unit 18 13
8. Su X et al (2017) Reconstitution of TCR signaling using supported lipid bilayers. Methods Mol Biol 1584:65–76
9. Su X et al (2016) Phase separation of signaling molecules promotes T cell receptor signal transduction. Science 352(6285):595–599
10. Huang WYC et al (2019) A molecular assembly phase transition and kinetic proofreading modulate Ras activation by SOS. Science 363 (6431):1098–1103
11. Dustin ML, Groves JT (2012) Receptor signaling clusters in the immune synapse. Annu Rev Biophys 41:543–556
12. Grakoui A et al (1999) The immunological synapse: a molecular machine controlling T cell activation. Science 285(5425):221–227

Chapter 14

Assessing the Impact of Phytochemicals on Immune Checkpoints: Implications for Cancer Immunotherapy

Melanie R. Power Coombs and David W. Hoskin

Abstract

Phytochemicals are the basis for many anticancer drugs currently in clinical use, as well as a potential source of future cancer treatments. Some phytochemicals have been found to modify the expression of checkpoint inhibitors of the immune response, as well as kill cancer cells. Cancer cells, in turn, may evade detection by the immune system by expressing molecules such as programmed death ligand 1 (PD-L1) that interacts with programmed cell death 1 (PD-1) on T cells to inhibit T cell activation and effector function. Phytochemicals have direct effects on cancer cells and/or T cells that may impact PD-L1/PD1 interactions, although this may vary depending on the phytochemical in question. Flow cytometric analysis of cancer cells stained with anti-PD-L1 antibodies following treatment with a given phytochemical enables the detection of any alteration in PD-L1 expression. The effect of the phytochemical on T cell function can be assessed using proliferation assays (e.g., tritiated thymidine incorporation, flow cytometric analysis of Oregon Green 488-stained cells) and enzyme-linked immunosorbent assay of interleukin-2 content in culture supernatants. Additional study is needed to better understand the impact of phytochemicals on cancer immunotherapy.

Key words Cancer, Co-cultures, Flow cytometry, Immune checkpoints, Immunotherapy, Interleukin-2, Proliferation, T cells

1 Introduction

1.1 Phytochemicals

Phytochemicals that naturally occur in fruits and vegetables have an impact on human health regarding immune cell function, microbes, and cancer [1–6]. Epidemiological studies have shown that a diet rich in fruits and vegetables positively correlates with a lower risk of various chronic diseases, including cancer [7–9]. Flavonoids are a class of phytochemicals that have the potential to serve as both preventative and therapeutic agents for the treatment of cancer, as well as other chronic diseases.

1.2 Cancer Treatment Strategies

Improved screening, diagnosis, and treatment are still required for cancer since, in spite of significant advances, 25% of cancer victims still die each year from the disease [10]. While treatment varies

Chaohong Liu (ed.), *T-Cell Receptor Signaling: Methods and Protocols*, Methods in Molecular Biology, vol. 2111,
https://doi.org/10.1007/978-1-0716-0266-9_14,

significantly between individual cancers, a combination of surgery, radiotherapy, hormonal therapy, chemotherapy, and receptor-targeted therapies is often employed [11, 12]. Surgery is effective in eradicating localized disease, whereas more advanced cancers require systemic hormonal therapy, chemotherapy, and/or radiotherapy. However, there is an urgent need for novel therapeutic agents with the ability to selectively kill cancer cells, irrespective of their growth rates and resistance mechanisms. Recently, immunotherapy has emerged as a promising approach to the eradication of metastatic disease [13–15].

1.3 Immune Checkpoints

Targeting the different mechanisms by which cancer cells evade immune responses is relevant for inhibiting cancer progression and has resulted in improved immune-based treatment strategies [16, 17]. Since cancer cells are self cells, they are, by nature, difficult for the immune response to detect as a threat. For example, some cancer cells decrease their expression of class I major histocompatibility complex (MHC) molecules, allowing them to evade cytotoxic T cells [18]. Cancer cells are also able to evade anti-tumor immune responses by expressing ligands that bind to inhibitory molecules on activated T cells that normally serve as checkpoints to regulate T cell immunity [16]. In this regard, cancer cells may upregulate their expression of ligands (PD-L1, PD-L2) for inhibitory PD-1 on T cells [6, 17, 19]. Blockade of these checkpoint-related inhibitory pathways is leading to improvements in cancer immunotherapy.

1.4 Impact of Natural Products on T Cell Checkpoints

Certain phytochemicals may potentiate immunotherapy since they are able to trigger a form of cancer cell death that may initiate an anti-tumor immune response, such as causing the release of the immune-modifying molecule high mobility group box 1 protein (HMGB1) [20]. Importantly, the phytochemical apigenin has been found to decrease the interferon (IFN) γ-induced expression of PD-L1 (*see* Fig. 1) by human and mouse mammary carcinoma cells [6]. Phytochemicals therefore have the potential to enhance the effectiveness of T cell-based cancer immunotherapy by prolonging T cell activation. The following details methods to determine the effect of a given phytochemical (and associated metabolites) on cancer cell expression of PD-L1/PD-L2, as well as T cell activation. The potential impact of phytochemicals on T cell function, and in particular immune checkpoints like PD-1, needs to be further explored in order to better understand their potential to enhance T cell-based cancer immunotherapies.

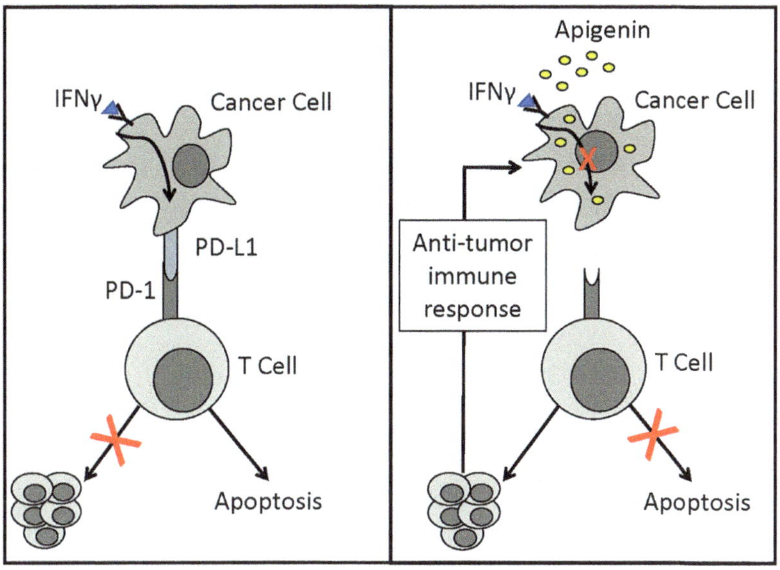

Fig. 1 Effect of apigenin on PD-L1 expression and T cell proliferation. IFNγ induces PD-L1 expression in some cancer cells, inhibiting T cell proliferation (left side). Apigenin inhibits IFNγ-induced PD-L1 expression promoting T cell proliferation (right side)

2 Materials

2.1 Cell Culture

1. T25 and T75 tissue culture flasks.
2. Dulbecco's Modified Eagle's Medium (DMEM).
3. Roswell Park Memorial Institute (RPMI)-1640 medium.
4. Fetal bovine serum (FBS), heat-inactivated at 56 °C for 30 min.
5. Penicillin-streptomycin (10,000 U/mL).
6. 4-(2-hydroxyethyl)-1-piperazineethanesulfonic acid (HEPES; 1M).
7. L-glutamine solution (200 mM).
8. Trypsin (0.25%) in 1 mM ethylenediaminetetraacetic acid (EDTA).
9. Polystyrene flat-bottom 6- or 24-well cell culture plates.

2.2 Determination of PD-L1 and/or PD-L2 Expression

1. Polystyrene round-bottom tubes, 5 mL (*see* **Note 1**).
2. Recombinant IFNγ.
3. TrypLE reagent.
4. Fluorochrome-conjugated (*see* **Note 2**) anti-PD-L1 antibodies or anti-PD-L2 antibodies and the appropriate fluorochrome-conjugated isotype control antibody.
5. Phosphate-buffered saline (PBS).

6. Flow cytometry buffer (0.2% NaN3, 1% BSA, in PBS).
7. 1% paraformaldehyde solution.
8. Aluminum foil (*see* **Note 3**).

2.3 T Cell Isolation from Mice

1. Pan T Cell Isolation MACS kit (Miltenyi Biotec).
2. C57BL/6 or other inbred strain of mice.
3. PBS.
4. Polystyrene round-bottom tubes, 13 mL.
5. Tissue homogenizer.
6. 0.2% NaCl solution.
7. 1.6% NaCl solution.
8. MACS buffer (2% bovine serum albumin (BSA) [w/v], 2 mM EDTA in PBS, pH 7.2).
9. Pre-separation filter (30 μm nylon mesh).
10. LS column.
11. MACS separator magnet.

2.4 T Cell Proliferation: Tritiated Thymidine Incorporation

1. Mouse T-activator CD3/CD28 antibody-coated beads (Dynabeads; *see* **Note 4**).
2. Tritiated thymidine stock solution (5 mCi/5 mL).
3. Titertek® Cell Harvester and fiberglass filter mats.

2.5 T Cell Proliferation: Oregon Green 488 Staining

1. Mouse T-activator CD3/CD28 antibody-coated beads (Dynabeads; *see* **Note 4**).
2. Oregon Green 488 dye.
3. Aluminum foil (*see* **Note 3**).
4. 1% paraformaldehyde solution.

2.6 Interleukin-2 (IL-2) Production

1. Enzyme-linked immunosorbent assay (ELISA) plate.
2. IL-2 capture antibody.
3. ELISA coating buffer (from BD Biosciences ELISA kit).
4. ELISA wash buffer (1× PBS, 0.05% Tween-20 (250 μL Tween in 500 mL PBS)).
5. Blocking buffer (10% FBS in PBS; *see* **Note 5**).
6. IL-2 standard (20 μg/mL stock), stored at −80 °C (*see* **Note 6**).
7. Biotin-conjugated anti-IL-2 antibody (0.5 mg/mL stock).
8. Streptavidin-horse radish peroxidase (HRP).
9. 3,3′,5,5′-Tetramethylbenzidine (TMB) substrate solution.
10. Stop solution (0.3 M H_2SO_4).

2.7 T Cell-Cancer Cell Co-cultures

1. Oregon Green 488 dye.
2. T25 and T75 tissue culture flasks.
3. 24-well flat-bottom tissue culture plates.
4. IFNγ (stock solution of IFNγ at 10 μg/mL in 0.1% BSA, stored at −80 °C).
5. Phytochemical such as apigenin (stock solution at 20 mM in dimethyl sulfoxide (DMSO) or other solvent, stored at 20 °C).
6. TrypLE reagent.
7. Polystyrene round-bottom tubes, 5 mL.
8. 1% paraformaldehyde solution.
9. Aluminum foil (*see* **Note 3**).

3 Methods

3.1 Cell Culture: Adherent Cancer Cells

1. Grow adherent cancer cells in a monolayer (*see* **Note 7**), using a T75 tissue culture flask, at 37 °C in a humidified 10% CO_2 incubator in DMEM containing 10% heat-inactivated FBS, 5 mM HEPES buffer (7.4 pH), 2 mM L-glutamine, and 100 U/mL penicillin-streptomycin (hereafter referred to as cDMEM) for 2–3 days.
2. Cells are lifted from flasks with 3 mL trypsin solution.
3. Once the cells appear rounded and lift from the flask, they are removed and placed in a tube containing 7 mL cDMEM.
4. Cells are ready to be counted and used in an experiment.

3.2 Cell Culture: Non-adherent Jurkat T Cells

1. Non-adherent Jurkat T cells are maintained at 37 °C in a humidified 5% CO_2 incubator in a T25 flask containing RPMI-1640 medium supplemented with 5% heat-inactivated FBS, 5 mM HEPES buffer (7.4 pH), 2 mM L-glutamine, and 100 U/mL penicillin-streptomycin (hereafter referred to as cRPMI).
2. Cells are split every 2–3 days and culture medium is replenished.

3.3 Cell Counting

1. Mix a 20 μL aliquot of the cell solution with 20 μL of 0.1% Trypan Blue dye (*see* **Note 8**).
2. Count the cells under a light microscope using a hemocytometer.
3. Calculate the number of cells needed for the experiment.

3.4 Determination of PD-L1 Expression

Some cancer cells, including breast cancer cells, increase their expression of PD-L1 in the tumor microenvironment, specifically in response to IFNγ [6]. The effect of phytochemicals on the upregulation of PD-L1 can be assessed by flow cytometry. Cancer cells are exposed to the desired phytochemical followed by treatment with IFNγ to determine whether the phytochemical impacts the expression of PD-L1. After a period of culture, cells are washed and stained with fluorochrome-conjugated anti-PD-L1 antibody and PD-L1 expression is determined by measuring fluorescence with a flow cytometer. Cells that are stained with an isotype control antibody serve as a control for background levels of fluorescence. Using the appropriate antibodies, this method can also be used to assess PD-L2 expression.

1. Place 50,000–150,000 cancer cells (e.g., mouse or human mammary carcinoma cells, respectively) per 1600 μL in each well of a 6-well flat-bottom plate. Allow cells to adhere overnight.
2. Treat cells with different concentrations of the desired phytochemical (10–100 μM) for 30 min.
3. Add mouse or human IFNγ (0.1, 1, and 10 ng/mL).
4. Culture treated cells for 24 h at 37 °C in a humidified 10% CO_2 atmosphere.
5. Remove culture supernatant and place in a 5 mL tube, add 1 mL TrypLE solution to each well for 5 min, then remove the lifted cells, and combine with the culture supernatant.
6. Wash each well with 1 mL cDMEM and combine with cells. Centrifuge tubes containing cells at 500 × *g* for 5 min at room temperature (RT). Discard supernatant and resuspend cell pellet in 50 μL flow cytometry buffer.
7. Resuspend cells in 50 μL of isotype control antibody or phycoerythrin (PE)-labeled anti-PD-L1 antibody (1.2 μg/mL) for 30 min at 4 °C.
8. Wash cells by adding 1 mL flow cytometry buffer to each tube and centrifuging at 500 × *g* for 5 min at 4 °C.
9. Discard supernatants and resuspend cell pellets in 400 μL of 1% paraformaldehyde.
10. Cover tubes containing labelled cells with aluminum foil and store for up to 48 h at 4 °C.
11. Read the samples on a flow cytometer. Measure fluorescence on the FL-2 channel if using PE-labeled antibodies to detect changes in PD-L1 expression on the surface of the cell (*see* Fig. 2).

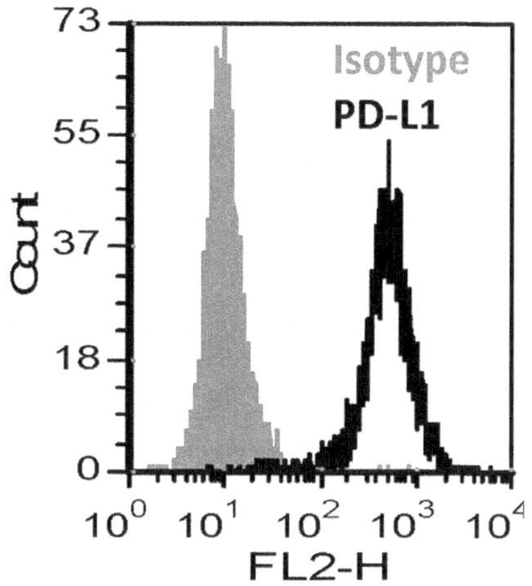

Fig. 2 Representative of flow cytometry histogram showing fluorescence of cells stained with phycoerythrin (PE)-labeled isotype (gray peak) or PD-L1 (open black peak) antibodies

3.5 T Cell Isolation from Mice

T cells can be obtained from the spleens of mice in order to examine the effect of a particular phytochemical on T cell activation, proliferation, and cytokine production. Companies such as Miltenyi Biotec make kits for mouse T cell isolation.

1. Collect spleens from mice, put 2 spleens into a 13 mL round-bottom tube containing 4 mL PBS, and place on ice (*see* **Note 9**).
2. Grind spleens with a tissue homogenizer to generate single-cell suspensions.
3. Pour cell suspension into a centrifuge tube.
4. Add PBS to bring volume up to 10 mL.
5. Centrifuge at 500 × *g* for 5 min.
6. Decant the supernatant.
7. Run the tube along an empty rack really well to lift and break up the cell pellet.
8. To lyse red blood cells, add 4 mL of 0.2% NaCl and pipette up and down once to mix. Leave for 20 s and then add 4 mL 1.6% NaCl. Add 2 mL of PBS to the tube.
9. Allow debris to settle for 5 min. Decant the supernatant that now contains only intact white blood cells into a new tube.
10. Centrifuge at 500 × *g* for 5 min.
11. Discard supernatant.
12. Run the tube along an empty rack to lift and break up the cell pellet.

13. Resuspend cells in 1 mL of MACS buffer.
14. Activate a pre-separation filter by running 500 μL buffer through it.
15. Add cells to the pre-separation filter and run them through the filter, followed by 6–8 mL MACS buffer to wash.
16. Collect cells and centrifuge at 500 × *g* for 5 min.
17. Resuspend cells in 2 mL MACS buffer and count.
18. Dilute to 125 million cells/mL and add 1 mL to as many tubes are needed (125 million cells/column).
19. Top tubes up to 10 mL with MACS buffer.
20. Centrifuge cells at 500 × *g* for 5 min, decant supernatant, and resuspend the cell pellet.
21. Add 450 μL MACS buffer and 50 μL of biotin-conjugated antibody cocktail to each tube.
22. Vortex and incubate for 10 min at 4 °C with loose cap.
23. Add 400 μL MACS buffer and 100 μL of anti-biotin microbeads to each tube.
24. Vortex and incubate for 15 min at 4 °C with loose cap.
25. Top up to 10 mL with MACS buffer.
26. Centrifuge at 500 × *g* for 5 min, decant supernatant, and resuspend cells in 500 μL MACS buffer.
27. Place LS column in the magnetic field of MACS separator. Rinse column with 3 mL MACS buffer.
28. Add cells to the column and collect the effluent. Wash column with 12 mL of MACS buffer (four washes of 3 mL each). Collect all of the effluent.
29. Centrifuge effluent containing T cells at 500 × *g* for 5 min, discard supernatant, and resuspend cell pellet in 10 mL medium.
30. Wash cells by centrifugation at 500 × *g* for 5 min, and then resuspend the cell pellet in 2 mL cRPMI medium.
31. Count cells and dilute to desired concentration prior to use in the experiment.

3.6 T Cell Proliferation: Tritiated Thymidine Incorporation

T cell proliferation following their activation can be assessed by measuring the incorporation of tritiated thymidine into newly synthesized DNA. T cells become activated in response to antibody-coated beads that bind to and crosslink cell-surface CD3 and CD28. Changes in T cell proliferation in the absence or presence of a given phytochemical are observed by detecting changes in thymidine incorporation.

1. Put 25,000–100,000 T cells per well in a final volume of 100–200 μL cRPMI medium in a 96-well round-bottom plate, in quadruplicate.
2. Use mouse T cell-activator anti-CD3 and anti-CD28-coated Dynabeads at a ratio of 1 bead to 2 cells (stock solution contains 40,000 beads/μL).
3. Culture T cells with or without different concentrations of the desired phytochemical (10–250 μM) with Dynabeads in a 37 °C humidified chamber in 10% CO_2 for 48, 72, or 96 h.
4. At 6 h before the end of culture, add tritiated thymidine to each well. Tritiated thymidine stock is 5 mCi/5 mL, and final working concentration is 0.25 μCi/well.
5. Harvest cells using onto fiberglass filter mats with a Titertek® Cell Harvester.
6. Tritiated thymidine incorporation into newly synthesized DNA is measured by liquid scintillation counting.

3.7 T Cell Proliferation: Oregon Green 488 Staining

T cells proliferation can also be assessed by flow cytometric analysis of cells stained with Oregon Green 488 dye prior to culture. Each time the T cell divides the fluorescence of the cell protein-associated Oregon Green 488 dye halves so the number of cell divisions can be quantified by measuring fluorescence with a flow cytometer. Measurement of cell proliferation by fluorescence avoids the potential hazards associated with the use of radioisotopes.

1. Pellet T cells by centrifugation at 500 × *g* for 5 min.
2. In the dark, resuspend the T cells in warm PBS containing Oregon Green 488 dye. Add 1.6 μL Oregon Green 488 dye (5 mM stock in DMSO) in 4 mL PBS to achieve a final concentration of 2 μM.
3. Incubate T cells for 10 min at RT on a rocker or plate shaker. Use aluminum foil to prevent exposing the T cell suspension to light.
4. Add 4 mL of warm FBS to bind excess Oregon Green 488 dye.
5. Centrifuge cells 500 × *g* for 5 min and resuspend in 10 mL warm cRPMI medium.
6. Incubate T cells for another 30 min in a 37 °C incubator (loosen the cap to allow CO_2 exchange).
7. Wash T cells by centrifugation at 500 × *g* for 5 min, resuspend in 1 mL cRPMI medium, count cells, and dilute to desired concentration for plating.
8. Treat T cells with different concentrations of the desired phytochemical and culture at 37 °C in a humidified 10% CO_2 atmosphere for 72 h. Retain a sample of stained T cells in 1% paraformaldehyde and stored at 4 °C for future use as a non-proliferative control.

9. At the end of culture, wash cells in PBS by centrifugation at 500 × *g* for 5 min and resuspend in 300 μL of 1% paraformaldehyde.
10. Read samples on FL-1 on the flow cytometer in order to detect green fluorescence. Cells that have proliferated will show a decrease in fluorescence, which is halved for each round of cell division. T cells that have not proliferated will have a fluorescence that is closer to the non-proliferative control.

3.8 IL-2 Production

Activated T cells secrete the cytokine IL-2, which is an important growth factor for T cells and also induces regulatory T cells [21]. Examining the impact of a given phytochemical on IL-2 synthesis by T cells can provide further information on the phytochemical's immune-modulating potential.

1. Coat a 96-well ELISA plate with capture antibody for IL-2 by adding 100 μL/well of antibody (1 μg/mL in coating buffer) to an ELISA plate. Wrap plate with parafilm and let the antibodies adhere to the plate by incubating overnight at 4 °C.
2. Wash the wells of the ELISA plate five times with ELISA wash buffer. Pat the plate dry on paper towel.
3. Block plates with by adding 200 μL assay diluent/well and incubate the plate at RT for 1 h or overnight at 4 °C.
4. Wash the wells of the plate fivetimes, as described in **step 2**.
5. Perform a twofold series dilution of IL-2 in assay diluent as a standard positive control, starting at 200 pg/mL and increasing to 3 pg/mL.
6. Add 100 μL standard, samples, or assay diluent as a negative control to appropriate wells, in duplicate, and incubate the ELISA plate overnight at 4 °C or incubate for 2 h at RT.
7. Wash the wells of the plate five times, as described in **step 2**.
8. Add 100 μL/well of the biotin-conjugated anti-IL-2 antibody (0.5 μg/mL) to the plate.
9. Incubate the plate at RT for 1 h.
10. Wash the wells of the plate five times, as described in **step 2**.
11. Add streptavidin-HRP at 100 μL/well to the ELISA plate.
12. Incubate the plate at RT for 30 min.
13. Wash the wells of the plate seven times, as described in **step 2**.
14. Add TMB substrate solution at 100 μL/well to the ELISA plate.
15. Incubate the ELISA plate in the dark for approximately 15 min or until the top standard is dark blue.
16. Add 50 μL stop solution (0.3 M H_2SO_4) to each well of the ELISA plate.

17. Determine the absorbance of the sampled by reading the plate on a microplate reader at a wavelength of 450 nm (*see* **Note 10**). The concentrations of IL-2 in samples can be determined by using SOFTmax PRO software and a standard curve generated with the IL-2 standards.

3.9 T Cell-Cancer Cell Co-cultures

Although effector T cells have the potential to selectively kill cancer cells, these malignant cells may evade cytotoxic T cells by expressing PD-L1, the ligand for inhibitory PD-1 expressed by activated T cells. Furthermore, cancer cells may express higher levels of PD-L1 as the result of exposure to IFNγ within the tumor microenvironment. The impact of tumor cell-associated PD-L1 in the absence or presence of a given phytochemical on T cell proliferation can be assessed using co-cultures of Oregon Green 488 dye-stained T cells and cancer cells, followed by flow cytometric analysis.

1. Seed adherent cancer cells (e.g., MDA-MB-468 human mammary carcinoma cells) into T75 tissue culture flasks at 1.0×10^6 cells/flask and culture for 24 h.
2. Treat cancer cells with DMSO as a vehicle control (0.15%), IFNγ (10 ng/mL), the desired phytochemical (10–250 μM) alone or with IFNγ (10 ng/mL) and culture for 24 h. Collect cancer cells by lifting with TrypLE, wash cells with cRPMI medium, and plate at 2×10^5 cells/well into a 24-well flat-bottom plate.
3. Incubate the plate containing the cancer cells for 4 h at 37 °C in a humidified 5% CO_2 incubator to allow cancer cells to adhere and form a monolayer.
4. Add Oregon Green 488-stained Jurkat T cells (5×10^4 cells/well) to the cancer cell monolayer and incubate the resulting co-culture for 48 or 72 h at 37 °C in a humidified 5% CO_2 incubator. Jurkat T cells are transformed and therefore proliferate in the absence of T cell receptor/CD28 stimulation.
5. After incubation, transfer Jurkat T cells to 5 mL round-bottom polystyrene tubes.
6. Wash wells once with PBS and combine the PBS with the fraction from **step 5**.
7. Centrifuge Jurkat T cells at $500 \times g$ for 5 min at 4 °C. Discard the cell-free supernatant.
8. Resuspend Jurkat T cells in 300 μL of 1% paraformaldehyde.
9. Assess T cell proliferation by flow cytometry on FL-1. The number of cell divisions is calculated using the formula $n = \ln \times (\text{mean fluorescence intensity (MFI)}_{control}/\text{MFI}_{sample})/\ln 2$, where n is the number of cell divisions and $\text{MFI}_{control}$ is the MFI of the non-proliferative control. Cell divisions are then normalized to the medium control.

4 Notes

1. In our experience, cell recovery is optimal in polystyrene tubes but poor in polypropylene tubes.
2. Many phytochemicals fluoresce at certain wavelengths. Curcumin, for example, fluoresces across a broad spectrum. It is therefore important to choose a fluorochrome that fluoresces at a different wavelength to avoid confounding fluorescence readings.
3. It is important to use aluminum foil or another opaque material to prevent quenching of the fluorescent dye or fluorochrome-conjugated antibodies by exposure to ambient light.
4. Titrate Dynabeads to determine an optimal concentration for T cell activation. In our experience, the concentration recommended by the manufacturer can be reduced while still providing optimal T cell activation. Dynabeads that activate human T cells are also available but are coated with antibodies that differ from those used to activate mouse T cells.
5. As an alternative to FBS, bovine serum albumin, gelatin, and skim milk powder can be used as blocking agents that may give a lower background signal, depending on the antibody that is being used.
6. It is important to store cytokine standards at −80 °C since activity can be lost when stored at −20 °C or 4 °C.
7. Do not let cell cultures become overgrown as this can change the phenotype and behavior of the cells. If culture medium becomes exhausted replace with fresh medium.
8. Trypan Blue is a viability dye that is excluded by live cells and taken up by dead cells; however, prolonged exposure to Trypan Blue dye is toxic to cells. It is therefore important to perform cell counting as soon as possible after the addition of Trypan Blue dye.
9. For T cell isolation by MACS, it is important to keep everything on ice, especially once cells have been labelled with biotin-conjugated antibodies in order to prevent endocytosis of surface-bound antibodies.
10. The inherent color of some phytochemicals may interfere with colorimetric assays such as ELISA. Design your experiment to allow for the subtraction of background values.

References

1. Forward NA, Conrad DM, Power Coombs MR, Doucette CD, Furlong SJ, Lin T-J, Hoskin DW (2011) Curcumin blocks interleukin (IL)-2 signaling in T-lymphocytes by inhibiting IL-2 synthesis, CD25 expression, and IL-2 receptor signaling. Biochem Biophys Res Commun 407:801–806
2. Yaffe PB, Power Coombs MR, Doucette CD, Walsh M, Hoskin DW (2015) Piperine, an alkaloid from black pepper, inhibits growth of human colon cancer cells via G1 arrest and apoptosis triggered by endoplasmic reticulum stress. Mol Carcinog 54:1070–1085
3. Harrison ME, Power Coombs MR, Delaney LM, Hoskin DW (2014) Exposure of breast cancer cells to a subcytotoxic dose of apigenin causes growth inhibition, oxidative stress, and hypophosphorylation of Akt. Exp Mol Pathol 97:211–217
4. Bernard M, Furlong SJ, Power Coombs MR, Hoskin DW (2015) Differential inhibition of T lymphocyte proliferation and cytokine synthesis by [6]-gingerol, [8]-gingerol, and [10]-gingerol. Phytother Res 29:1707–1713
5. Fernando W, Coombs MRP, Hoskin DW, Rupasinghe HPV (2016) Docosahexaenoic acid-acylated phloridzin, a novel polyphenol fatty acid ester derivative, is cytotoxic to breast cancer cells. Carcinogenesis 37:1004–1013
6. Coombs MRP, Harrison ME, Hoskin DW (2016) Apigenin inhibits the inducible expression of programmed death ligand 1 by human and mouse mammary carcinoma cells. Cancer Lett 380:424–433
7. Knekt P, Järvinen R, Seppänen R, Heliövaara M, Teppo L, Pukkala E, Aromaa A (1997) Dietary flavonoids and the risk of lung cancer and other malignant neoplasms. Am J Epidemiol 146:223–230
8. Hertog MG, Feskens EJ, Hollman PC, Katan MB, Kromhout D (1994) Dietary flavonoids and cancer risk in the Zutphen elderly study. Nutr Cancer 22:175–184
9. Bosetti C, Spertini L, Parpinel M, Gnagnarella P, Lagiou P, Negri E, Franceschi S, Montella M, Peterson J, Dwyer J, Giacosa A, Vecchia CL (2005) Flavonoids and breast cancer risk in Italy. Cancer Epidemiol Prev Biomark 14:805–808
10. Canadian Cancer Society's Advisory Committee on Cancer Statistics (2017) Canadian Cancer Statistics 2017. Toronto, ON: Canadian Cancer Society
11. Pedro C, Mira B, Silva P, Netto E, Pocinho R, Mota A, Labareda M, Magalhães M, Esteves S, Santos F (2018) Surgery vs. primary radiotherapy in early-stage oropharyngeal cancer. Clin Transl Radiat Oncol 9:18–22
12. Maughan KL, Lutterbie MA, Ham PS (2010) Treatment of breast cancer. Am Fam Physician 81:1339–1346
13. Yousefi H, Yuan J, Keshavarz-Fathi M, Murphy JF, Rezaei N (2017) Immunotherapy of cancers comes of age. Expert Rev Clin Immunol 13:1001–1015
14. Peters S, Reck M, Smit EF, Mok T, Hellmann MD (2019) How to make the best use of immunotherapy as first-line treatment for advanced/metastatic non-small-cell lung cancer. Ann Oncol 30:884–996
15. Esteva FJ, Hubbard-Lucey VM, Tang J, Pusztai L (2019) Immunotherapy and targeted therapy combinations in metastatic breast cancer. Lancet Oncol 20:e175–e186
16. Donini C, D'Ambrosio L, Grignani G, Aglietta M, Sangiolo D (2018) Next generation immune-checkpoints for cancer therapy. J Thorac Dis 10:S1581–S1601
17. Alsaab HO, Sau S, Alzhrani R, Tatiparti K, Bhise K, Kashaw SK, Iyer AK (2017) PD-1 and PD-L1 checkpoint signaling inhibition for cancer immunotherapy: mechanism, combinations, and clinical outcome. Front Pharmacol 8:561
18. Garcia-Lora A, Martinez M, Algarra I, Gaforio JJ, Garrido F (2003) MHC class I-deficient metastatic tumor variants immunoselected by T lymphocytes originate from the coordinated downregulation of APM components. Int J Cancer 106:521–527
19. Aguiar PN, De Mello RA, Hall P, Tadokoro H, de Lima Lopes G (2017) PD-L1 expression as a predictive biomarker in advanced non-small-cell lung cancer: updated survival data. Immunotherapy 9:499–506
20. Yin S-Y, Yang N-S, Lin T-J (2017) Phytochemicals approach for developing cancer immunotherapeutics. Front Pharmacol 8:386
21. Apert C, Romagnoli P, van Meerwijk JPM (2018) IL-2 and IL-15 dependent thymic development of Foxp3-expressing regulatory T lymphocytes. Protein Cell 9:322–332

Chapter 15

Assessment of Immune Protective T Cell Repertoire in Humans Immunized with Novel Tuberculosis Vaccines

Mangalakumari Jeyanathan and Zhou Xing

Abstract

Tuberculosis (TB) is one of the major global health concerns. There has been a lack of an effective vaccine strategy. The Bacillus Calmette-Guerin (BCG), the only licensed vaccine against TB, is not effective against adult pulmonary TB, the highly contagious form of TB. In the past two decades or so, many novel TB vaccines have been developed, and some of them were evaluated in clinical trials. However, the lack of validated immune correlates to assess the clinical relevance of novel TB vaccines before their entry into costly efficacy trials is a huge challenge to the field of TB vaccine development. Here we describe a general protocol for the procedure of a systematic immunological approach that can be utilized to better assess the clinical relevance of TB vaccine-activated T cells in early phases of clinical studies.

Key words Tuberculosis, Mycobacteria, Ad5Ag85A, Peptide, Epitope, PBMC, Ag-specific T cell line, MGIA

1 Introduction

Tuberculosis (TB) is the leading infectious killer in the world [1]. Global TB control program is challenged by the unavailability of effective vaccines and diagnostic tools [2]. Adding to this is the emergence of multi-resistant to total drug-resistant strains of *Mycobacterium tuberculosis*, the causative agent of TB. The only currently available licensed vaccine is the Bacillus Calmette-Guerin (BCG), a century-old vaccine that protects against childhood disseminated TB but not adult pulmonary TB, the highly contagious form. With the vision of World Health Organization (WHO) and partners to end TB by 2035 (defined as less than one incident case per million), scientists have focused the efforts to develop novel TB vaccines that can either boost or replace BCG [3]. Up to now, more than ten novel TB vaccines are being evaluated in clinical trials [4]. However, those vaccines with demonstrated safety and immunogenicity failed to enhance protective efficacy [5, 6]. Currently, the lack of a validated correlate to assess the

Chaohong Liu (ed.), *T-Cell Receptor Signaling: Methods and Protocols*, Methods in Molecular Biology, vol. 2111,
https://doi.org/10.1007/978-1-0716-0266-9_15,

clinical relevance of novel tuberculosis vaccines before their entry into costly efficacy trials thus signifies a huge challenge to the field of TB vaccine development [7, 8]. Mycobacterial growth inhibition assays (MGIA) and intradermal BCG challenge approach have been used to inform the correlates of protection [9–11]. However, there is a need for additional approaches to evaluate whether novel TB vaccine activate the type of T cells capable of recognizing antigenic epitope found in genetically diverse, latently infected but otherwise healthy humans or those with active disease. We have recently reported on an integrated immunological approach that enables identification of immune protective T cell repertoires in humans vaccinated with a novel adenovirus (Ad)-based tuberculosis vaccine (Ad5Ag85A) [12]. Here we describe a general protocol for the procedure of a systematic immunological approach that can be utilized to assess the immune protective T cell repertoire in humans immunized with novel tuberculosis vaccines.

The systematic approach consists of seven steps:

Step 1: Determination of T cell reactivity to peptide pools spanning the protein expressed by the vaccine (each pool containing 7–10 overlapping peptides). This allows identification of maximum T cell-reactive peptide pools for each participant in the clinical vaccine study.

Step 2: Generation of new peptide pools spanning the maximum T cell-reactive peptide pools identified in Step 1.

Step 3: Identification of maximum T cell-reactive single peptides for each participant using the new peptide pools generated in Step 2.

Step 4: Generation of Ag-specific T cell lines and cryopreservation of expanded T cell lines.

Step 5: Identification of single peptide-specific CD4 and CD8 T cells within expanded Ag-specific T cell lines for each participant.

Step 6: Evaluation of the Ag-specific T cells inhibiting mycobacterial outgrowth in autologous monocytes.

Step 7: Determination of clinical relevance of identified epitopes in Step 5 and human population coverage.

2 Basic Methods

Prepare all media using cell-culture-grade components, and filter using 0.45 μm pore size filters. Store all media at 4 °C. Avoid prolonged exposure of media to light and air. Avoid using cell culture medium which has changed color. Also, avoid using cell culture media if it is old (optimal if used within a month of preparation unless stated otherwise). Always use warm media to resuspend cells undergoing culture or immediately after thawing a frozen vial of cells. Follow approved methods of waste disposal for cell culture wastes and BCG-infected cultures.

2.1 Media, Buffers, and Other Solutions

1. Complete RPMI (cRPMI), mix RPMI 1640, 10% fetal bovine serum, 20 mM glutamine, 1000 U/mL penicillin, and 1 mg/mL streptomycin, and filter using 0.2 μM pore size filter.
2. 12.5% Human serum albumin (HSA), mix 2.5 g HSA in 20 mL RPMI. Knock on table to wet the powder, and leave overnight in fridge or for a few hours at room temperature (RT). When dissolved, filter using 0.2 μM pore size filter (*see* **Note 1**).
3. Freezing medium (12.5% HSA and 25% DMSO), mix 10 mL 12.5% HAS and 2.5 mL DMSO. Drip DMSO into 12.5% HSA while swirling the conical tube. Clumps will disappear with sufficient swirling (*see* **Note 2**).
4. T cell expansion medium, supplement cRPMI with 100 U/mL rIL-2 (human recombinant IL-2).
5. Mycobacterial growth inhibition assay medium (MGIA medium), mix RPMI 1640, 10% HSA, 20 mM glutamine, 1000 U/mL penicillin, and 100 U/mL human recombinant IL-2, and filter using 0.2 μM pore size filter.
6. FACS buffer, 0.5% bovine serum albumin (BSA) in PBS.
7. Fixative solution, 2% methanol-free formaldehyde in PBS.

2.2 Peptide Pool Preparation

1. For preparation of peptide stocks, use 50 μL DMSO to dissolve each peptide (2 mg); this leads to a concentration of 40 μg/μL (*see* **Note 3**).
2. Aliquot each peptide to ten vials, 5 μL/vial (each aliquot contains 200 μg of peptide), and store in a −70 °C freezer (labeled as p1, p2, p3...etc.).
3. For preparation of peptide pools of 7–10 peptides, establish six peptide pools by combining p1–10, p11–20, p21–30, p31–40, p41–50, and p51–57 by combining 5 μL of each ten consecutive peptides; each pool thus has a final volume of 50 μL and contains 200 μg of each peptide except the last pool that has only 35 μL.
4. To pools P1, P2, P3, P4, and P5, add 950 μL of RPMI culture media, and this brings the final volume of each peptide pool to a total of 1000 μL (0.2 μg each peptide/μL).
5. To P6, add 665 μL of RPMI culture media, and bring the final volume of this pool to 700 μL (0.2 μg each peptide/μL).
6. Aliquot each peptide pool to 50 μL of aliquots, and store in −70 °C freezer (labeled as P1–10, P11–20, P21–30, etc.).

3 Methods

3.1 Human IFN-γ ELISpot Assay by Using Fresh PBMC to Determine T Cell Reactivity to Peptide Pools Spanning the Protein Expressed by the Vaccine

1. Prepare peptide pools consisting of the 15-mer peptides with a 10-mer overlap spanning the protein expressed in the vaccine. For instance, there are a total of 57 of 15-mer peptides with 10-mer overlap of Ag85A protein (complete peptide library) expressed by Ad5Ag85A. Six pools, each comprising 7–10 overlapping peptides (2 μg of each peptide), may be prepared using this complete peptide library.
2. Perform ELISpot with human IFN-γ ELISpot set according to manufacturer's instruction using the fresh PBMC isolated from the whole blood collected before (baseline) and at various timepoints after vaccination that are stimulated with the peptide pools prepared. Count spot-forming units (SFU) in each well. A study participant is considered responding to the vaccine if the spot-forming units (SFUs)/million PBMC is two times more than the spots recorded at baseline.
3. Adopt a ranking system to categorize vaccine responders to each of peptide pools as highly reactive, >500 SFUs/million PBMC; medium highly reactive, >100 SFUs/million PBMC; and poorly reactive, 0–99 SFUs/million PBMC. For this, use the peak response by each participant. For instance, after Ad5Ag85A vaccination, peak responses by each participant were categorized as shown in Fig. 1.
4. Select the peptide pools to which most of the study participants have reactive T cells. These peptides will be used in the next step to prepare a new set of peptide pools. This allows to narrow down the number of peptides to include in the identification of maximum T cell-reactive single peptide. For instance, after Ad5Ag85A vaccination, as shown in Fig. 1, most of the participants highly reacted to first three pools (first 30 peptides of Ag85A peptide library).

3.2 Three-Dimensional Matrix Design for the Preparation of New Peptide Pools to Identify T Cell-Reactive Single Peptide

1. Design a 3D matrix so that each peptide is present exactly in three pools testing positive [13]. For instance, 3 T cell-reactive peptide pools comprising 10 overlapping peptides in each pool (30 peptides in total) can be distributed into 12 pools with 7 or 8 peptides per pool as shown in Fig. 2a. This results in, for example, peptide 1 being present in p1, p5, and p9 (Fig. 2b).
2. Prepare new peptide pools based on the 3D matrix as shown in Fig. 2b to have 0.25 μg of each peptide/μL.

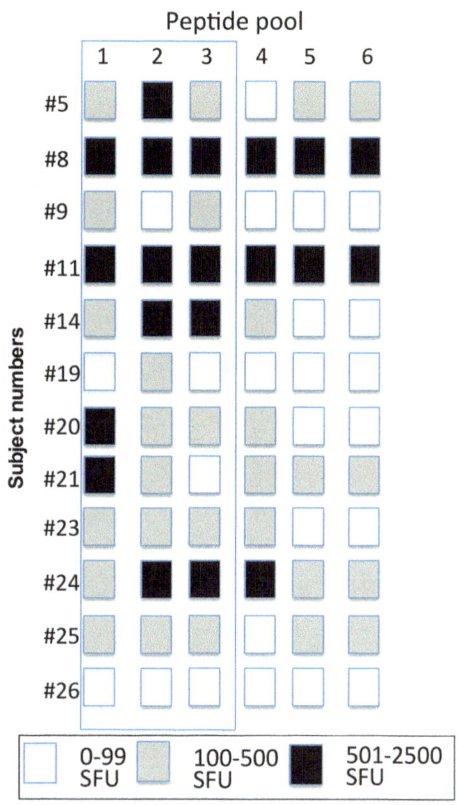

Fig. 1 Extent of PBMC reactivity in 12 study participants. ELISPOT data for six peptide pools identified as p1–p6 (top panel) using fresh PBMC from participants vaccinated with Ad5Ag85A identified by numerical numbers (left side). Reactivity was classified as poorly (0–99 spot-forming units [SFUs]/million PBMC), medium-highly (100–500 SFUs/million PBMC), and highly (501–2500 SFUs/million PBMC) reactive

3.3 Human IFN-g ELISpot Assay by Using Frozen PBMC to Determine Maximum T Cell-Reactive Single Peptide

1. Revive a vial of frozen PBMC collected at baseline and at peak response visit for 4–6 h.
2. For revival, prepare an appropriate number of 15 mL polypropylene tubes for your samples by adding 10 mL of cRPMI medium to each tube and placing in the incubator at 37 °C. Separately warm an aliquot of 30 mL of cRPMI per sample at 37 °C.
3. Bring a bucket of liquid N_2 to sample storage room. Wear protective clothing, and remove the cryo-samples, and place them into liquid N_2. Remove the samples from liquid N_2, and thaw them in the 37 °C water bath (*see* **Note 4**).
4. Allow time for a small amount of ice to remain in the sample, and immediately process the cells in the laminar flow hood as described below.

	Pool 5	Pool 6	Pool 7	Pool 8	
Pool 1	1	9			**Pool 9**
		10	17		**Pool 10**
			18	24	**Pool 11**
	2			25	**Pool 12**
Pool 2	3	11			**Pool 9**
		12	19		**Pool 10**
			20	26	**Pool 11**
	4			27	**Pool 12**
Pool 3	5	13			**Pool 9**
		14	21		**Pool 10**
			22	28	**Pool 11**
	6				**Pool 12**
Pool 4	7	15			**Pool 9**
		16	23		**Pool 10**
				29	**Pool 11**
	8			30	**Pool 12**

Pool	Peptides
p1	1,2,9,10,17,18,24,25
p2	3,4,11,12,19,20,26,27
p3	5,6,13,14,21,22,28
p4	7,8,15,16,23,29,30
p5	1,2,3,4,5,6,7,8
p6	9,10,11,12,13,14,15,16
p7	17,18,19,20,21,22,23
p8	24,25,26,27,28,29,30
p9	1,3,5,7,9,11,13,15
p10	10,12,14,16,17,19,21,23
p11	18,20,22,24,26,28,29
p12	2,4,6,8,25,27,30

Fig. 2 Design of 3D matrix using 30 peptides. (**a**) Twelve new peptide pools of N-terminal 30 peptides of Ag85A protein were constructed by using a 3D matrix approach. (**b**) Specific peptides included in each of the 12 new pools. By this design, each peptide is present in three unique pools (an example is highlighted in red)

5. Remove the required number of 15 mL polypropylene tubes containing culture medium from the incubator. Wipe the sides of the cryovials with an alcohol swab as to avoid contaminating the cells with water from the water bath. Using a sterile 1 mL pipette, transfer the contents of the cryovial by dripping into the pre-warmed culture medium in each 15 mL polypropylene tube.

6. Bring the volume in the tubes to 15 mL adding pre-warmed cRPMI (*see* **Note 5**). Once samples have been resuspended in cRPMI, centrifuge samples in swing-out buckets at 21 °C at 447 × *g* for 10 min. Decant the supernatant, and loosen the cell pellet by gently tapping the tube, and then add 15 mL of pre-warmed cRPMI to the tube, and mix by slowly inverting the tubes two times.

7. Centrifuge in a swing-out bucket at 21 °C at 447 × *g* for 10 min. Decant the supernatant. Then remove all supernatant by tilting the tube so that remaining supernatant runs down to the end of the tube. Remove it carefully with a light-duty tissue wipe. Loosen the cell pellet by gently tapping the tube. Top up to 5 mL with cRPMI.

8. Add 25 μL of DNase 1 (5 μL per 1 mL cell suspension (10 U/mL)) to the sample from **step 7**. Gently mix by inverting the tubes twice. Place tubes with cells in the incubator (37 °C with 5% CO_2) leaving cap of tubes loose (*see* **Note 6**).

9. Perform ELISpot with human IFN-γ ELISpot set according to manufacturer's instruction using the revived PBMC stimulated with new peptide pools prepared in Step 2. Count the SFUs in each well, and subtract the number of SFUs at the baseline visit

from the SFUs at the peak response for each participant. Next, identify individual peptides with strong T cell reactivity in each study participant by using the 3D matrix. For example, a participant in Ad5Ag85A trial study reacted to new peptide pools (Fig. 2b) 2, 3, 6, 7, 8, 9, 10, 11, and 12. Based on the 3D matrix criteria that a single reactive peptide should be present in three T cell-reactive peptide pools, this participant was identified as having high T cell reaction to single peptides 12, 14, 27, and 28 out of 30 single peptides.

3.4 Ag-Specific T Cell Expansion in PBMC to Identify Single Peptide Reactive CD4 and CD8 T Cells

1. Revive a vial of frozen PBMC collected at baseline and at memory response visit (when responses declined after vaccination) for 4–6 h in PBMC-reviving medium.
2. After reviving PBMC, use a T cell enrichment kit to purify Pan T cells from the PBMC collected at the memory response visit according to the manufacturer's instruction.
3. Resuspend purified Pan T cells and revived baseline PBMC in T cell expansion medium to a concentration of 2×10^6 cells/mL.
4. Co-culture 100,000 Pan T cells (50 μL) from memory response visit and 100,000 revived autologous PBMC (50 μL) from baseline visit per well in a U-bottom 96 well plate. Total volume per well is 200 μL. Set up 4–6 wells to generate enough Ag-specific T cell lines.
5. To each well, add the protein/peptide pool of the antigen expressed by the vaccine. For instance, to expand Ag85A-specific T cells in PBMC from Ad5Ag85A-vaccinated participants, rAg85A (2.5 μg) and single peptide pool containing all 57 peptides spanning the Ag85A (2 μg of each peptide) were added to the co-culture.
6. Incubate the plate in 5% CO_2, 37 °C incubator, and examine the plate every day under a dissection microscope for heaps of cells at the bottom of the well with aggregates/clumps of cells. Examine for change in the media color. At ~6 days, media will change color from pink to yellowish orange. *Note*: From this point on, always use T cell expansion medium.

   Day 6: Replenish media by removing ~80 μL from top of the well without disturbing the cells and adding fresh 130 μL media to make a total volume of 200 μL. Continue to incubate the cells.

   Day 9: Count cells from one well by preparing 10× dilution of the cells. The cell number increased 4× from the initial number. Transfer cells to a 24 well plate using P1000 pipette tip, and try not to disturb cell clumps (200 μL from 96 well plate + 100 μL used to wash well + 700 μL media in 24 well plate = 1 mL total media). The clumps do break up a bit but are still visible. *Note*: confluence 50%. Continue to incubate the cells (*see* **Note 7**).

Day 12: At this point, cell confluence is 90%. Count cells by resuspending one well (pipetting up and down in the well, prepare 50× dilution). Cell number would have increased 30× from the initial number. Transfer the cells from each well to a 25 cm^2 flask in a total volume of 5 mL media/flask. Continue to incubate the cells.

Day 15: Transfer cells to 150 cm^2 flasks by combining cells from three 25 cm^2 flasks. Make the total volume to 30 mL with warm media. Count cells in the 150 cm^2 flask by preparing 2× dilution. The cell number would have increased 130× from the initial number. (Should be ~15 × 10^6 cells—can expand to 60 × 10^6 cells before splitting again.)

Day 19: Count cells by preparing 10× dilution. The cell number increased 250× from the initial number. Transfer the cells to 50 mL Falcon tubes (one per flask), and spin down the cells at 447 × *g* for 10 min at RT. Resuspend pellet by gently shaking back and forth, and add 30 mL warm media, and transfer to 150 cm^2 flask, and continue to incubate the cells.

Day 21: Count cells in each flask by preparing 2× dilution. By now, the cells should have stopped growing.

Day 22: Transfer the cells to 50 mL Falcon tube, and spin down the cells at 447 × *g* for 10 min at room temperature. Resuspend the cells in cold 12.5% HSA to prepare 14 × 10^6 cells/mL. To this, add an equal volume of cold freezing media yielding a 7 × 10^6 cells/mL cell suspension. Mix gently, aliquot 1 mL per cryovial, and place into Mr. Frosty. Leave Mr. Frosty in −70 °C O/N. Transfer to liquid N_2 after 24 h (Fig. 3).

3.5 Intracellular Cytokine Staining and Flow Cytometry to Identify Single Peptide Reactive CD4 and CD8 T Cells in Expanded Ag-Specific T Cell Lines

1. Revive the frozen PBMC collected at the baseline visit to use as autologous Ag presenters and the frozen expanded Ag-specific T cell line from the same participant overnight.
2. After overnight revival, count live cells in both PBMC and Ag-specific T cell line preparations using Trypan blue (1:3 dilution).
3. Adjust the cell concentration to 1 × 10^6 cells/mL using warm cRPMI. Irradiate the PBMC with gamma radiation (1000–5000 rad) for 40 min.
4. Add 0.5 mL (500,000 cells) of each cell suspension (irradiated PBMC + Ag-specific T cell line) to each flow cytometry tube (FACS tubes), and stimulate the cells with single peptides highly reactive to T cells identified in Step 3 in the presence of co-stimulatory molecules (anti-CD28 and anti-CD49d). Incubate the FACS tubes with caps loosened in 5% CO_2, 37 °C incubator (*see* **Note 8**).

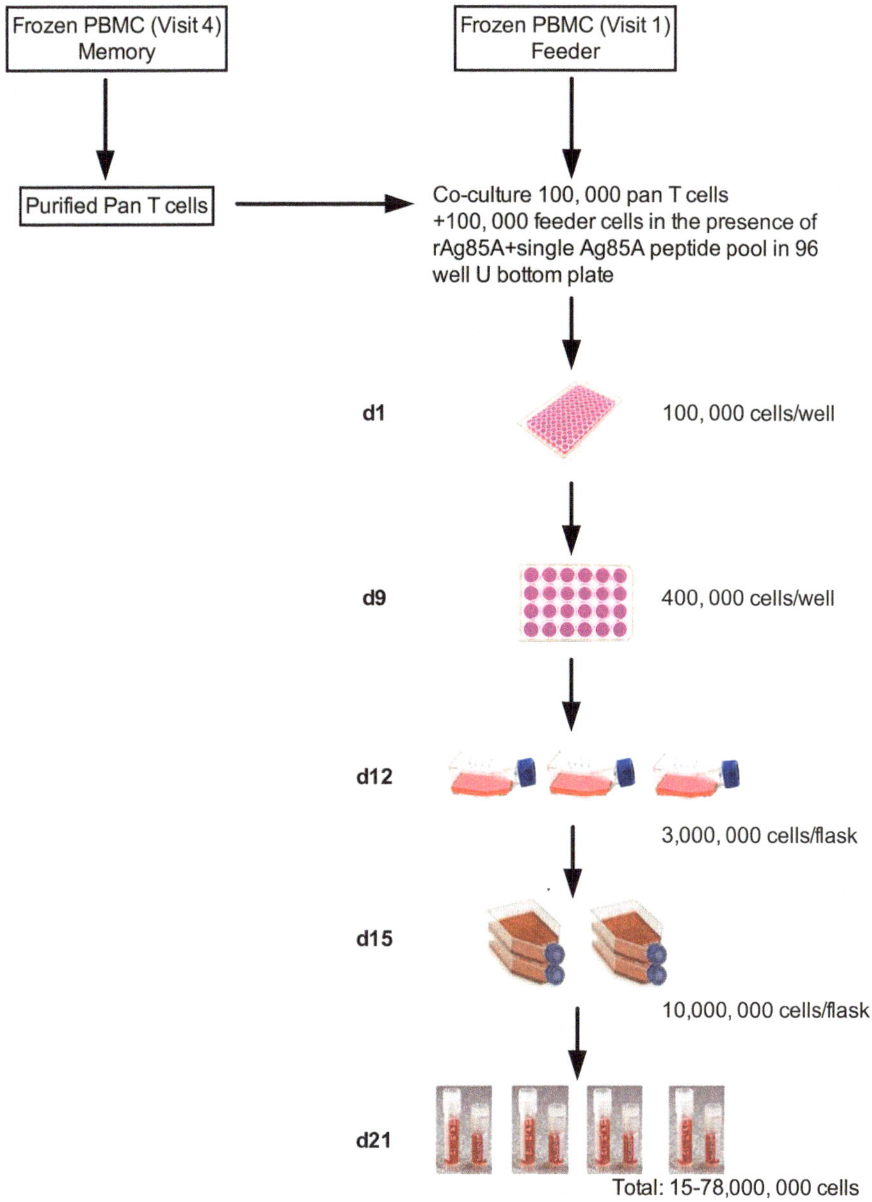

Fig. 3 Step-by-step illustration of expanding and establishing Ag85A-specific memory T cell lines from Frozen PBMC of study participants

Note: Table 1 depicts an example of stimulation conditions employed to identify single peptide reactive CD4 and CD8 T cells following Ad5Ag85A vaccination. Use the single peptide pool of the antigen expressed by the vaccine as a positive control. Set up a separate tube with only co-stimulatory molecules as a negative control.

Table 1
Stimulation conditions for identification of immunodominant CD4 and CD8 memory T cell epitopes in study participants following Ad5Ag85A vaccination

Antigen	Stock concentration	Amount per mL (working concentration)	Amount per 1 mL cell suspension
Negative control (unstimulated)			
Ag85A peptide (1–57) (single peptide pool)	0.7 μg Each peptide/μL	2 μg/mL Each peptide	2.8 μL (undiluted; can refreeze at −70 °C)
Single peptide (identified at Step 3)	40 μg/μL	25 μg/mL Each peptide	6.3 μL (1:10 of stock)—dilute by adding 45 μL RPMI, and refreeze at −70 °C
Anti-CD28 Anti-CD49d	1 mg/mL 1 mg/mL	0.5 μg/mL 0.5 μg/mL	10 μL of 1:20 dilution of both (20 μL of CD28 and 20 μL CD49d in 360 μL PBS)
Brefeldin A	5 mg/mL	10 μg/mL	20 μL 1:10 Dilution (PBS)

5. After 1 h of incubation, add freshly prepared Brefeldin A to all tubes. Mix well with P1000 pipette, and return to incubator, and continue to incubate for further 5 h (*see* **Note 9**).
6. At the end of incubation, remove tubes from the incubator; spin them at 447 × *g* for 10 min at RT. Stain immediately at the end of stimulation.
7. For intracellular staining, wash the cells once with 2 mL PBS, and spin for 5 min at 447 × *g* at RT.
8. After discarding the supernatant, gently tap the tubes to resuspend the cells. Then add 1 mL PBS at RT. To this, add 1 μL reconstituted Near Red live/dead stain. Incubate at RT for 30 min in the dark (follow manufacturer's instructions for reconstitution of Near Red).
9. At the end of incubation, spin the tubes for 5 min at 447 × *g* at RT. Decant the supernatant, and wash once with 2 mL FACS buffer at RT.
10. Decant the FACS buffer, and remove the last drop by blotting on a tissue. 50 μL of volume remains following this procedure. Stain the cells with surface receptor antibody cocktail prepared as CD14-V450 (Clone M5E2) 2.5 μL per test; CD19-V450 (Clone SJ25C1) 2.5 μL per test; CD4 AF700 (Clone RPA-T4) 1 μL of ¼ dilution per test. Add a total of 6 μL of surface antibody cocktail to each tube (*see* **Note 10**). Vortex gently. Incubate at RT for 15 min in the dark.

11. Wash the cells once with 2 mL FACS buffer, and spin for 5 min at 447 × *g* at RT. Discard the supernatant, and vortex pellet gently to resuspend cells.
12. Add 250 μL Cytofix/Cytoperm from BD Biosciences to all tubes. Vortex gently, and incubate at 4 °C (fridge) for 20 min in the dark. After fix/perm treatment, spin the tubes at 739 × *g* for 5 min at RT. Discard the supernatant. Wash once with 2 mL Perm/Wash from BD Biosciences, and spin for 4 min at 739 × *g* at RT (*see* **Note 11**). Discard the supernatant, and vortex gently to resuspend cells.
13. Stain the cells with intracellular cytokine staining antibodies (ICCS). Prepare the antibody cocktail as CD3-PerCPcy5.5 (Clone SK7) 10 μL per test; CD8-PeCy7 (Clone RPA-T8) 1 μL of 1/14 dilution per test; IFN-γ PE (Clone 4S.B3) 10 μL per test. Add a total of 21 μL of ICCS antibody mix per test. Vortex. Incubate at RT for 30 min in dark.
14. Wash once with 2 mL of Perm/Wash from BD Biosciences, and spin for 5 min at 739 × *g* at RT. Discard the supernatant, and vortex gently to resuspend cells.
15. Resuspend the pellet in fixative solution, and incubate for 10 min at RT. Spin the tubes at 739 × *g* for 5 min. Discard the supernatant, and vortex gently to resuspend cells. Add 200 μL of FACS buffer. Tap to mix. Wrap in aluminum foil. Store at 4 °C (fridge) in dark, and acquire the data within 24 h in a flow cytometry. Collect at least 200,000 events of CD3+ cells. Analyze the data using FlowJo software.
16. Prepare compensation controls for flow cytometry, setting up LSRII for data acquisition and gating strategy for data analysis in FlowJo.
17. For preparation of compensation controls, use the same ratio of irradiated PBMCs plus expanded T cells as for stimulations but a total of 500,000 cells (or less) in a total volume of 50 μL FACS buffer (*see* **Note 12**).
18. Add 100 μL FACS buffer for AF700, V450, and PE-Cy7 compensation color control tubes, and add 50 μL FACS buffer for PE and PerCPCy5.5 compensation color tubes.
19. Use the following antibody conjugates to prepare compensation controls. Incubate for 30 min at RT in the dark. CD4-AF700 (Clone RPA-T4) 1 μL; CD4-PB (Clone RPA-T4) 5 μL; CD8-PE Cy7 (Clone RPA-T8) 1 μL of 1/14 dilution; CD3-PE (Clone SK7) 10 μL; CD3-PerCPCy5.5 (Clone SK7) 10 μL. Set up a separate tube for unstained/ unstimulated cells.
20. Add 2 mL FACS buffer to each tube, and spin at 447 × *g* for 5 min at RT. Discard supernatant, and resuspend the pellet in fixative solution, and incubate for 10 min at RT. Set up a

separate tube for Near Red live/dead stain. Add 1 mL PBS at room temperature (RT) to this, and add 1 μL reconstituted Near Red live/dead stain. Incubate at RT for 30 min in the dark. At the end of incubation, spin the tubes at 739 × *g* for 5 min. Discard the supernatant, and vortex gently to resuspend cells. Add 200 μL of FACS buffer. Tap to mix. Wrap in aluminum foil. Store at 4 °C (fridge) in dark.

21. Set up a separate tube for unstained/unstimulated cells.
22. For compensation setup, while creating compensation, tick FSC-A, FSC-H, and FSC-W to allow doublet elimination during analysis.
23. Set the FSC voltage to 480 and SSC voltage to 380 for unstained cells. Run one of the stained samples to make sure all the events are falling within the range. If not, adjust the voltages for the marker.
24. When analyzing the data on FlowJo, first, gate on FSC-A vs. FSC-W. Then gate on SSC-A vs. SSC-W. Now from this population, gate on live cells by excluding Near Red positive dead cells. From the live cells, gate on CD14/CD19 negative population by excluding CD14/CD19 positive population. Now gate on FSC-A vs. SSC-A for lymphocyte population. From lymphocyte population, gate for CD3+ cells and from CD3+ for CD4+ and CD8+ cell population.

Figure 4 depicts an example of CD4 and CD8 T cell epitope identification for a participant vaccinated with

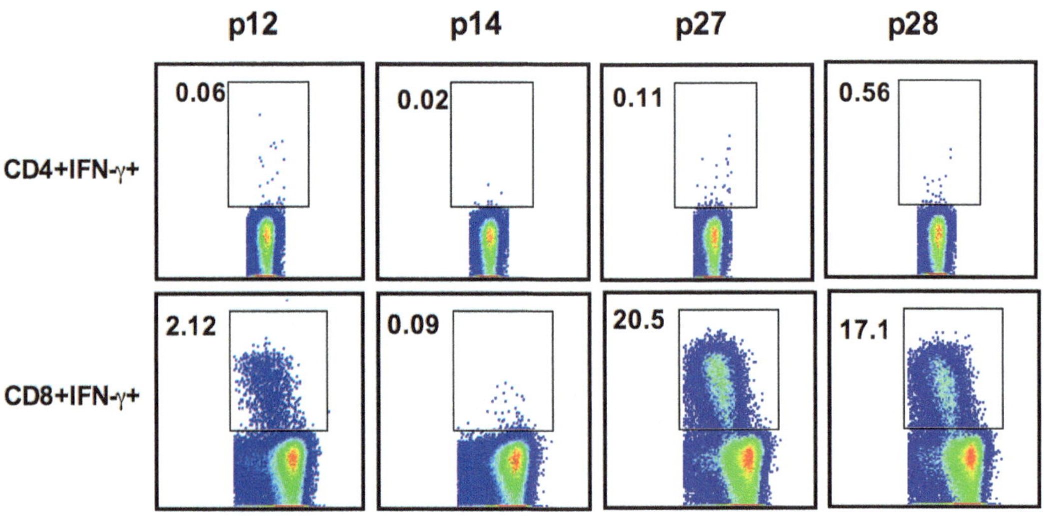

Fig. 4 Ag-specific T cell line responses to single peptide stimulation. Dotplots showing IFNγ-producing CD4 and CD8 T cells in the expanded T cell line stimulated with single peptides p12, p14, p27, and p28 from a participant vaccinated with Ad5Ag85A. Numbers in the plots represent frequencies of CD4/CD8 T cells producing IFNγ

Ad5Ag85A. In Step 3, four single peptides (p12, p14, p27, and p28) were identified as highly T cell-reactive peptides for this participant. Stimulation of established T cell lines with these peptides and IFNγ intracellular cytokine staining identified three single peptides to be immunodominant (p12, p27, p28) that were highly recognized by CD8 T cells.

3.6 Mycobacterial Growth Inhibition Assay (MGIA) Using BCG to Evaluate Mycobacterial Growth Inhibition by Vaccine-Induced T Cells Compared to T Cells Before Vaccination

1. Revive frozen PBMC from baseline visit (V1) and 8 weeks after vaccination visit (V4) for 4 h.
2. Purify Pan T cells from revived PBMC using a T cell enrichment kit to enrich Pan T cells according to manufacturer's instructions. Resuspend the cells to 2×10^6 cells/mL in MGIA medium (*see* **Note 13**).
3. Purify autologous monocytes from PBMC from baseline visit (V1) using a negative monocyte purification kit according to manufactures' instruction, and use as targets for pan T cells (*see* **Note 13**). Resuspend the cells to 2×10^5 cells/mL in MGIA medium. Hereafter, use MGIA medium in all steps.
4. Co-culture 50 μL of purified autologous monocytes (10,000 monocytes) and 50 μL of Pan T cells (100,000 cells) in a 96 well U-bottom plate containing total volume of 100 μL. Set up separate wells for baseline-visit Pan T cells and autologous monocytes and 8 weeks after vaccination-visit Pan T cells and autologous monocytes. Set up triplicates for each. Keep the plate in 5% CO_2, 37 °C incubator until the infection with BCG.
5. For preparation of BCG, grow BCG AERAS in Middlebrook 7H9 broth supplemented with OADC to mid-log phase; harvest, aliquot, and freeze the stock vials at −80 °C. Titrate a vial, and record the titer. Generally, mid-log-phase culture yields 100,000 colony-forming units (CFU) per μL. Each frozen vial contains 500 μL total volume.
6. Thaw a vial just before setting up the infection of cells. Spin the thawed vial at $13792 \times g$ for 5 min. Discard the supernatant, and wash the pellet ×2 with PBS with 0.05% tween 80. Resuspend the pellet in 500 μL MGIA media.
7. Prepare serial dilutions, and plate on 7H10 agar plate to check the titer.
8. Prepare a separate supply for infection at a concentration of 600 CFU/mL, considering the titer recorded during freezing of stock vial.
9. Take the plate containing the cells from the incubator, and add 100 μL of infection supply (60 BCG/well) and a single peptide pool of the antigen expressed by the vaccine (2 μg of each peptide) to each well (*see* **Note 14**). Place the plates back in 5% CO_2, 37 °C incubator for 3 days.

10. For assaying CFU to evaluate inhibition of BCG growth, after the 3-day incubation, remove the supernatant from each well into separate Eppendorf tubes to recover extracellular bacteria. Spin the tubes at 14,000 rpm for 5 min in a microcentrifuge, and discard the supernatant by carefully pipetting out. Resuspend the pellet in each tube in 50 μL of sterile (autoclaved) water.
11. Add 100 μL of sterile (autoclaved) water to each well, and leave for 10 min to burst the cells to release BCG from the cells.
12. Pool the 100 μL from the wells (pipette up and down and scrape the bottom of the well with the pipette tip) with the corresponding 50 μL from the Eppendorf tubes (lysed pellet). Also, wash the wells with 90 μL water, and pool as well. This yields a total volume of ~240 μL/Eppendorf tube BCG suspension.
13. Plate 100 μL of BCG suspension on each 7H10 agar plate quadrant.
14. From the leftover 140 μL suspension, make two separate 1/10 serial dilutions by adding 50 μL to 450 μL of water. Plate each dilution in triplicate (three quadrants).
15. Place the plates in a polythene bag, tie the bag loosely, and incubate the plates in a dry 37 °C incubator.
16. Next day, invert the plates, and wait for 14 days to count colonies.
17. Calculate the percentages of BCG growth inhibition by using the following formula:

$$\%\text{Inhibition} = 100 - 100 \times \frac{\text{cfu from V4}}{\text{cfu from V1}}$$

Figure 5 shows an example of bacterial burden in the cultures of baseline and 8-week post-Ad5Ag85A vaccination in three participants.

3.7 HLA-Associated Recognition of Identified Single Peptides by Using Immune Epitope Database (IEDB) and Clinical Relevance of Immunodominant T Cell Epitopes in Study Participants

1. Genotype HLA for major histocompatibility complex (MHC) class I and class II haplotypes at an intermediate level of resolution using PBMC. This can be outsourced to a genotyping service. This allows to determine the extent of promiscuity of the HLA-associated T cell epitope recognition identified among vaccinated participants.
2. Next, deduce the CD4 and CD8 T cell epitopes present in the protein expressed by the vaccine by considering their relative high affinity (half-maximal inhibitory concentration, <50 nM) using IEDB analysis resource consensus tool [14].
3. Select most common class I and class II HLA alleles. For instance, after Ad5Ag85A vaccination, p12, p14, p27, and

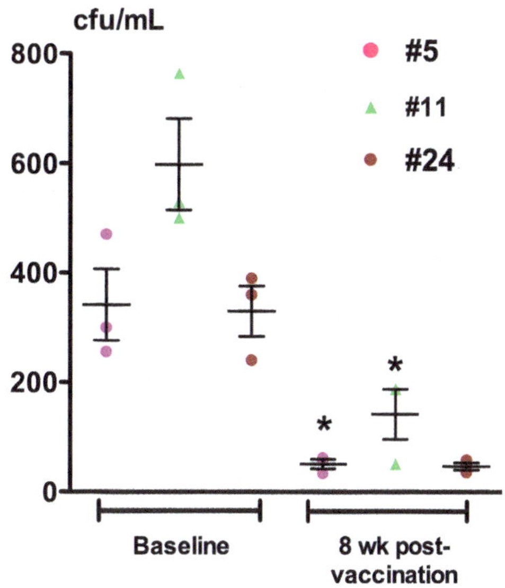

Fig. 5 Mycobacterial growth inhibition in autologous monocytes by Ag-specific T cells. Scatterplot showing the BCG burden in the co-culture of Pan T cells from the baseline and week 8 after Ad5Ag85A vaccination with BCG-infected baseline autologous monocytes in the presence of single Ag85A peptide pool. Data are expressed as the mean number (± standard error of the mean) of colony-forming units (CFU) of three independent co-cultures of pan-T cells and infected autologous monocytes from each indicated study participant

p28 single peptides were identified as highly CD8 T cell-reactive peptides among study participants.

4. Thus, HLA restriction was identified against the most common class I HLA alleles (A*0101, A*0201, A*2402, A*0301, A*1101, B*0702, B*0801, B*1501) and B*3501 and B*3503 (B*35 is one of the most common HLA-B type in West Africa). These alleles are shared by 85–90% of the world population [15].
5. Next, submit the single peptide as one continuous sequence against each HLA allele. The predicted output will be given in IC50 nM. Peptides <IC50 have high affinity according to IEDB analysis resource.
6. Consider epitopes of lengths 8–11 as potential candidate CD8 T cell epitopes. For instance, four single peptides identified among Ad5Ag85A vaccine study participants were restricted to HLA-B type with epitopes located at 58–80 aa and 135–149 aa (Table 2).
7. Use this output to determine the HLA-associated T cell epitope recognition among the study participants.
8. Carry out thorough literature search to identify the epitopes of the protein expressed by the vaccine previously recognized by

Table 2
Deduced Ag85A CD8 T cell epitopes based on identified immunodominant peptides and common class I HLA alleles in humans

Peptides	Alleles	Epitopes	Epitope location (aa)
p12	B*40:02	FEWYDQSGLSV	58–68
p14	B*07:02	MPVGGQSSF	70–78
	B*15:01	VVMPVGGQSSF	68–78
	B*15:01	VMPVGGQSSF	69–78
	B*35:01	MPVGGQSSF	70–78
	B*35:01	MPVGGQSSFY	70–79
	B*35:01	MPVGGQSSFYS	70–80
p27	B*35:01	LAIYHPQQF	135–143
	B*35:01	LAIYHPQQFVY	135–145
p28	B*07:02	HPQQFVYAGAM	139–149
	B*15:01	QQFVYAGAM	141–149
	B*35:01	HPQQFVYAGAM	139–149

the healthy people positive to tuberculin skin test (latent infected) and people with active tuberculosis.

9. Compare the epitopes recognized by the study participants with those recognized by the latently TB infected and active tuberculosis patients to understand the clinical relevance of the epitope recognition following vaccination.

4 Notes

1. When preparing 12.5% HSA, avoid excess agitation which can cause the HSA buffer to become foamy. Do not use 12.5% HSA past 7 days after initial preparation.
2. Do not pipette DMSO down the side of polypropylene tube; this will cause clumps that are difficult to dissolve. Prepare it freshly just before use.
3. The pellet may be tough, requiring repeated pipetting or even breaking with pipette tip.
4. Thaw only two cryo-samples at a time. This is very important in order to retain good viability.
5. If processing many cryovials, add warm cRPMI to a few vials, and place them into the incubator before processing additional vials, to ensure cell viability.

6. We find that adding DNase 1 during revival is critical, which reduces clumping of cells and the background staining during immunostaining.
7. It is very important to keep the cells clumped. Too much disturbance to the clumps will affect the expansion.
8. Protocol described above is optimized for stimulation done in FACS tubes. Optimization is required to adapt this protocol for stimulation and staining done in a 96 well plate.
9. Brefeldin A stock is supplied in 5 mg vials (powder). Prepare stock solution by reconstituting the powder with 1 mL DMSO (5 mg/mL). Store the stock solution in 20 μL aliquots at −20 °C. For every use, take out a vial, and prepare 1/10 dilution using PBS. Do not use vials that are repeatedly frozen and thawed.
10. Antibody dilutions provided are for antibodies bought from BD Biosciences. For those bought from a different supplier, titration of antibodies is required.
11. Upon permeabilization and fixation, since cells become very light, centrifuging at higher speed is critical to avoid cell losses.
12. For color controls, always use the same cell suspension as the ones undergoing staining. This will allow to set up cell population voltages in the LSR II that will give proper separation between cell populations of interest.
13. Generally, 40% of the total frozen PBMC are Pan T cells, and 10% are monocytes.
14. We find that adding a single peptide pool of antigen expressed by the vaccine to cultures enhances the differences in the bacterial burden between baseline and 8 weeks post vaccination. We also find that heaps of cells at the bottom of the well with aggregates/clumps of cells in the cultures with 8-week postvaccination cells are signs of proliferating memory Ag-specific T cells.

Acknowledgments

The work is supported by funds from the Foundation Program of the Canadian Institutes of Health Research (CIHR) and the Collaborative Health Research Program of CIHR and the Natural Sciences and Engineering Research Council of Canada.

References

1. WHO (2018) Global tuberculosis report. WHO, Geneva
2. Voss G, Casimiro D, Neyrolles O et al (2018) Progress and challenges in TB vaccine development. F1000Res 7:199

3. Kaufmann SHE, Dockrell HM, Drager N et al (2017) TBVAC2020: advancing tuberculosis vaccines from discovery to clinical development. Front Immunol 8:1203. https://doi.org/10.3389/fimmu.2017.01203
4. Xing Z, Jeyanathan M, Smaill F (2014) New approaches to TB vaccination. Chest 146:804–812
5. Tameris MD, Hatherill M, Landry BS et al (2013) Safety and efficacy of MVA85A, a new tuberculosis vaccine, in infants previously vaccinated with BCG: a randomised, placebo-controlled phase 2b trial. Lancet 381:1021–1028
6. Nemes E, Geldenhuys H, Rozot V et al (2018) Prevention of M. tuberculosis infection with H4:IC31 vaccine or BCG revaccination. N Engl J Med 379:138–149
7. Kaufmann SHE, Weiner J, von Reyn CF (2017) Novel approaches to tuberculosis vaccine development. Int J Infect Dis 56:263–267
8. Jeyanathan M, Yao Y, Afkhami S et al (2018) New tuberculosis vaccine strategies: taking aim at un-natural immunity. Trends Immunol 39:419–433
9. Li Q, Song H-Y, Ellner JJ et al (2002) Bactericidal activity in whole blood as a potential surrogate marker of immunity after vaccination against tuberculosis. Clin Vaccine Immunol 9:901–907
10. Li Q, Hoft DF, Kampmann B et al (2002) Investigation of the relationships between immune-mediated inhibition of Mycobacterial growth and other potential surrogate markers of protective Mycobacterium tuberculosis immunity. J Infect Dis 186:1448–1457
11. Minassian AM, Satti I, Poulton ID et al (2012) A human challenge model for Mycobacterium tuberculosis using Mycobacterium bovis bacille Calmette-Guérin. J Infect Dis 205:1035–1042
12. Jeyanathan M, Damjanovic D, Yao Y et al (2016) Induction of an immune-protective t-cell repertoire with diverse genetic coverage by a novel viral-vectored tuberculosis vaccine in humans. J Infect Dis 214:1996–2005
13. Hoffmeister B, Kiecker F, Tesfa L et al (2003) Mapping T cell epitopes by flow cytometry. Methods 29:270–281
14. Peters B, Sette A, Kim Y et al (2012) Immune epitope database analysis resource. Nucleic Acids Res 40:W525–W530
15. Weichold FF, Mueller S, Kortsik C et al (2007) Impact of MHC class I alleles on the M. tuberculosis antigen-specific CD8+ T-cell response in patients with pulmonary tuberculosis. Genes Immun 8:334–343

Chapter 16

Retroviral Gene Transduction into T Cell Progenitors for Analysis of T Cell Development in the Thymus

Ryunosuke Muro, Hiroshi Takayanagi, and Takeshi Nitta

Abstract

The thymus is an organ where T cells develop throughout life. Using mice as a model animal, molecular mechanisms of intrathymic T cell development have been studied. Fetal thymus organ culture technique enables ex vivo reconstitution of fetal-specific T cell development, while bone marrow chimera technique allows in vivo reconstitution of T cell development in adult thymus. These techniques can be combined with retroviral gene transduction into the T cell progenitors to evaluate the function of genes of interest in developing T cells. Here, we describe the basic protocols for retrovirus gene transduction into fetal or adult T cell progenitors and reconstitution of thymic T cell development including experimental tips such as using cryopreserved fetal liver or bone marrow cells as sources of T cell progenitors.

Key words Thymus, T cells, Retrovirus, Fetal thymus organ culture, Bone marrow chimera

1 Introduction

T cells are central to the immune system in vertebrates. Two types of T cells, αβT cells and γδT cells, both require the thymus for their development and T cell receptor (TCR) repertoire formation. Early T cell progenitors derived from fetal liver or adult bone marrow enter the thymus where through interaction with thymic stromal cells they undergo differentiation and repertoire selection, to give rise to mature T cells with diverse yet self-tolerant TCR repertoire [1]. Understanding molecular mechanisms regulating intercellular and intracellular signals for T cell development still remains challenging.

This chapter describes basic protocols for retroviral gene transduction into T cell progenitors and reconstitution of T cell development in fetal thymus as well as adult mice. We suggest using cryopreserved fetal liver or bone marrow cells as sources of T cell progenitors. For reconstitution of fetal T cell development, we use the fetal thymus organ culture (FTOC) technique that allows the gene transduction into fetal-specific T cell subsets such as IL-17-

Chaohong Liu (ed.), *T-Cell Receptor Signaling: Methods and Protocols*, Methods in Molecular Biology, vol. 2111,
https://doi.org/10.1007/978-1-0716-0266-9_16,

producing γδT (γδT17) cells [2]. Using bone marrow chimera technique, developing T cells in adult thymus can be transduced to express any genes of interest at high efficiency and reproducibility, in order to evaluate its in vivo function.

2 Materials

2.1 Preparation of Retroviral Supernatant

1. 100 mm Sterile plastic dishes for cell culture.
2. 100 μg/ml Poly-L-lysine.
3. Plat-E cells [3].
4. Retrovirus vector pMSCV-IRES-EGFP (*see* **Note 1**).
5. The culture medium for Plat-E cells is DMEM supplemented with 10% fetal calf serum (FCS), 100 U/ml penicillin, 100 μg/ml streptomycin. Plat-E cells are maintained in the culture medium in a 37 °C, 5% CO_2 incubator.
6. Lipofectamine 3000 (Invitrogen).
7. 10 ml Sterile syringes and syringe-driven filters (0.45 μm pore size, 16 mm diameter, sterile).

2.2 Gene Transduction into Fetal T Cell Progenitors

1. Timed pregnant C57BL/6 mice (gestational days 14.5–15.5) (*see* **Note 2**).
2. 13.5 mM 2-Deoxyguanosine (dGuo) in sterile PBS.
3. 5 μg/ml Anti-Ly-6G/Ly-6C (Gr-1, clone RB6-8C5) and anti-TER119 (clone TER-119) conjugated with anti-phycoerythrin (PE), diluted with MACS buffer.
4. Anti-PE antibody conjugated with magnetic beads (Miltenyi Biotec).
5. MACS MS column (Miltenyi Biotec) (*see* **Note 3**).
6. MACS separator system (Miltenyi Biotec).
7. MACS buffer: Sterile PBS containing 0.5% bovine serum albumin (BSA) and 2 mM EDTA.
8. IMDM medium: IMDM supplemented with 20% FCS, 4 mM L-glutamine, 1 mM sodium pyruvate, 0.1 mM nonessential amino acids, 100 U/ml penicillin, and 100 U/ml streptomycin.
9. Cytokines: SCF and IL-7.
10. Terasaki 96-well plates.
11. Sterile Gelfoam gelatin sponges (Pfizer). Cut into small pieces (e.g., 1 cm square), and store dry at room temperature.
12. 13 mm Polycarbonate (PC) filter membranes (Whatman). Autoclave to sterilize, and store dry at room temperature.
13. Twenty-four-well plates (16 mm diameter, sterile).

14. RPMI medium: RPMI1640 supplemented with 10% FCS, 50 μM 2-mercaptoethanol, 10 mM HEPES, 2 mM L-glutamine, 1× nonessential amino acids, 1 mM sodium pyruvate, 100 U/ml penicillin, and 100 μg/ml streptomycin.
15. 10 mg/ml Polybrene (hexadimethrine bromide) in sterile PBS.

2.3 Gene Transduction into Adult T Cell Progenitors

1. C57BL/6 mice (10–20 weeks old, sex-matched).
2. 10 mg/ml 5-Fluorouracil (5-FU) in sterile PBS.
3. PBS + Gm: Sterile PBS containing 50 μg/ml gentamycin.
4. Cellbanker 1 (ZENOAQ).
5. 5 μg/ml Anti-Sca1 antibody (clone D7) conjugated with PE, diluted with MACS buffer.
6. Anti-PE antibody conjugated with magnetic beads, MACS MS column, MACS separator system, and MACS buffer; *see* Subheading 2.2, **items 4–7**.
7. IMDM medium; *see* Subheading 2.2, **item 8**.
8. Cytokines: SCF, IL-6, and IL-3.
9. X-ray irradiator.
10. 10 mg/ml Polybrene in sterile PBS.

3 Methods

3.1 Preparation of Retroviral Supernatant

1. Prepare the poly-L-lysine-coated 100 mm dish. Fill a 100 mm dish with 5 ml of 20 μg/ml poly-L-lysine (diluted with sterile PBS), remove the solution immediately, and let the dish dry for 1 h in the tissue culture hood. Dried dishes may be kept at 4 °C overnight. Immediately prior to plating the cells, rinse the dishes with culture medium two times.
2. Set up the Plat-E cell culture. In a poly-L-lysine-treated 100 mm dish, 4×10^6 cells are seeded with 10 ml of culture medium. Cells are cultured in a 37 °C, 5% CO_2 incubator for 12–18 h.
3. Transfect Plat-E cells with retroviral plasmid DNA by use of Lipofectamine 3000 reagent, according to the manufacturer's protocol. We usually use 17 μg of plasmid DNA per 100 mm dish. Culture cells in a 37 °C, 5% CO_2 incubator.
4. Twelve hours after the transfection, remove the supernatant, and add 5 ml of pre-warmed culture medium to wash the cells. Remove supernatant, and add 10 ml of fresh, pre-warmed culture medium. Culture cells in a 37 °C, 5% CO_2 incubator.
5. Forty-eight hours after the transfection, collect culture supernatant containing retroviruses, and filter it through 0.45 μm syringe filters. The supernatant can be used immediately or may

be aliquoted into 1.5 ml microtubes and then stored at −80 °C until use. After collecting the supernatant, the cells can be used for further retroviral production. To do so, gently add 10 ml of fresh, pre-warmed culture medium to the plate, and continue culture in a 37 °C, 5% CO_2 incubator. Retroviral supernatants can be collected every 12 h between 48 and 72 h after transfection (*see* **Note 4**).

3.2 Gene Transduction into Fetal T Cell Progenitors

Fetal liver contains unique progenitors of T cell subsets that develop only early in ontogeny. γδT17 cells are a representative fetal-specific T cell subset whose development is dependent on fetal liver-derived progenitors [4] and fetal thymic environment. Introduction of the progenitor cells into mouse fetal thymus cultured in vitro reconstitutes fetal T cell development, and this technique can be combined with retroviral gene transduction into the progenitor cells, to express any genes of interest in fetal-specific T cells such as γδT17. Fig. 1a shows the scheme of experimental schedule:

1. Sacrifice timed pregnant mice to remove fetus-filled uteri. Isolate fetal thymus lobes as well as fetal livers from fetuses under a stereomicroscope. Place the isolated thymus in a 60 mm dish filled with RPMI medium at 4 °C. Put the fetal livers in a 1.5 ml microtube filled with 1 ml of RPMI medium, dissociate them by gently pipetting, and centrifuge at 600 × *g* for 3 min. Discard the supernatant. Suspend the cells with 500 μl of Cellbanker 1, and store them at −80 °C. The cryopreserved cells can be stored at least for several months (*see* **Note 5**).
2. Place one piece of the gelatin sponge in a culture well of a 24-well plate, and fill the culture well with 1 ml of RPMI medium. Press the gelatin sponge to transfuse the RPMI medium. Discard 500 μl of the RPMI medium from the culture well. Wet both sides of a PC membrane with RPMI medium, and place it onto each gelatin sponge.
3. Place up to five thymus lobes onto the PC membrane, and drop 55 μl of 13.5 mM dGuo to the thymus lobes directly (Fig. 2a). The final concentration of dGuo is 1.35 mM. Incubate the fetal thymus lobes in 37 °C, 5% CO_2 for 5 days.
4. After 4 days of dGuo treatment, thaw the cryopreserved fetal liver cells in a 37 °C water bath. Wash the cells by adding MACS buffer, centrifuge at 600 × *g* for 3 min, and discard the supernatant. Add 5 μg/ml anti-Ly-6G/Ly-6C and anti-TER119 antibody conjugated with PE (100 μl per 1 × 10^7 leukocytes) onto the cell pellet, resuspend gently by pipetting, and incubate at 4 °C for 30 min.

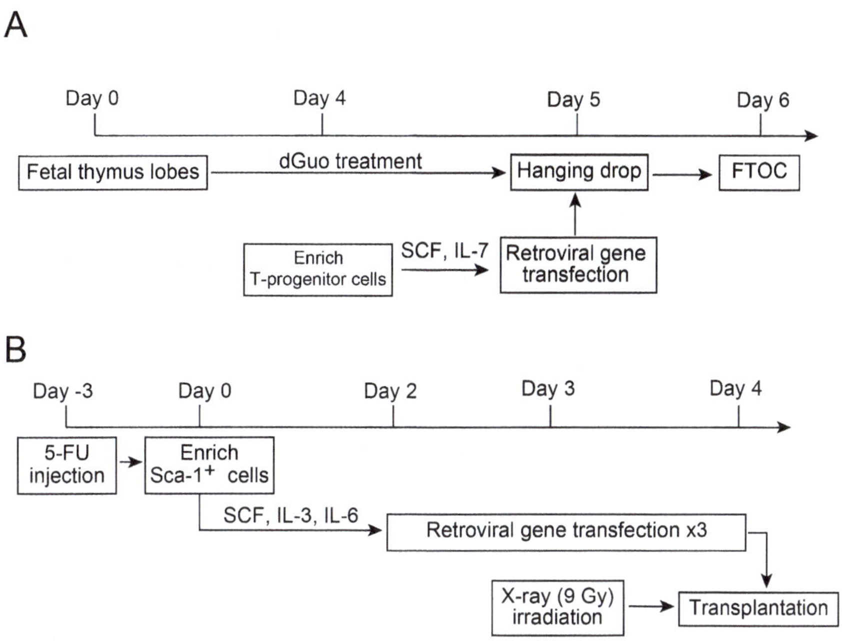

Fig. 1 Scheme of experimental schedules. (**a**) Scheme of experimental schedule of FTOC. (**b**) Scheme of generation of retrogenic mice

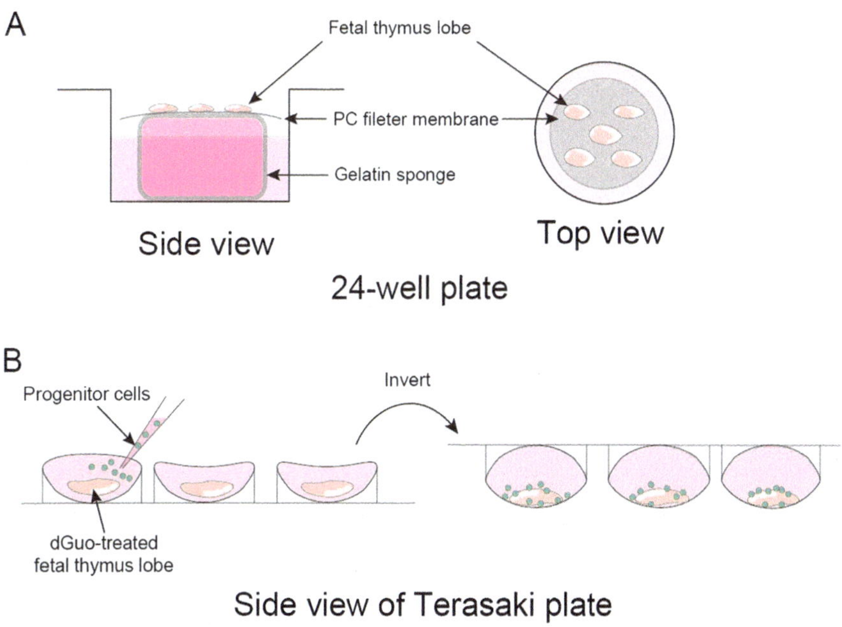

Fig. 2 Diagram of fetal thymus organ culture. (**a**) Side and top views of culture well for FTOC. (**b**) Scheme of hanging drop culture

5. Add 1 ml of cold MACS buffer to the cells, and centrifuge at 600 × *g* for 3 min at 4 °C. Remove supernatant, resuspend cell pellet in cold MACS buffer (70 μl per 1×10^7 leukocytes), add anti-PE microbeads (20 μl per 1×10^7 leukocytes), and incubate at 4 °C for 20 min.
6. Add 1 ml of cold MACS buffer to the cells, and centrifuge at 600 × *g* for 1 min at 4°. Remove supernatant, resuspend cell pellet in 1 ml of cold MACS buffer, centrifuge at 600 × *g* for 1 min at 4 °C. Remove supernatant, and resuspend cell pellet in cold MACS buffer (100 μl per 1×10^7 leukocytes).
7. Pass the cells through MS column using magnetic separator, and collect flow through containing unlabeled cells, according to the manufacturer's protocol. T cell progenitors are enriched in the Ly-6G/6C$^-$ TER119$^-$ fraction (*see* **Note 6**).
8. Collect cells by centrifugation, and wash cells with IMDM medium two times. Resuspend cells in IMDM medium, and count the number of cells. Transfer cell suspension into 48-well plate (approximately 1×10^6 viable cells/500 μl of culture medium/well), and add SCF (final 50 ng/ml) and IL-7 (final 25 ng/ml). Culture cells in a 37 °C, 5% CO_2 incubator for 24 h.
9. After 5 days of dGuo treatment, thaw the frozen retroviral supernatant quickly in a 37 °C water bath.
10. Remove the top 80% of the culture supernatant from the wells of 48-well plate in which fetal liver cells are cultured. Be careful not to take the cells at the bottom of the wells. Add 1 ml of retroviral supernatant containing 10 μg/ml polybrene to the cells, and seal the plate with Parafilm, and spin the plate at 1000 × *g* for 90 min at room temperature.
11. Collect the cells from plate into a sterile 1.5 ml microtube, and centrifuge at 600 × *g* for 3 min. Resuspend the cells in RPMI medium. Count the number of cells, and adjust cell concentration to 1×10^4 cells per 10 μl of RPMI medium.
12. Transfer the dGuo-treated fetal thymus lobes to the 100 mm dish filled with 10 ml of RPMI medium. Incubate in a 37 °C, 5% CO_2 incubator for 20 min to diffuse away dGuo.
13. Transfer 20 μl of culture medium containing one dGuo-treated fetal thymus lobe per well of a Terasaki plate, by pipetting.
14. Add 10 μl of cell suspension from **step 11** into each well of the Terasaki plate containing the dGuo-treated fetal thymus lobe.
15. Place the lid on the plate, and gently invert. Make sure that the thymus lobes are located at the bottom of the drop (Fig. 2b). If not, gently pipette the well. This "hanging drop culture" technique enables seeding of fetal thymus by fetal liver-derived progenitor cells. Culture it in a 37 °C, 5% CO_2 incubator for 24 h.

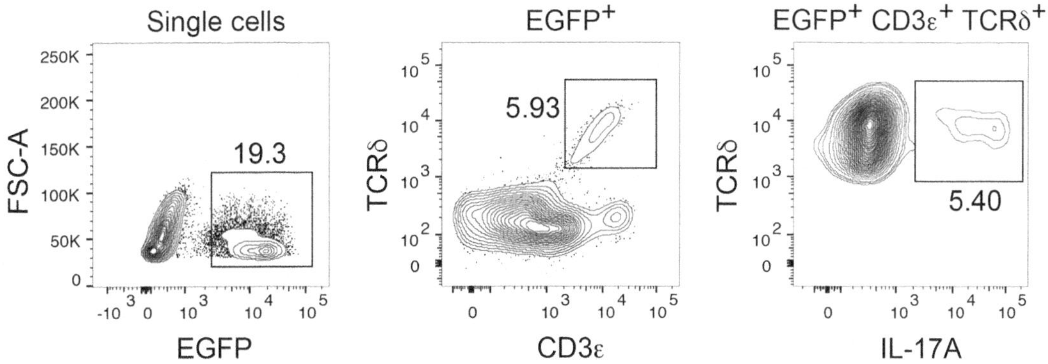

Fig. 3 Flow cytometry analysis of thymocyte from FTOC. TER119⁻ Ly-6G/Ly-6C⁻ cells were infected with retroviruses expressing EGFP and reconstituted in dGuo-treated fetal thymus in vitro. Fourteen days after FTOC, thymocytes following stimulation of phorbol myristate acetate (PMA) and ionomycin were analyzed by flow cytometry

16. Prepare gelatin sponge and PC filter membrane in a 24-well plate as described in **step 2**.
17. Rinse the thymus lobes containing infected T cell progenitors in 100 mm dish filled with 10 ml of the RPMI medium.
18. Place the thymus lobes onto the PC filer membrane by pipetting, and culture them in a 37 °C, 5% CO_2 incubator. Change medium once every 3 days by removing 300 μl of medium from each well and adding back 300 μl of pre-warmed fresh RPMI medium. If necessary, add cytokines and inhibitors.
19. T cell development in the cultured thymic lobes can be evaluated by the flow cytometry analysis. Representative results of gene transduction into γδT17 cells are shown in Fig. 3 and [5].

3.3 Gene Transduction into Adult T Cell Progenitors

Hematopoietic progenitor cells can be enriched in the bone marrow of adult mice treated with 5-FU, which ablates proliferating mature hematopoietic cells. Sca1 is a cell surface marker widely used to enrich hematopoietic stem cells and progenitor cells from the 5-FU-treated bone marrow. Since retroviruses are only able to infect dividing cells, Sca1$^+$ cells have to be stimulated with cytokines prior to infection. Cell culture procedures including retrovirus infection must be completed within 4 days, because prolonged culture period (longer than 5 days) results in the loss of potential of progenitor cells to differentiate into T-lineage cells. Figure 1b shows the scheme of experimental schedule:

1. Inject 5-FU (0.15 mg per gram body weight) intraperitoneally into donor mice using a 27-G needle.
2. Three to four days after 5-FU injection, sacrifice the mice. Cut out the femur and tibia, cut both ends off each bone, and keep

bones in ice-cold PBS + Gm. Flush the bone marrow into a 100 mm plastic dish using an appropriate volume of ice-cold PBS + Gm and 10 ml syringe attached with a 25-G needle. Collect the bone marrow cells from the dish, and filter them through a 70 μm cell strainer. Put the filtered cells in 50 ml tube, and centrifuge the cells at 400 × *g* for 5 min at 4 °C. Remove supernatant.

3. At this step, bone marrow cells can be cryopreserved until use. Suspend the cells in Cellbanker 1 (500 μl of Cellbanker 1 per 1×10^7 leukocytes), and store them at −80 °C. The cells can be stored at least for several months. Upon use, thaw the frozen cells quickly in a 37 °C water bath, add 10 ml of IMDM medium, and then centrifuge at 400 × g for 5 min at 4 °C. Remove supernatant.
4. Add 5 μg/ml anti-Sca1 antibody conjugated with PE (100 μl per 1×10^7 leukocytes) onto the cell pellet, resuspend gently by pipetting, and incubate at 4 °C for 30 min.
5. Add 1 ml of cold MACS buffer to the cells, transfer the cell suspension to 1.5 ml microtube, and centrifuge at 600 × *g* for 1 min at 4 °C. Remove supernatant, resuspend cell pellet in 1 ml of cold MACS buffer, and centrifuge at 600 × *g* for 1 min at 4 °C.
6. Remove supernatant, resuspend cell pellet in cold MACS buffer (70 μl per 1×10^7 leukocytes), add anti-PE microbeads (20 μl per 1×10^7 leukocytes), and incubate at 4 °C for 15 min.
7. Add 1 ml of cold MACS buffer to the cells, and centrifuge at 600 × *g* for 1 min at 4 °C. Remove supernatant, resuspend cell pellet in 1 ml of cold MACS buffer, and centrifuge at 600 × *g* for 1 min at 4 °C. Remove supernatant, and resuspend cell pellet in cold MACS buffer (100 μl per 1×10^7 leukocytes).
8. Enrich $Sca1^+$ cells using MS column and magnetic separator, according to the manufacturer's protocol for positive selection of magnetically labeled cells (*see* **Note 6**).
9. Collect cells by centrifugation, and wash cells with IMDM medium two times.
10. Resuspend cells in IMDM medium, and count the number of cells. Transfer cell suspension into 24-well plate (approximately 3×10^6 viable cells/1 ml of IMDM medium/well), and add SCF (final 50 ng/ml), IL-6 (final 50 ng/ml), and IL-3 (final 10 ng/ml). Culture cells in a 37 °C, 5% CO_2 incubator.
11. During the culture, a fraction of $Sca1^+$ cells becomes enlarged and begins to proliferate while other cells die. Check the cells daily under the microscope (Fig. 4). If the cells are confluent, split the cells into multiple wells with the IMDM medium containing the cytokines.

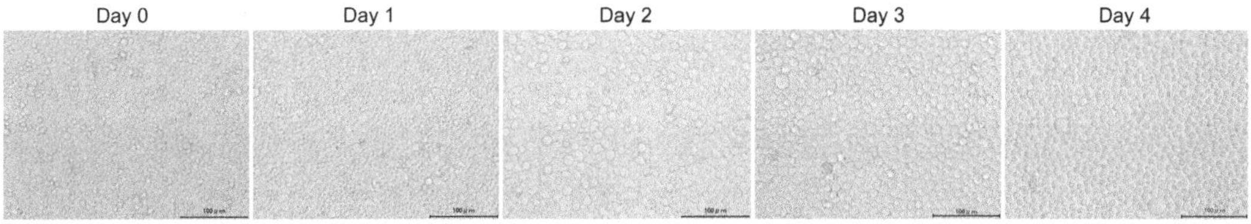

Fig. 4 Cultured bone marrow cells. Sca1$^+$ bone marrow cells were cultured and observed under microscope at the indicated days after culture. Scale bar, 100 μm

12. After 40 h, thaw the frozen retroviral supernatant quickly in a 37 °C water bath.
13. Carefully collect the top 80% of the culture supernatant from the wells into a sterile 1.5 ml microtube, centrifuge at 600 × *g* for 1 min to collect any cells, and remove supernatant.
14. Resuspend the cells with 1 ml of retroviral supernatant containing 10 μg/ml polybrene, and transfer it back to the culture well in the 24-well plate. Seal the plate with Parafilm, and spin the plate at 1000 × *g* for 90 min at room temperature.
15. Carefully collect the top 80% of the retroviral supernatant from the wells into a sterile 1.5 ml microtube, spin at 600 × *g* for 1 min to collect any cells, remove supernatant, and resuspend the cells in the same volume of the IMDM medium containing the cytokines. Transfer the cell suspension back to the culture well in the 24-well plate, and continue to culture in a 37 °C, 5% CO_2 incubator.
16. After 24 h (64 h after the culture), thaw the frozen retroviral supernatant, and repeat **steps 13–15**.
17. Irradiate recipient mice the day before injection. Regular irradiation dose for C57BL/6 mice (and other wild-type mice) is 9.0 Gy. Optimal dose has to be established empirically, according to the status of the irradiator and mouse facility.
18. After 24 h (88 h after the culture), thaw the frozen retroviral supernatant, and repeat **steps 13–15**.
19. Ninety-six hours after the culture, collect cells by pipetting and centrifugation, and wash cells with ice-cold PBS + Gm two times (*see* **Note 7**).
20. Inject the cells into the irradiated mice intravenously. We routinely inject 1 × 10^6 cells per recipient mouse (approximately one donor mouse to one recipient mouse).
21. Check the mice weekly to ensure that they remain healthy.
22. It takes 4–5 weeks to reconstitute T cell development in the thymus (Fig. 5 and ref. 6) and 5–8 weeks to fully reconstitute the peripheral T cell populations in the recipient mice (*see* **Note 8**).

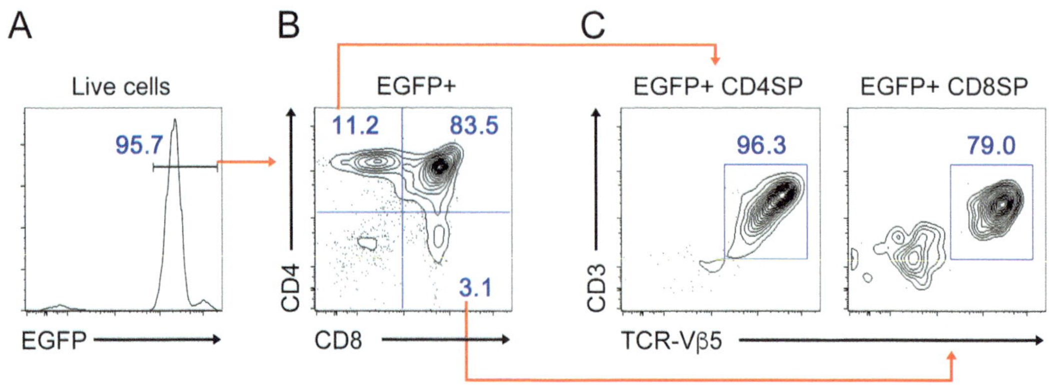

Fig. 5 Flow cytometry analysis of thymocytes from retrogenic mice. $Sca1^+$ bone marrow cells were infected with retroviruses expressing a rearranged TCRβ chain (TCR-Vβ5) and reconstituted in irradiate wild-type mice. Five weeks later, thymocytes were analyzed by flow cytometry. (**a**) EGFP expression in total thymocytes. (**b**) CD4 and CD8 expression profile of gated $EGFP^+$ cells. (**c**) Expression of CD3 and TCR-Vβ5 in gated $EGFP^+$ CD4 single positive (SP) or CD8SP cells. Virtually all of the $EGFP^+$ $CD3^+$ mature T cells expressed the retrovirally transduced TCR-Vβ5, indicating successful gene transfer into developing T cells. Number indicates percentage of cells within indicted areas

4 Notes

1. For gene transduction into developing T cells, the murine stem cell virus (MSCV-based retrovirus vectors) [7] should be used to prevent transcriptional suppression during differentiation processes. Various types of MSCV-based retrovirus vectors are available from Addgene (https://www.addgene.org/).
2. Timed pregnant mice of commonly used strains such as C57BL/6 are commercially available from various mouse suppliers. Generally, one pregnant C57BL/6 mouse has eight to ten fetuses.
3. Practical binding capacity of the MS column is 5×10^6 magnetically labeled cells. Larger size of column (LS column) may be used. Select proper number and size of column according to the number of cells. The frequency of $Sca1^+$ cells in 5-FU-treated bone marrow leukocytes is usually 20–40%.
4. Twenty-four hours after transfection, GFP expression in transfected Plat-E cells can be monitored under fluorescent microscope. The efficiency of the transfection may be evaluated after the collection of retroviral supernatant, by trypsinizing the transfected Plat-E cells and analyzing GFP expression by flow cytometry. The transfection efficiency using this protocol ranges 50–80%.
5. For detailed instructions for the isolation of fetal thymus, *see* ref. 8. Video instructions are also available online [9].

6. The efficiency of magnetic cell depletion or enrichment may be evaluated by flow cytometric analysis of PE fluorescence.
7. Cell culture can be prolonged up to 4.5 days after the cell seeding. After day 5, T cell reconstitution potential is drastically reduced.
8. In our regular experiments, the frequency of retrovirally transduced ($EGFP^+$) cells in reconstituted thymocytes is 74% on average (ranges from 33 to 97%).

Acknowledgments

We thank S. Nitta and M. Tsutsumi for the technical assistance. This study was supported by Grants-in-Aid for Research from the Japan Society for the Promotion of Science (JSPS) (KAKENHI 15H05703, 16H05202, 16K19102, and 17H05788), the Kanehara-Ichiro Foundation (grants 29–23), and the Kanae Foundation for the promotion of medical science (grant 47th).

References

1. Abramson J, Anderson G (2017) Thymic epithelial cells. Annu Rev Immunol 35:85–118. https://doi.org/10.1146/annurev-immunol-051116-052320
2. Shibata K, Yamada H, Nakamura R et al (2008) Identification of CD25+ gamma delta T cells as fetal thymus-derived naturally occurring IL-17 producers. J Immunol 181(9):5940–5947
3. Morita S, Kojima T, Kitamura T (2000) Plat-E: an efficient and stable system for transient packaging of retroviruses. Gene Ther 7:1063–1066. https://doi.org/10.1038/sj.gt.3301206
4. Spidale NA, Sylvia K, Narayan K et al (2018) Interleukin-17-producing γδ T cells originate from SOX13. Immunity 49(5):857–872.e5. https://doi.org/10.1016/j.immuni.2018.09.010
5. Muro R, Nitta T, Nakano K et al (2018) γδTCR recruits the Syk/PI3K axis to drive proinflammatory differentiation program. J Clin Invest 128(1):415–426. https://doi.org/10.1172/JCI95837
6. Nitta T, Kochi Y, Muro R et al (2017) Human thymoproteasome variations influence CD8 T cell selection. Sci Immunol 2:eaan5165. https://doi.org/10.1126/sciimmunol.aan5165
7. Hawley RG, Fong AZC, Burns BF et al (1992) Transplantable myeloproliferative disease induced in mice by interleukin 6 retrovirus. J Exp Med 176:1149–1163
8. Nitta T, Ohigashi I, Takahama Y (2013) The development of T lymphocytes in fetal thymus organ culture. Methods Mol Biol 946:85–102. https://doi.org/10.1007/978-1-62703-128-8_6
9. Jenkinson W, Jenkinson E, Anderson G (2008) Preparation of 2-dGuo-treated thymus organ cultures. J Vis Exp. https://doi.org/10.3791/906

Chapter 17

Testing the Efficiency and Kinetics of Negative Selection Using Thymic Slices

Tyng-An Zhou, Chia-Lin Hsu, and Ivan Lilyanov Dzhagalov

Abstract

Central tolerance is an efficient barrier to autoimmunity and negative selection of self-reactive thymocytes is one of its major manifestations. Because of its importance, negative selection has been studied extensively through numerous in vitro and in vivo approaches that have tremendously increased our understanding of the process. Recently, in situ experimental systems using thymus slices have been developed that combine some of the advantages of in vitro assays such as ease of manipulation and high throughput with the existence of three dimensional mature thymus microenvironment. These approaches offer unprecedented opportunity to study negative selection. Here, we describe how thymic slices can be used to measure the kinetics and magnitude of negative selection. Taking the OT-1/Ova system as an example, we provide detailed guidance on cutting thymic slices, labeling and overlaying thymocytes on them and reading out the extent of negative selection by flow cytometry. The system can easily be adapted to evaluate the effects of various mutations or treatments on negative selection or to study the behavior of different cells in the thymus through time-lapse imaging.

Key words Thymus, Thymocytes, Negative selection, Central tolerance, Flow cytometry, Thymic slices, Vibratome, OT-1, Ova

1 Introduction

The random nature of T cell receptor (TCR) gene rearrangement allows the generation of T cells recognizing essentially all pathogens, but also poses the threat of autoreactivity. A major mechanism upholding self-tolerance is negative selection in the thymus, usually in the form of clonal deletion. Since its first description more than 30 years ago [1], negative selection has been extensively studied by a variety of in vivo and in vitro approaches. The current gold standard for assessment of negative selection is the cross of TCR transgenic mice with mice expressing that TCR's ligand in the thymus (e.g., OT-1 × RIPmOva or OT-2 × RIPmOva) [2–4]. These models are invaluable in testing the importance of individual molecules for the ultimate success of negative selection; however, obtaining kinetics

Chaohong Liu (ed.), *T-Cell Receptor Signaling: Methods and Protocols*, Methods in Molecular Biology, vol. 2111, https://doi.org/10.1007/978-1-0716-0266-9_17,

information from them is not straightforward. As a typical endpoint assay, one mouse provides a single data point and obtaining data on multiple time points and different treatments require large number of experimental animals.

Complementing the in vivo methods, a number of in vitro models have been used to study negative selection [5, 6]. These methods are usually easier to set up and multiplex. However, the absence of three-dimensional environment of the thymus with all supporting cell types as well as of key features of thymocyte behavior, such as motility, typically limits the utility of these models to assessment of cell death or signaling. Moreover, the lack of physiological survival factors results in high background level of apoptosis that can obscure small differences in the extent of negative selection.

A third group of methods, bridging the advantages of in vivo and in vitro assays, are the in situ models that use tissue explants or organotypic culture. Just like in the in vivo methods, negative selection takes place in thymocytes' native three-dimensional environment that allows normal behavior and motility and is similar to the in vitro methods they are amenable to high-throughput analysis. Fetal thymic organ cultures (FTOCs) and re-aggregate thymic organ cultures (RTOCs) have been used to assess the contribution of different molecules or cell types to negative selection [6, 7]. Recently, a new in situ method is gaining popularity—the thymic slice. Vibratome-cut slices have all the elements of adult thymus environment, including fully developed cortex and medulla and are superior to FTOCs that represent fetal environment with underdeveloped medulla or RTOCs that do not have separate medulla and cortex. Additional important advantage of the slices is that they can easily be populated with a labeled population of interest by simply overlaying the cells for a couple of hours that allows them to migrate inside the tissue and behave in a very similar way to cells that have developed within the organ [3, 8]. Traditionally, marked cells could only be introduced into the thymus through the creation of bone marrow chimeric mice that usually requires at least 6 weeks. Moreover, it is not difficult to generate ~20 slices from a single thymus, which is enough to set up an experiment with six treatments or time points in triplicates using only one mouse. Thymic slices can easily be imaged and time-lapse two-photon microscopy is commonly used to observe the behavior of thymocytes and their supporting cell in the thymus and particularly in the medulla, which is difficult to access through the capsule of the organ due to its deeper position [3]. Similar to FTOCs and RTOCs, the viability of the thymocytes in the slice cultures is difficult to maintain for longer than 5 days, limiting long-term experiments. Thymic slices have been used to study positive selection [8–10], negative selection [3, 11–13], signaling in thymocytes [8, 14], regulatory T cell generation [15], and thymocyte motility

[4, 16–18]. Several methods describing the use of thymic slices for observing the behavior of cells in the thymus by time-lapse two-photon microscopy have recently been published [19–21]. This chapter provides a detailed protocol to measure the kinetics and magnitude of negative selection using thymic slices and the OT-1/$Ova_{257-264}$ system. OT-1 transgenic thymocytes can be stimulated to undergo negative selection by the encounter with their cognate peptide–$Ova_{257-264}$ presented by H2-K^b [22]. The presence of a polyconal thymocyte population coming from C57BL/6 controls for any factors not related to negative selection and allows comparison between different experiments. Here, we describe how to prepare the thymic slices; isolate and label thymocytes and introduce them into the slice; and analyze the results by flow cytometry. The assay can be used to measure how a mutation or a particular treatment of thymocytes or thymus stroma affects the extent or kinetics of negative selection in situ.

2 Materials

1. 70% alcohol in water (v/v).
2. Styrofoam dissection board.
3. Erlenmeyer flasks.
4. 500 mL beakers.
5. 10 cm forceps.
6. 11 cm Iris scissors.
7. One-sided razor blades.
8. Metal spatula.
9. Paper towels.
10. Kimwipes.
11. 70 μm nylon mesh (e.g., Small Parts, part # B0043D1SZG).
12. 3 mL syringes.
13. 6 cm Petri dishes.
14. 6-well plates.
15. 1.5 mL microcentrifuge tubes.
16. 15 mL conical tubes.
17. 5 mL polystyrene round bottom tubes (Falcon, cat. #352052).
18. Phosphate-buffered saline (PBS).
19. Hanks buffered salt solution (HBSS).
20. Complete Dulbecco's modified Eagle medium (cDMEM): high-glucose DMEM supplemented with 10% fetal bovine serum, 2 mM L-glutamine, 100 U/L penicillin, 100 μg/L streptomycin sulfate and 0.05 mM 2-mercaptoethanol.

21. Low-melting point agarose (e.g., NuSieve GTG Agarose, Lonza).
22. Disposable 22 mm embedding molds, square top tapered to 12 mm bottom (Polysciences, cat #18986).
23. Tissue glue (e.g., Vetbond, 3M).
24. Cell Proliferation Dye eFluor 450, stock solution at 5 mM in DMSO (eBioscience, ThermoFisher).
25. Cell Proliferation Dye eFluor 670 stock, solution at 5 mM in DMSO (eBioscience, ThermoFisher).
26. Ovalbumin ($Ova_{257-264}$) peptide (Anaspec), stock solution: 1 μM $Ova_{257-264}$ in DMEM, stored at −80 °C.
27. FACS buffer: PBS supplemented with 0.5% bovine serum albumin (BSA, w/v), 1 mM EDTA, and 0.1% sodium azide (NaN_3, w/v).
28. Propidium iodide (PI) stock solution at 1 mg/mL.
29. 0.4 μm pore size PET Cell Culture Inserts for 6-well plates (e.g., Falcon cat. #353090).
30. C57BL/6 mice (Jackson Labs, stock #000664).
31. OT-1 mice (NARLabs, Taipei, stock #RMRC13211 or Jackson Labs, stock #003831).
32. Microwave.
33. 37 °C and 55 °C water baths.
34. Hemocytometer.
35. Refrigerating centrifuge (e.g., Eppendorf 5810R).
36. Vibratome (e.g., Leica VT 1000S).
37. CO_2 incubator.
38. Flow cytometer equipped with violet (405 nm), blue (488 nm), and red (633 nm) lasers (e.g., LSR-Fortessa, BD Biosciences).
39. FlowJo flow cytometry analyzing software (BD Biosciences).
40. Microsoft Excel.

3 Methods

Note: All procedures used in animal experiments described here have been approved by the Institutional Animal Care and Use Committee (IACUC) of National Yang-Ming University, Taipei.

3.1 Preparation of Low-Melting Point (LMP) Agarose

1. Dissolve 2 g low-melting point agarose in 50 mL HBSS in a 250 mL Erlenmeyer flask for 4% final concentration. Add the agarose to the buffer, otherwise clumping might occur.

2. Boil in a microwave oven to completely dissolve the agarose. Wear heat-protecting gloves and avoid boiling over. In practice, we heat the mixture until it starts boiling and swirl to mix. We repeat these two steps until no trace of the agarose could be observed and the liquid is homogeneous and transparent.
3. Cover the flask with aluminum foil and put in a 56 °C water bath to cool down. Just before the dissection of the mouse, move the flask with the agarose to a 37 °C water bath (*see* **Note 1**).

3.2 Dissection of Thymus for Slices Preparation

1. Euthanize a C57BL/6 mouse by an institutionally approved method. We use CO_2 suffocation followed by neck dislocation.
2. Place and secure the mouse on a dissection board on its back and saturate the abdomen and chest with 70% alcohol.
3. Using forceps and sharp scissors open up the peritoneal cavity and cut vena cava (*see* **Note 2**).
4. Pierce the diaphragm with scissors and cut it from the rib cage. Piercing of the diaphragm leads to collapse of the lungs.
5. Cut the side of the rib cage all the way to the armpits. Lift the sternum and the ribs and secure them with pins above the head to expose the mediastinum.
6. Carefully remove the thymus without squeezing or pulling it and place it in PBS on ice. We usually pull the heart that is attached to the thymus to separate the thymus from surrounding tissues. Then using sharp scissors, we cut around the thymus (*see* **Note 3**).
7. Clean the thymus of connective tissue: Place the thymus on wet paper towels and using small forceps and sharp scissors cut out all connective tissue. Keep the thymus always wet by spraying it occasionally with PBS. A good test for the thoroughness of cleaning is dipping the thymus in buffer. If there is any connective tissue left, it is usually clearly visible inside the buffer (*see* **Note 4**).
8. Separate the two lobes of the thymus and remove the connective tissue between them. Once the lobes are cleaned and separated, place them in a 6 cm dish with 5 mL of PBS on ice.

3.3 Embedding the Thymus in LMP Agarose

1. Place a 22 mm square tissue embedding mold in a beaker filled with water and crushed ice.
2. Fill the mold with LMP agarose to the point where the side walls become straight. Work quickly as the agarose will start solidifying soon (*see* **Note 5**).
3. Gently pick one thymus lobe for the capsule from the side that will be its top. Quickly dry the thymus by touching it with a Kimwipe. Submerge the thymus lobe in the agarose. Push it

Fig. 1 Preparation of a mouse thymus lobe for cutting with Vibratome. (**a**) Top view of a thymus lobe embedded in agarose. (**b**) The agarose block taken out of the mold and ready for trimming. Note that there are 2–3 mm of agarose between the thymus and the edge of the block. (**c**) Trimming of the agarose block in preparation for cutting with the Vibratome

with the forceps until it sinks below the surface; otherwise the surface tension might prevent it from sinking (*see* **Note 6**).

4. Wait until the agarose solidifies and becomes translucent—around 1–2 min (Fig. 1a).

3.4 Cutting Slices with Vibratome

1. Put the mold with the thymus on a solid flat surface and press the bottom down with your thumb for several seconds. That should be enough for the agarose block to separate from the mold (*see* **Note 7**).
2. Put the agarose block on an elevated platform (the lid of a pipette tip box works well) and trim into a rectangular prism with a one-sided razor blade. Leave at least 2–3 mm on each side of the thymus and ~5 mm on the bottom side (*see* **Note 8**) (Fig. 1b and c).
3. Dry the bottom side of the agarose block with a Kimwipe.
4. Place a drop of tissue glue or superglue on the stage of the Vibratome and mount the agarose block on the glue. Press gently to ensure good adhesion. Secure the Vibratome stage in the cutting chamber.
5. Lower the blade just above the agarose block. Adjust the starting and finishing positions of the blade. Retract the blade to the starting position (*see* **Note 9**).
6. Fill the Vibratome cutting chamber with ice-cold PBS and the area around it with crushed ice.
7. Start cutting slices from the agarose block (Fig. 2a). We typically use the following settings on Leica 1000S: speed, 482 (~0.2 mm/s); amplitude, 1 mm; angle, 5°; and frequency, 8 (~80 Hz). Most often we cut slices that are 400 μm thick. We prefer to do single-slice cuts rather than continuous cutting, so that we can inspect each slice and make adjustments if necessary.

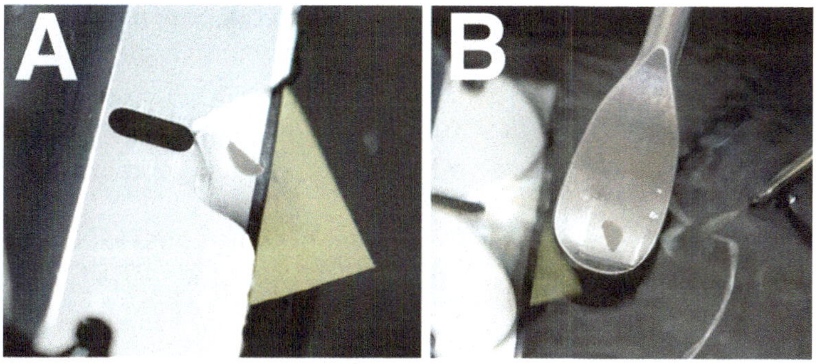

Fig. 2 Cutting slices with the Vibratome. (**a**) Cutting of a 400-μm-thick thymic slice. (**b**) Retrieval of a cut slice with a spatula

8. Once a slice is cleanly cut, carefully transfer it to a 6 cm dish with cDMEM on ice using a bent spatula (*see* **Note 10**) (Fig. 2b).
9. Keep the slices on ice until ready to overlay the thymocytes.

3.5 Harvesting and Labeling of Thymocytes

1. Euthanize a C57BL/6 mouse and an OT-1 mouse by an institutionally approved method (*see* **Note 11**).
2. Harvest the thymuses and clean them well according to Subheading 3.2 of the protocol (*see* **Note 12**).
3. Create single cell thymocyte suspension by smashing the thymus of each mouse with the back side of a plunger of a 3 mL syringe in 5 mL of PBS in a 6 cm dish.
4. Filter through 70 μm filter in a 15 mL conical tube (*see* **Note 13**).
5. Dilute the cell suspension to 10 mL with PBS.
6. Count the cells with a hemocytometer.
7. Take 10×10^6 cells of each cell suspension in fresh 15 mL conical tubes (*see* **Note 14**).
8. Spin down at $450 \times g$, 4 °C for 5 min. Aspirate carefully the supernatant.
9. Resuspend the cells from OT-1 mouse in 1 mL of pre-warmed to 37 °C PBS containing 1 μM Cell Proliferation Dye eFluor 450 (*see* **Note 15**). Vortex. Incubate at 37 °C for 15 min. Add 10 mL cDMEM medium and spin down at $450 \times g$, 4 °C for 5 min. Aspirate carefully the supernatant.
10. Resuspend the cells from C57BL/6 mouse in 1 mL of pre-warmed to 37 °C PBS containing 5 μM Cell Proliferation Dye eFluor 670. Vortex. Incubate at 37 °C for 15 min. Add 10 mL cDMEM medium and spin down at $450 \times g$ 4 °C for 5 min. Aspirate carefully the supernatant.

11. Resuspend each cell suspension in 200 μL cDMEM medium so that the concentration is 50×10^6 cells/mL.
12. Mix the cell suspensions together so that the final volume is 400 μL and the concentration of each cell type—25×10^6/mL. Vortex to ensure good mixing of the cells.

3.6 Overlaying Thymocytes on Slices

1. In a biosafety cabinet, put Cell Culture Inserts into the wells of a 6-well plate.
2. Add 1 mL of cDMEM medium to the bottom of each well. This volume is enough to reach the membrane of the insert and keep the slices moist.
3. Carefully add three slices into one Cell Culture Insert with a bent spatula (Fig. 3a). Dry the slices before putting them inside by touching their edges to a Kimwipe (*see* **Note 16**). Make sure the slices do not touch each other or the walls of the Cell Culture Inserts.
4. Vortex the labeled cell suspension and carefully add 10–20 μL on top of each slice (*see* **Note 17**) (Fig. 3b).
5. Once all the slices have been covered with cell suspension, the 6-well plate should be covered with its lid and carefully moved to a CO_2 incubator.
6. After 2 h, take the 6-well plate out and gently wash the cell suspension off the top of the slice with 1 mL of cDMEM medium (*see* **Note 18**). Remove the medium from the Cell Culture Insert by aspiration with a pipette.
7. Add 1 mL of cDMEM medium containing 10 nM $Ova_{257-264}$ peptide to induce negative selection or 10 nM control peptide (*see* **Note 19**). Return back to the CO_2 incubator.
8. After 30 min remove the peptide containing suspension by aspiration with a pipette. Add 10 μL of cDMEM medium on top of each slice to prevent them from drying during the continued incubation. Incubate for the desired time (*see* **Note 20**).

3.7 Slice Dissociation and Flow Cytometry Analysis

1. After the end of the incubation period, take out the 6-well plate and add 1 mL of FACS buffer to the respective Cell Culture Inserts to facilitate the collection of the slices.
2. Use a bent spatula to move each slice to a 6 cm dish with 5 mL of ice-cold FACS buffer inside (*see* **Note 21**).
3. Create single cell suspension from the slice with the back side of a plunger of a 3 mL syringe. Make sure there are no pieces of intact tissue remaining.
4. Filter the suspension into 5 mL FACS tubes using pre-cut autoclaved 70 μm filters or cell strainer caps (*see* **Note 22**).

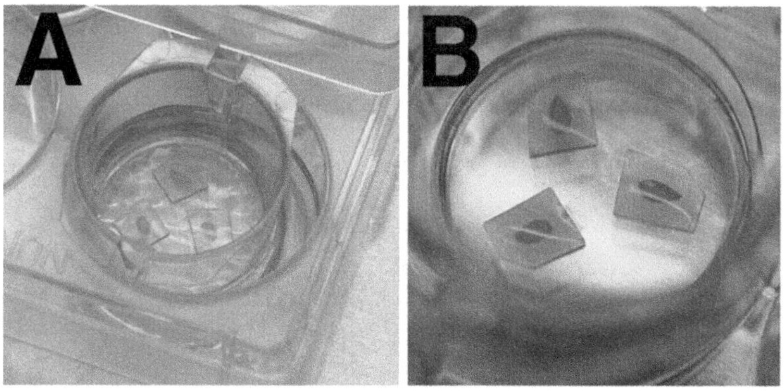

Fig. 3 Overlaying of labeled thymocytes on cut thymic slices. (**a**) Arrangement of slices in a Cell Culture Insert. Note that the slices do not touch each other or the walls of the Cell Culture Insert. (**b**) Thymic slices overlaid with 10–20 μL labeled cell suspension that is retained on top of them

5. Move 1 mL of cell suspension to fresh FACS tubes (*see* **Note 23**).
6. Spin down at 450 × *g* 4 °C for 5 min. Aspirate carefully the supernatant.
7. Resuspend in 300 μL FACS buffer containing PI at 1:1000 dilution.
8. Proceed with the flow cytometry analysis using standard techniques. Acquire at least 200,000 cells, preferably 500,000 cells. This basic procedure requires flow cytometer with violet (405 nm), blue (488 nm), and red (633 nm) lasers. No compensation is required.

3.8 Analyzing the Flow Cytometry Data and Calculating the Cell Loss

1. We use FlowJo software for our flow cytometry analysis, so this workflow follows FlowJo's conventions, but it can easily be adapted to any other software.
2. First gate on the live cells in FSC-A vs. PI plots and then gate on single cells in FSC-A vs. FSC-W plots.
3. Gate on the C57BL/6 cells and OT-1 cells in BV421 vs. APC plots (Fig. 4a).
4. Obtain the number of cells in each gate. For multiple samples this is easily done by exporting the cell counts to a table.
5. Copy the numbers to Microsoft Excel and divide the number of OT-1 cells by the number of C57BL/6 cells for each sample. This is the raw ratio.
6. Normalize the ratio to your control (usually control peptide stimulation)—obtain the average of the raw ratios for the control and divide all raw ratios by this number (*see* **Note 24**).
7. Plot the ratios for different treatments or time points and evaluate by an appropriate statistical method (Fig. 4b).

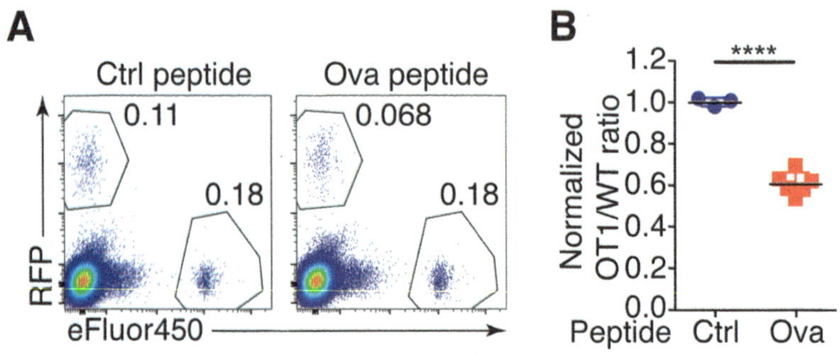

Fig. 4 Representative data from an in situ negative selection experiment. (**a**) Example plots of OT-1thymocytes expressing red fluorescent protein (RFP) overlaid together with eFluor450 labeled C57BL/6 thymocytes (WT) on a C57BL/6 thymic slice and treated with control peptide (Ctrl) or $Ova_{257-264}$ (Ova) peptide for 30 min. The slices were incubated overnight in a CO_2 incubator, mechanically dissociated and analyzed by flow cytometry. The numbers are the proportions of cells in each gate out of total live, single cells. (**b**) Statistical analysis of the normalized ratio of live OT-1/WT thymocytes recovered from the slices. Error bars are standard error of the mean. $N = 3$ for control peptide and $N = 6$ for Ova peptide ($****p < 0.0001$, unpaired t test)

4 Notes

1. The agarose should be ~37 °C when the thymus is added in it. Keeping low melting point agarose at 37 °C for prolonged time can lead to solidification. The agarose can be stored and reused within a week. In this case it should be boiled again. However, we avoid boiling it more than two times, because the loss of water leads to increased agarose concentration and viscosity and, ultimately, difficulty in embedding the thymus.
2. The opening of vena cava decreases the chance of heavy bleeding during the dissection of the thymus.
3. Physical damage to the thymus such as cuts or squeezing of the parenchyma decrease the overall health of the tissue and can cause problems with migration or excessive cell death later and should be avoided as much as possible.
4. Thorough cleaning is critical for successful cutting of the thymus. Any remaining connective tissue on top of the capsule can interfere with separation of the slices by the Vibratome blade.
5. We suggest that every lab establishes optimal conditions for embedding the thymus. In our lab, we wait for ~20 s after pouring the agarose and then submerge the thymus in it. If you dip the thymus too soon, it will sink to the bottom and there will not be enough agarose on that side. If you wait for too long, the agarose will solidify and the thymus will not sink at all.

6. We typically submerge the thymus along its longest axis, but other orientations are possible. The guiding principle is that there should be enough agarose on each side once the agarose block is trimmed for slicing.
7. The separation of the agarose block can be facilitated if the corners of the mold are cut with a blade.
8. The agarose block can be trimmed in different shapes, but we prefer the rectangular prism. Each researcher should carefully evaluate the position of the thymus in the block and decide in advance which side should be bottom and which side should be top. The bottom side will be parallel to the cutting plane and should be smoothly cut. It should also have large enough surface to hold the whole block when glued to the Vibratome stage. Too small bottom sides detach easily during cutting, especially if the agarose block is tall. Because the Vibratome blade cannot reach the bottom of the stage, leave at least 5 mm between the bottom side of the agarose block and the bottom of the thymus. Do not leave too much agarose around the thymus, because the slices will be very big, difficult to cut cleanly, and difficult to position in a tissue culture insert later.
9. It is much easier to determine the starting and finishing position of the blade before the cutting chamber is filled with PBS.
10. Common problems during slice cutting are as follows: (1) The whole agarose block detaches from the cutting stage. In this case, the cutting chamber needs to be emptied of PBS and dried. Then the agarose block can be re-glued to the stage. (2) The blade cannot cut through the thymus and the slice remains attached to the agarose block through a thin piece of connective tissue. This is usually caused by suboptimal cleaning of the thymus lobes from connective tissue. The slice can be separated manually with scissors from the block; however, this usually results in defects in the slice. That is why it is advisable to prepare at least two thymuses in agarose blocks and very carefully clean the thymus from connective tissue. (3) The thymus slice detaches from the agarose. This could result from moving the thymus lobe once the agarose around it has started to solidify.
11. If you are doing an experiment with few slices (e.g., <8), it might be possible to use one lobe of the C57BL/6 thymus for slices and the other lobe for single cell suspension. In this protocol, we use OT-1 mice, because these are one of the most common TCR transgenic mice that have extensively been used to study negative selection. However, the method can be used with any TCR transgenic mice with a known cognate peptide.

12. The cleaning of the thymus for single cell suspension does not have to be as rigorous as for slice preparation. Leaving some connective tissue is acceptable, because it will be filtered out in subsequent steps. However, it is still advisable to remove any blood from the capsule by rolling the thymus on wet paper towels.
13. The cheapest option is to use pre-cut autoclaved filter membranes (e.g., Small Parts, part #B0043D1SZG or similar from Amazon). Alternatives include filtering into 5 mL polystyrene round-bottom (FACS) tubes with filter caps (Falcon, cat #352235) or into 50 mL conical tubes using 70 μm cell strainer (e.g., Falcon cat. #352350).
14. That number of cells will be enough for at least 20 slices. If the experiment needs to be scaled up, increase the number of cells accordingly.
15. Cell Proliferation Dyes 450 and 670 bind to free amino groups, so the buffer should not contain protein or TRIS, hence the use of PBS. Each lab is advised to titrate both dyes to find the optimal concentration for their purposes.
16. Careful drying is critical for the success of overlaying cells on the slice. If the slice is not dry, the surface tension cannot be maintained on its top and the cell suspension will leak to the membrane leaving no cells on top of the slice. Alternatively, a hydrophobic barrier such as vacuum grease silicon (e.g., Beckman, cat#335148) or Teflon O-ring (The O-Ring Store) can be used to make sure that the cell suspension will stay in place. The vacuum grease silicon can be applied with a syringe and a plastic needle. The O-rings can be bought in different sizes and one that fits the particular slice (surrounds all of the thymus tissue, but does not go outside of the agarose) can be put on top.
17. If the slice is dried well, the cell suspension will hold as a small drop on top of the slice. Avoid adding too much of the cell suspension as this will make it difficult for the drop to stay on top.
18. Careful, but thorough, washing is critical for the success of the experiment. Too little washing and many cells will be stuck on the top of the slice where their apoptosis will proceed with different kinetics compared to the cells inside the slice. The cells on top will likely be much more numerous than the ones inside and will dilute the effect of peptide or any other treatment. If there is no cell loss after specific peptide addition, this is the most likely step that needs to be optimized. Too much washing and the top layers of the slice will be washed away leaving very few labeled cells inside. We have found out that drop-wise addition of medium just above the thymus tissue while holding the plate tilted at 45° works best. Two or three rounds of such washing are usually sufficient.

19. The minimum concentration of agonist peptide can be as little as 1 nM [11], but more consistent results are obtained with 10 nM.
20. Depending of the purpose of the experiment, the incubation time can be different. In general, activation of OT-1 thymocytes could be seen as soon as 30 min after the peptide addition, signs of apoptosis (e.g., caspase activation, phosphatidylserine exposure) can be seen within 2 h of peptide addition, and cell loss can be observed starting after 3 h [11]. Approximately 40–60% of all OT I cells will be lost within the first 10 h. If the experiment is done with OT-1 × $RAG1^{-/-}$ mice, the cell loss will be even greater [23].
21. Make sure to label all dishes and tubes in advance. The incubation is good time to prepare them. Always confirm that you are using the correct containers.
22. The suspension will inevitably contain pieces of agarose, but we have found that they are readily filtered out and, typically, do not affect the flow cytometry analysis.
23. In our experience, the average slice has around 5×10^6 cells. In this case 1 mL of cells suspension will have ~10^6 cells. If the slice is very small or unusually big, it might be worth counting the cells to make sure that at least 10^6 cells are transferred for flow cytometry analysis. The entire cell suspension can also be used for flow cytometry. However, if further staining with antibodies is required, we recommend counting the cells in all samples and using equal number of cells for the staining.
24. The normalization is done to facilitate comparison between different experiments. Although every effort is made to make sure that the two cell populations are at 1:1 ratio, often times this is not the case, which can lead to raw ratios varying considerably between experiments and obscuring the majority of the differences.

Acknowledgments

We would like to thank Hsing-Kai Feng and Yun-Tzu Chen for help with taking the photos and carefully reviewing the manuscript. This work was supported by grants from MOST, 106-2320-B-010-026-MY3 (ILD), 107-2320-B-010-016-MY3 (ILD), 104-2628-B-010-002-MY4 (CLH), and 107-2320-B-010-020 (CLH), and a grant from the Yen Tjing Ling Medical Foundation, CI-108-5 (ILD).

References

1. Kappler JW, Roehm N, Marrack P (1987) T cell tolerance by clonal elimination in the thymus. Cell 49:273–280
2. Anderson MS, Venanzi ES, Chen Z et al (2005) The cellular mechanism of Aire control of T cell tolerance. Immunity 23:227–239. https://doi.org/10.1016/j.immuni.2005.07.005
3. Le Borgne M, Ladi E, Dzhagalov IL et al (2009) The impact of negative selection on thymocyte migration in the medulla. Nat Immunol 10:823–830. https://doi.org/10.1038/ni.1761
4. Hu Z, Lancaster JN, Sasiponganan C, Ehrlich LIR (2015) CCR4 promotes medullary entry and thymocyte-dendritic cell interactions required for central tolerance. J Exp Med 212:1947–1965. https://doi.org/10.1084/jem.20150178
5. Ferrero I, Anjuère F, MacDonald HR, Ardavín C (1997) In vitro negative selection of viral superantigen-reactive thymocytes by thymic dendritic cells. Blood 90:1943–1951
6. Mcgargill MA, Hogquist KA (1999) Antigen-induced coreceptor down-regulation on thymocytes is not a result of apoptosis. J Immunol 162:1237–1245
7. Guerri L, Peguillet I, Geraldo Y et al (2013) Analysis of APC types involved in CD4 tolerance and regulatory T cell generation using reaggregated thymic organ cultures. J Immunol 190:2102–2110. https://doi.org/10.4049/jimmunol.1202883
8. Bhakta NR, Oh DY, Lewis RS (2005) Calcium oscillations regulate thymocyte motility during positive selection in the three-dimensional thymic environment. Nat Immunol 6:143–151. https://doi.org/10.1038/ni1161
9. Melichar HJ, Ross JO, Herzmark P et al (2013) Distinct temporal patterns of T cell receptor signaling during positive versus negative selection in situ. Sci Signal 6:ra92–ra92. https://doi.org/10.1126/scisignal.2004400
10. Ross JO, Melichar HJ, Au-Yeung BB et al (2014) Distinct phases in the positive selection of CD8+ T cells distinguished by intrathymic migration and T-cell receptor signaling patterns. Proc Natl Acad Sci 111:E2550–E2558. https://doi.org/10.1073/pnas.1408482111
11. Dzhagalov IL, Chen KG, Herzmark P, Robey EA (2013) Elimination of self-reactive T cells in the thymus: a timeline for negative selection. PLoS Biol 11:e1001566. https://doi.org/10.1371/journal.pbio.1001566
12. Melichar HJ, Ross JO, Taylor KT, Robey EA (2015) Stable interactions and sustained TCR signaling characterize thymocyte-thymocyte interactions that support negative selection. J Immunol 194:1057–1061. https://doi.org/10.4049/jimmunol.1400169
13. Ki S, Thyagarajan HM, Hu Z et al (2017) EBI2 contributes to the induction of thymic central tolerance in mice by promoting rapid motility of medullary thymocytes. Eur J Immunol 14:377–312. https://doi.org/10.1002/eji.201747020
14. Au-Yeung BB, Melichar HJ, Ross JO et al (2014) Quantitative and temporal requirements revealed for Zap70 catalytic activity during T cell development. Nat Immunol 15:687–694. https://doi.org/10.1038/ni.2918
15. Weist BM, Kurd N, Boussier J et al (2015) Thymic regulatory T cell niche size is dictated by limiting IL-2 from antigen-bearing dendritic cells and feedback competition. Nat Immunol 16:635–641. https://doi.org/10.1038/ni.3171
16. Ueda Y, Katagiri K, Tomiyama T et al (2012) Mst1 regulates integrin-dependent thymocyte trafficking and antigen recognition in the thymus. Nat Commun 3:1098–1013. https://doi.org/10.1038/ncomms2105
17. Ehrlich LIR, Oh DY, Weissman IL, Lewis RS (2009) Differential contribution of chemotaxis and substrate restriction to segregation of immature and mature thymocytes. Immunity 31:986–998. https://doi.org/10.1016/j.immuni.2009.09.020
18. Halkias J, Melichar HJ, Taylor KT et al (2013) Opposing chemokine gradients control human thymocyte migration in situ. J Clin Invest 123:2131–2142. https://doi.org/10.1172/JCI67175DS1
19. Dzhagalov IL, Melichar HJ, Ross JO, et al (2012) Two-photon imaging of the immune system. Curr Protoc Cytom Chapter 12: Unit12.26. https://doi.org/10.1002/0471142956.cy1226s60
20. Ross JO, Melichar HJ, Halkias J, Robey EA (2016) Studying T cell development in thymic slices. Methods Mol Biol 1323:131–140. https://doi.org/10.1007/978-1-4939-2809-5_11
21. Lancaster JN, Ehrlich LIR (2017) Analysis of thymocyte migration, cellular interactions, and activation by multiphoton fluorescence microscopy of live thymic slices. In: Wells CM,

Parsons M (eds) Cell migration. Springer New York, New York, NY, pp 9–25

22. Hogquist KA, Jameson SC, Heath WR et al (1994) T cell receptor antagonist peptides induce positive selection. Cell 76:17–27. https://doi.org/10.1016/0092-8674(94)90169-4
23. Sood A, Dong M, Melichar HJ (2016) Preparation and applications of organotypic thymic slice cultures. J Vis Exp 114:e54355. https://doi.org/10.3791/54355

Chapter 18

Investigating T Cell Receptor Signals In Situ by Two-Photon Microscopy of Thymocytes Expressing Genetic Reporters in Low-Density Chimeras

Marilaine Fournier, Mengqi Dong, and Heather J. Melichar

Abstract

T cell development is a dynamic process accompanied by extensive thymocyte migration, cellular interactions, and T cell receptor (TCR) signaling. In particular, thymic selection processes that ensure a functional, self-tolerant repertoire require TCR interactions with self-peptide presented by major histocompatibility complex molecules expressed by specialized thymic antigen-presenting cells. The quantity and quality of these TCR signals influence T cell fate. Two-photon microscopy, which enables live imaging of cells in intact tissue, has emerged as a powerful tool to gain insights into thymocyte migration and TCR signaling during T cell development in situ. Here we describe the generation of non-irradiated, low-density chimeric mice by neonatal injection of adult bone marrow engineered to express fluorescent TCR signaling reporters for imaging by two-photon microscopy. We also describe how the thymic lobes from chimeric mice are prepared for live imaging of thymocyte behavior and TCR signaling events. While we focus on imaging TCR signals associated with T cell development in the thymus, these techniques can also be adapted to study TCR signaling in mature T cells in peripheral lymphoid organs.

Key words Neonatal chimera, Thymus, Thymocyte behavior, Genetic reporters, T cell receptor signaling, Thymic slices, Two-photon microscopy

1 Introduction

Two-photon microscopy has revolutionized our understanding of T cell behavior in situ. The speed, migration patterns, and cellular interactions of T cells at different developmental stages in the thymus as well as at steady state or during antigen challenge in secondary lymphoid organs and tissues can now be interrogated from multiple perspectives given access to appropriate infrastructure [1]. Advances in two-photon microscopy have also enabled visualization of T cell receptor (TCR) signal dynamics [2]. These studies have provided valuable insights into the TCR signals that

Marilaine Fournier and Mengqi Dong contributed equally to this work.

Chaohong Liu (ed.), *T-Cell Receptor Signaling: Methods and Protocols*, Methods in Molecular Biology, vol. 2111, https://doi.org/10.1007/978-1-0716-0266-9_18,

accompany thymic selection and peripheral T cell activation in physiological contexts.

The quality and quantity of TCR signals influence T cell fate during development in the thymus as well as T cell function after activation in the periphery. In particular, developing T cells in the thymus express a diverse repertoire of TCRs that undergo selection processes to test each antigen receptor for functionality and autoreactivity through interactions with self-peptide (self-p) presented by major histocompatibility complex (MHC) molecules on thymic epithelial cells and other antigen-presenting cells (APC). In general, low to moderate affinity TCR and self-pMHCs interactions lead to positive selection (survival and differentiation), whereas higher affinity interactions lead to clonal deletion (apoptosis). The differences in TCR signaling that accompany these processes and generate such distinct outcomes are only starting to be understood. When the TCR/CD3 complex and co-receptors are brought into proximity by engagement with self-pMHC on an APC, the co-receptor-associated tyrosine kinase Lck phosphorylates the intracellular domain of CD3 and the ζ chains, recruiting ζ chain-associated protein kinase of 70 kD (Zap70) which subsequently propagates the signal via linker for activation of T cells (LAT) [3]. It has only recently been shown that these TCR proximal signals are quantitatively, rather than qualitatively, distinct in low and high affinity interactions associated with positive versus negative selection [4]. At later stages of TCR signal propagation, qualitative differences begin to emerge. Negative selection is thought to be associated with higher levels of phosphorylated Zap70 recruitment and subsequent phosphorylation of LAT leading to activation of Ras and mitogen-activated protein kinase (MAPK) pathways at the plasma membrane, whereas weaker TCR signals associated with positive selection lead to Ras-MAPK pathway activation in the Golgi apparatus [3, 5–11]. In addition, distinct intracellular events downstream of TCR signaling lead to differences in calcium influx, re-localization of nuclear factor of activated T cell (NFAT) from the cytosol to the nucleus, and distinct transcriptional programs that lead to positive versus negative selection as well as CD4 versus CD8 T cell lineage commitment [5, 9, 12–19]. Indeed, in situ analyses of MHC class I restricted thymocytes by two-photon microscopy suggest that positive selection is associated with serial, transient increases in intracellular calcium ($[Ca^{2+}]_i$) that accumulates in the cell over time to reach a threshold prior to differentiation [15–17]. These low affinity TCR signals are associated with the preferential localization of NFAT in the cytoplasm [15]. In contrast, negative selection of MHC class I restricted thymocytes is associated with prolonged, stable increases in $[Ca^{2+}]_i$, translocation of NFAT to the nucleus, and migratory arrest [15]. In situ analysis of MHC class II restricted thymocytes, however, demonstrated robust relocalization of NFAT to the nucleus and sustained increases in $[Ca^{2+}]_i$ during positive selection [18, 19].

In situ visualization of the TCR signaling processes noted above typically relies on real-time induction of fluorescence using calcium indicators or intracellular re-localization of TCR signaling intermediates. Tools such as calcium indicator dyes or genetically encoded reporters allow visualization of differences in $[Ca^{2+}]_i$ during TCR signaling events [15–17, 19–26]. Ratiometric calcium indicator dyes, such as Indo-1, are particularly sensitive for detecting small fluctuations in $[Ca^{2+}]_i$; this is important for the quantification of TCR signals associated with positive selection [15–17, 19]. However, the half-life of the dyes limits their use in longer-term studies [27]. The sensitivity of genetically encoded calcium sensors is continually improving and is, at present, sufficient to detect changes in $[Ca^{2+}]_i$ in activated T cells [22, 23, 25]. Despite challenges in detecting changes to protein localization with two-photon microscopy, owing to the limited resolution of this approach, the technique has been used to show that fluorescent proteins coupled to TCR signaling molecules (e.g., TCR, LAT, PKCζ, NFAT) are differentially distributed during T cell migration and TCR signaling in situ [15, 18, 28–31].

Real-time two-photon microscopy studies of TCR signaling in the thymus have largely been done using addition of thymocytes loaded with the ratiometric calcium dyes or engineered to express fluorescent fusion proteins to reaggregate thymic organ cultures or thymic slices. The addition of thymocyte subsets to thymic tissue slices (described in Chap. 17 of this book as well as in other recent works [32–34]) is a powerful and flexible model to study the behavior of thymocytes and allows two-photon microscopy of cells in the cortex and medulla. Some differences, however, may exist in the behavior of acutely overlaid thymocyte populations that engage self-pMHC for the first time as compared to those that are endogenous to the tissue. Thus, we present here a complementary model system that permits analysis of TCR signaling in tissue endogenous cells during T cell development in the thymus and beyond.

In this chapter, we describe, in detail, the generation of neonatal chimeric mice with retrovirally transduced adult bone marrow (BM) for the development of adult thymocytes with TCR signal reporters for direct in situ imaging by two-photon microscopy (Fig. 1). The injection of BM into neonates does not require irradiation of recipient mice that can damage thymic epithelial cells necessary to support T cell development and generates low levels of chimerism (~0.1–2.0%), a density appropriate for two-photon imaging (Fig. 2) [35–41]. The neonatal chimera model also allows flexibility in that both the donors and recipients can be of any background (wild type, transgenic, or knockout) depending on the experimental question. Using low-density chimeric mice, we can visualize thymocyte behavior in the thymic cortex by directly imaging the thymic lobe ex vivo. In addition,

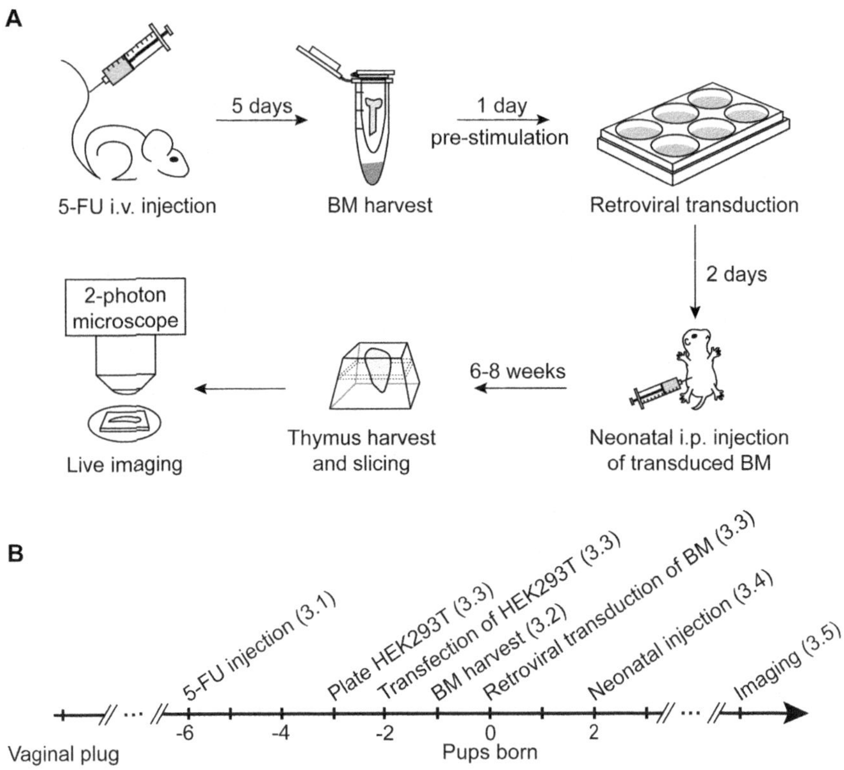

Fig. 1 Schematic overview of procedures described in this chapter. Depiction of the major steps (**A**) and timing (**B**) of generating BM expressing genetic reporters of TCR signaling for neonatal injection and, ultimately, imaging of the behavior of developing T cells in thymic slices by two-photon microscopy

tissue slices can be generated from the thymus for imaging thymocyte behavior in the cortex and medulla. We briefly discuss strategies for analyzing TCR signaling data after acquisition by two-photon microscopy (Fig. 3). Although we focus on imaging of TCR signaling in the thymus, these models can be easily adapted for experiments examining TCR signaling under various conditions in the secondary lymphoid organs.

2 Materials

2.1 5-Fluorouracil (5-FU) Injection

1. 50 mg/mL 5-FU solution.
2. Phosphate-buffered saline (PBS).
3. Adult (6–12 weeks old) female mice (*see* **Notes 1–3**).
4. Balance.
5. Heat lamp.
6. Mouse restrainer.
7. 70% ethanol pads.
8. 1 mL *tuberculin* syringe with 27G × ½ needle.

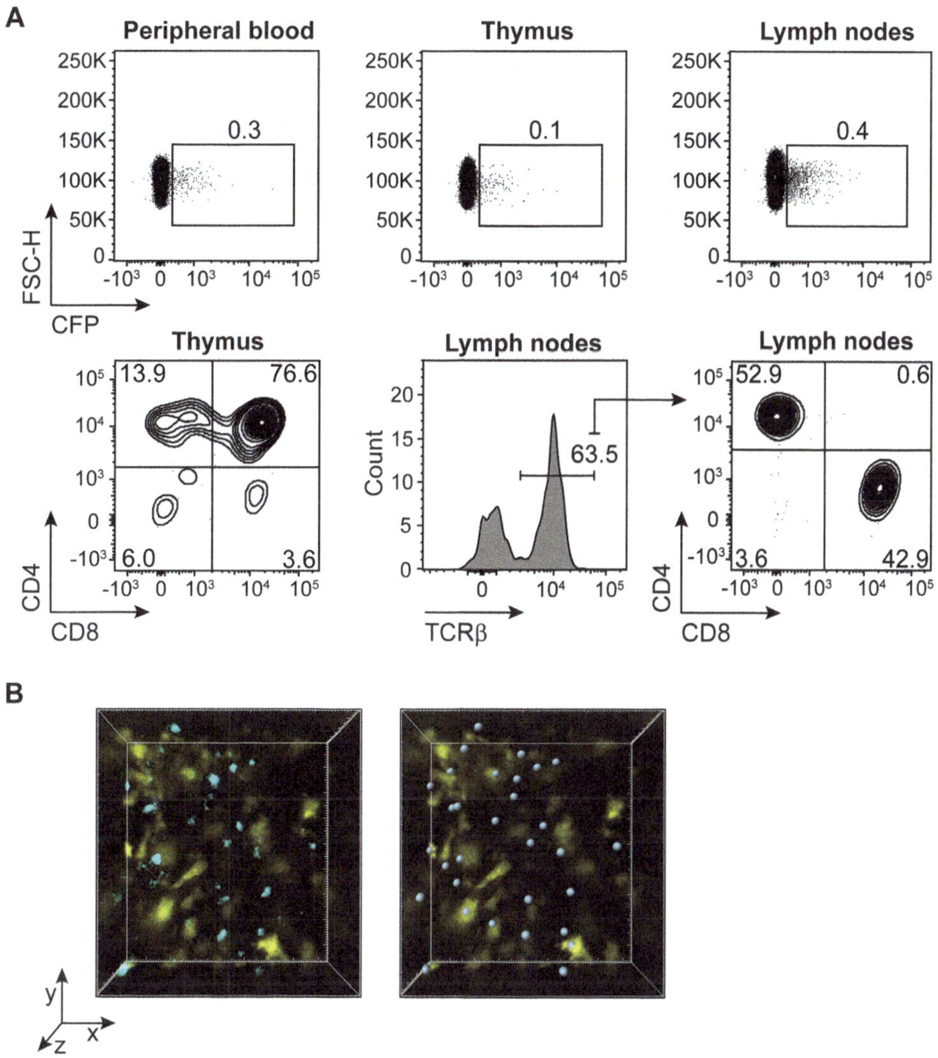

Fig. 2 Evaluation of chimerism in the blood, thymus, and secondary lymphoid organs by flow cytometry and two-photon microscopy. (**A**) A single dose of CAG-ECFP BM was injected into CD11c-YFP transgenic neonates. Six weeks after injection, peripheral blood and organs were harvested, and CFP^+ cells were detected (top panel) by flow cytometry. The T cell profile of CFP^+ cells in the thymus and lymph nodes was also determined using labeled antibody against TCRβ, CD4, and CD8 (bottom panel). (**B**) Representative two-photon microscopy image of CFP^+ T cells (cyan) and the CD11c-YFP^+ dendritic cells (yellow) in the medulla of a thymic slice without (left panel) or with (right panel) cell tracking spots (162.5 × 172.3 × 72.0 μm)

2.2 BM Harvest

1. 5-FU-injected adult female mice.
2. CO_2 chamber.
3. Dissection board with tissue pins.
4. 70% ethanol.
5. Broad-tipped forceps.
6. 11.5 cm small straight scissors.

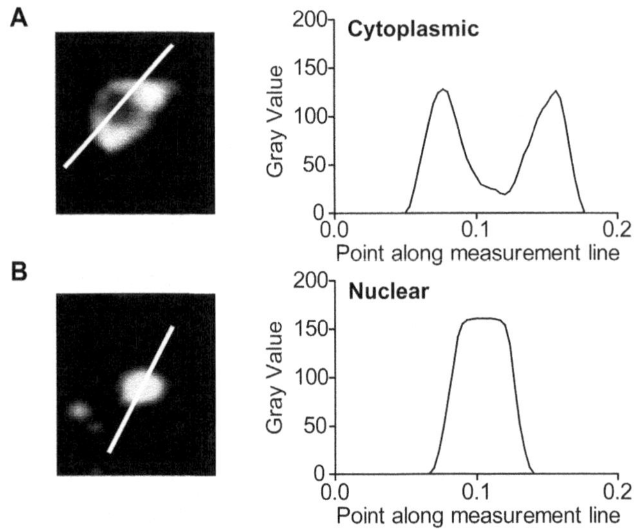

Fig. 3 Quantification of NFAT-GFP localization in individual cells. Representative images (left panels) of cytoplasmic (**A**) or nuclear (**B**) NFAT-GFP localization were extracted from two-photon time-lapse movie of NFAT-GFP expressing OT-I TCR transgenic cells in the cortex of a thymic slice incubated in the presence of OVA peptide. The intensity of the GFP signal based on gray value (0–255) along a measurement line was determined using ImageJ software (National Institutes of Health) (right panels)

7. 6 cm tissue culture dish.
8. PBS.
9. 0.5 mL microcentrifuge tube pierced at the bottom with a 21G needle.
10. 1.5 mL microcentrifuge tubes.
11. PBS supplemented with 2% FBS (PBS 2% FBS).
12. Microcentrifuge.
13. 70 μm cell strainer.
14. ACK lysis buffer: 150 mM NH_4Cl, 10 mM $KHCO_3$, 0.1 mM Na_2EDTA; adjust pH to 7.2–7.4.
15. 0.4% trypan blue in PBS.
16. Hemocytometer.
17. Hematopoietic progenitor cell (HPC) media: DMEM, 15% heat-inactivated (HI) fetal bovine serum (FBS), 6 ng/mL mouse interleukin-3 (mIL-3), 40 ng/mL mIL-6, 100 ng/mL mouse stem cell factor (mSCF), 1000 IU penicillin, 1000 μg/mL streptomycin and 10^{-5} M 2-mercapthoethanol.
18. 6-well plate for suspension cells.

2.3 Retroviral Transduction of BM Cells

1. Packaging cells (e.g., HEK293T).
2. Complete HEK293T cell media: DMEM, 10% HI FBS, 1000 IU penicillin and 1000 μg/mL streptomycin.
3. 10 cm tissue culture-treated dish.
4. Packaging vector expressing *gag-pol* and *env* genes (e.g., pCL-Eco).
5. Retroviral vector expressing the gene of interest (e.g., pMSCV-NFAT-GFP).
6. Serum-free media (e.g., Opti-MEM I).
7. Transfection reagent (e.g., Lipofectamine 2000).
8. 1.5 mL microcentrifuge tubes.
9. 0.45 μm PVDF syringe filter.
10. 10 mL syringe.
11. HI FBS.
12. Cytokines to complement viral supernatant (mIL-3, mIL-6, mSCF).
13. 4 μg/μL polybrene (hexadimethrine bromide) solution in H_2O.
14. 15 mL conical tubes.
15. 6-well plate for suspension cells.
16. Tabletop centrifuge.

2.4 Neonatal Injection

1. Donor BM cell suspension.
2. PBS.
3. Insulin syringe with 28G × ½ needle.
4. Litter of pups as recipients (2–6 days old) (*see* **Note 4**).
5. Empty cage.

2.5 Thymic Lobe Harvesting, Slicing, and Preparation for Imaging

1. 6–8-week-old chimeric mice.
2. CO_2 chamber.
3. Dissection board with tissue pins.
4. 70% ethanol.
5. 11.5 cm small straight scissors.
6. Curved fine forceps (e.g., Dumont #7 Forceps Inox Tip Size 0.17 × 0.10 mm).
7. Ultra-fine-tipped forceps.
8. 6 cm tissue culture dish.
9. PBS.
10. Tissue wipes.

11. Low-melting-temperature agarose (e.g., NuSieve GTG Agarose, Lonza).
12. 250 mL glass Erlenmeyer flask with aluminum foil cover.
13. Microwave.
14. 50 °C water bath.
15. 12 × 12 × 20 mm square disposable plastic tissue embedding mold (e.g., PEEL-A-WAY).
16. Slushy ice basin.
17. Stainless steel double-edged razor blades.
18. Vibratome (e.g., Leica VT1000S).
19. 200 μL pipet tip.
20. Tissue glue (e.g., 3M Vetbond).
21. Bent spatula.
22. Round microslides.

2.6 Two-Photon Imaging of Thymic Tissue and Analysis

1. Upright multi-photon microscope system (*see* **Note 5**).
2. Band pass filter and dichroic mirror sets.
3. Vacuum grease.
4. Micro-incubation perfusion system including a perfusion chamber with heater, peristaltic pump, bubbling stone and tubing, as well as 95% O_2 and 5% CO_2 gas mixture to allow perfusion of tissue with warmed (37 °C), oxygenated media during imaging.
5. Perfusion media: DMEM with 4.5 g/L glucose and sodium pyruvate, without L-glutamine and phenol red.
6. Computers equipped with acquisition (e.g., Zeiss Zen black) and analysis (e.g., Imaris-Bitplane, ImageJ-National Institutes of Health, Matlab-Mathworks) software.

3 Methods

Procedures described in Subheadings 3.1–3.4 should be performed under sterile conditions. All animal experimentation must be approved by the appropriate animal care committee. Careful planning is required as the protocols described below overlap in time (Fig. 1). In order to prepare donor cells and viral supernatant prior to the birth of recipient pups, we suggest verifying the presence of a vaginal plug in the breeding cage of interest to determine the date of birth of the neonates to be injected.

3.1 5-FU Injection

Plan to perform 5-FU injections of donor mice 8 days prior to neonatal injection.

1. Dilute 5-FU (*see* **Note 6**) to a concentration of 15 mg/mL in PBS.
2. Weigh the donor mice to inject.
3. Warm mice under heat lamp or with other heating source for ~3–5 min taking care not to overheat the animals.
4. Place a mouse in a commercial restraining device.
5. Wipe tail with a 70% ethanol pad.
6. Inject each adult donor mouse intravenously (i.v.) with 10 μL of 5-FU per gram of body weight in the lateral tail vein.
7. Return mice to cage and observe for several minutes to ensure return to normal activity.

3.2 BM Harvest

1. Euthanize donor mice 5 days post-5-FU injection according to your local animal care committee guidelines (e.g., CO_2 inhalation followed by cervical dislocation).
2. Pin each mouse on its back to a dissection board and spray lightly with 70% ethanol.
3. Lift the abdominal skin with forceps and cut the skin from the abdomen toward each of the lower limbs to the paws with scissors to expose the legs.
4. Dissect muscle tissue around the tibia, femur, and iliac bones from both legs and put them in a 6 cm dish containing PBS (*see* **Note 7**).
5. Remove all the connective tissue and muscle from the bones.
6. Place a pierced 0.5 mL polypropylene tube in a 1.5 mL tube containing 200 μL of PBS 2% FBS.
7. Carefully cut the femoral and iliac bones in the middle and put in the 0.5 mL tube with the cut edge towards the bottom. Cut the tibiae at the smaller end and place them similarly in the 0.5 mL tube. All bones from a single mouse can be put together in the same tube.
8. Close the 0.5 mL tube containing the bones and transfer both tubes into a microcentrifuge. Spin at 16,000 × *g* for 10 s. The BM cells should transfer into the 1.5 mL tube in the PBS 2% FBS and the remaining bones (now white) can be discarded with the 0.5 mL tube.
9. Aspirate the PBS 2% FBS, and add 1 mL of fresh PBS 2% FBS in the 1.5 mL tube, resuspend the BM cells, and pass through a 70 μm cell strainer into a new 1.5 mL tube. Spin the cells at 500 × *g* for 5 min and remove supernatant.

10. Resuspend the cells in 1 mL of ACK lysis buffer and incubate for 3 min at room temperature (RT). Fill the tube with PBS 2% FBS and centrifuge the cells at 500 × *g* for 5 min.
11. Resuspend the cells in PBS 2% FBS and count an aliquot of cells mixed with trypan blue using a hemocytometer.
12. Pellet cells by centrifugation and resuspend in HPC media at a concentration of 1 × 10^6 cells/mL. Place in 6-well plate for suspension cells and incubate at 37 °C for 24 h prior to transduction.

3.3 Retroviral Transduction of BM Cells

Plan to transfect HEK293T cells for retroviral production 4 days prior to neonatal injection.

1. One day before transfection, plate HEK293T cells (*see* **Note 8**) in a 10 cm dish to achieve 80–90% confluence the following day in complete HEK293T cell media.
2. On the day of transfection, prepare the retroviral packaging and expression plasmid mix by diluting 5 μg of pCL-Eco and 10 μg of retroviral vector in 500 μL of Opti-MEM I media. Separately, dilute 25 μL of Lipofectamine 2000 in 500 μL of Opti-MEM I media.
3. Mix the diluted DNA and Lipofectamine 2000 by flicking the tube with your finger, and incubate for 15 min at RT for DNA-liposome complexes to form.
4. During the incubation, delicately remove HEK293T cell media and add 4 mL of Opti-MEM I media by slowly pipetting against the side of the dish.
5. Add 1 mL of DNA-liposome complexes dropwise to the cells, swirl the plate gently, and incubate at 37 °C for 6 h.
6. Remove the transfection media and delicately replace with 8–10 mL of fresh complete HEK293T cell media.
7. Forty eight hours post-transfection, harvest the supernatant containing viral particles (*see* **Note 9**), and, using a 10 mL syringe and 0.45 μm syringe filter, filter the viral supernatant into a 15 mL conical tube. Supplement the viral supernatant with fresh HI FBS and cytokines to match the composition of the HPC media. In addition, add polybrene to a final concentration of 4 μg/mL.
8. Collect BM cells (from Subheading 3.2, **step 12**) in a 15 mL conical tube, centrifuge at 500 × *g* for 5 min, and remove media.
9. Resuspend the BM cells in the viral supernatant described in **step 7** at a concentration of 10^6 cells/mL and transfer to 6-well plates for suspension cultures.

10. Centrifuge the plates at 1000 × *g* for 90 min at RT. Transfer the plates to incubate at 37 °C for 48 h before injection into recipient mice.

3.4 Neonatal Injection

1. Prepare single doses to inject 1×10^6 donor BM cells in 70 μL of PBS in an insulin syringe (*see* **Note 10**).
2. Transfer some of the nesting material from the breeding cage along with the pups to inject into a new cage.
3. Loosely restrain a pup with the index finger and thumb. Inject BM cells intraperitoneally (i.p.) below the sternum but above the gut on the right side of the abdomen.
4. Put the pup back in the original cage with the mother.

3.5 Thymic Lobe Harvesting, Slicing, and Preparation for Imaging

1. 6–8 weeks after neonatal injection, euthanize the recipient experimental mice according to your local animal care committee guidelines (*see* **Note 11**).
2. Pin each mouse on its back on a dissection board and lightly spray with 70% ethanol.
3. Lift the abdominal skin with forceps and make longitudinal incisions toward the upper limbs with scissors to expose the ribcage.
4. Open the ribcage by cutting the diaphragm and then by cutting the ribcage on both sides to the upper limbs. Flip the ribcage towards the head of the mouse and pin it to expose the thymus located just above the heart.
5. Using pairs of ultra-fine- and curved fine-tipped forceps, delicately remove connective tissue around the thymus lobes to release them from the thoracic cavity and separate two lobes without directly touching the thymus. Carefully transfer the individual lobes to a 6 cm dish containing PBS.
6. Transfer the thymic lobes to a tissue wipe soaked with PBS and remove any residual connective tissue. Return the lobes to the dish containing PBS until ready to slice (*see* **Note 12**).
7. Prepare 50 mL of 4% low melting temperature agarose solution in PBS in a 250 mL Erlenmeyer flask. Heat at low power setting in a microwave to dissolve the agarose. Remove flask from microwave, cover with aluminum foil, and maintain in 50 °C water bath until needed.
8. When ready to embed the tissue, cool the agarose solution to just below 40 °C, and pour in the lower half of a tissue embedding mold. Gently pick up a thymic lobe and dry it on a fresh tissue wipe to remove residual PBS. Insert the lobe in the agarose, place the mold into a slushy ice basin, and position the thymus vertically in the mold. Let sit in the ice until the agarose has set (~5–10 min).

9. Once the agarose has solidified, turn the embedding mold over and press on the bottom of the mold to liberate the agarose block containing the thymic lobe. Trim the edges with a razor blade to have 1–3 mm of agarose around the lobe and with the base of the block slightly larger than the top.
10. Set up the vibratome by inserting half of a double-edged razor blade at a 5° cutting angle.
11. Using a 200 μL pipet tip, transfer a drop of tissue glue on the vibratome stage and put the agarose block on it. Fill the stage with PBS.
12. Proceed to cut 500–1000 μm slices at maximum amplitude and minimum speed. Collect the slices with a bent spatula when they are released in the PBS and transfer them into a 6 cm dish with PBS.
13. Using a 200 μL pipet tip, transfer a drop of tissue glue on a round coverslip. With a bent spatula, transfer the thymic lobe or slice on the glue, wait 10–20 s, and maintain in a dish with PBS on ice until ready to image.

3.6 Two-Photon Imaging of Thymic Tissue and Analysis

1. Set up the two-photon microscope by warming up the laser and identifying the appropriate band pass filters and dichroic mirrors based on the emission spectrum for each of the fluorescent populations within the sample to collect the broadest fluorescent spectrum in each channel while limiting spectral overlap.
2. Set up the perfusion system to ensure that the sample tissue will be bathed in warm (37 °C), oxygenated media during the imaging period, with a perfusion speed of ~2 mL/min.
3. Add vacuum grease to the bottom of the coverslip to which the thymic tissue is adhered and transfer to the perfusion chamber on the microscope. Allow tissue to warm for ~15 min prior to imaging.
4. Use the eyepiece and transmitted light to locate the tissue and then switch to scanning mode.
5. Identify a region of interest for subsequent imaging based on thymic region (e.g., cortex versus medulla), labeled thymocyte density, and any additional fluorescent cells or landmarks important to the experimental question.
6. Adjust the two-photon microscope and acquisition parameters including excitation wavelength (typically 740–920 nm) and power, photomultiplier tube (PMT) sensitivity, acquisition speed and averaging, scan area, depth of imaging (typically 40–100 μm starting at least 10 μm below the cut surface), *z* step (typically 3–5 μm), imaging frequency (typically 20–30 s), and duration (typically 20–30 min) in the acquisition software (e.g., Zen black) based on the experimental question, and proceed with time-lapse image acquisition (*see* **Note 13**).

7. For analysis, open the time-lapse image series in analysis software (e.g., Imaris) that allows tracking of individual fluorescent cells in three dimensions over time, provides measurements of the intensity of fluorescence in individual channels in individual cells, and allows for calculations of the distance between cells. These data can be imported into programs (e.g., Matlab or ImageJ) for further analysis to calculate migration parameters such as instantaneous and average speed, straightness/chemotactic index, persistence length, mean square displacement versus time lag, and cell-cell interactions. In addition, calculation of the fluorescence emission ratio in an individual cell (e.g., FRET-based calcium biosensors) or the distribution of fluorescence under a line drawn through the cell (e.g., NFAT-GFP) (Fig. 3) or around the perimeter of the cell (e.g., LAT-GFP) can be used to assess reporter activation or localization as surrogates for TCR signaling [15, 22, 30].

4 Notes

1. Donor strain depends on the experimental question and setup. It may be useful to use TCR transgenic RAG-deficient donor mice. If studying thymic selection, this will increase the frequency of visualized cells undergoing positive selection or negative selection (in the presence of cognate antigen). TCR transgenic RAG-deficient donors will also be useful to study antigen-specific activation of mature T cells in the periphery in the absence of labeled B cells.
2. The number of mice to be treated depends on the number of pups to be injected. We typically recover ~0.5×10^6 cells from the femurs and tibiae of one 5-FU-treated mouse.
3. If using a reporter transgenic mouse model (e.g., *tetO-mCameleon; Rosa26*$^{\text{M2-rtTA}}$ mice [22]) to study TCR signaling or ubiquitously labeled cells (e.g., CAG-ECFP mice [42]) to study thymocyte or T cell behavior, 5-FU injection and retroviral transduction are not required prior to BM harvest and neonatal injection, respectively. The BM can be harvested as described in Subheading 3.2 and injected as described in Subheading 3.4. We suggest depleting T cells via i.p. injection of anti-mouse Thy1 antibody or magnetic bead depletion from harvested BM.
4. The recipient mouse strain depends on the experimental question. For example, if looking to distinguish the thymic cortex and medulla for two-photon imaging, CD11c-YFP and Act-EGFP recipients will allow the identification of different thymic microenvironments based on the density of labeled cells [41, 43–45]. In addition, it is possible to study TCR signaling

from OT-I or OT-II TCR transgenic mice undergoing thymic selection by transferring them into recipients in which cognate antigen (OVA) is expressed in a subset of cells (e.g., RIP-mOVA in which OVA is restricted to medullary thymic epithelial cells [46]).

5. We use a Zeiss Axio Examiner Z1 motorized upright microscope equipped with a Zeiss LSM880 scan head with spectral detector, a Spectra-Physics MaiTai DeepSee Ti:Saphire femtosecond pulsed laser, four non-descanned multialkali PMT detectors a water dipping W plan-APO 20×/1.0 WD = 1.8 objective.
6. 5-FU is a chemotherapeutic agent that should be opened only in a chemical fume hood or appropriate biosafety cabinet. It depletes lineage committed cells and leads to hematopoietic stem/progenitor cell cycling [47]. We describe i.v. injection of 5-FU. However, 5-FU can also be injected i.p. with similar results. Local animal facility guidelines should be followed to identify cages in which mice have been treated with 5-FU. As an alternative to 5-FU injection, it is possible to perform a magnetic bead depletion of lineage committed cells from harvested BM before infection to enrich for HPCs.
7. Depending on the number of cells needed and any limitations in the number of donor mice, it is also possible to collect BM from the upper limbs, sternum, and spinal cord. For the latter two, cut the bones into small pieces after removing other tissue and extract the cells using a mortar and pestle.
8. Phoenix cells (available from the ATCC) are a packaging cell line that stably express the *gag-pol* and *env* retroviral genes [48]. It is possible to use Phoenix cells transfected only with the expression plasmid instead of HEK293T cells.
9. It is possible to add fresh media at this point to continue production of viral particles for an additional 24 h. Alternatively, fresh viral particles can be produced in advance and stored at −80 °C until needed for BM infection.
10. A transduction rate of at least 25% of the BM cells should be achieved or the chimerism of fluorescent cells in this model will likely be too low for imaging. It is possible to sort transduced BM cells prior to injection to ensure a proper density of the cells of interest. To achieve low-level chimerism, a single injection is sufficient, but it is also possible to inject pups 2 or 3 times at 2–3 days intervals (e.g., 2, 4, and 6 days after birth) to increase chimerism levels. Alternatively, transduced cells can be diluted with untransduced BM cells to the desired level and i.v. injected into lethally irradiated WT or sublethally irradiated RAG-deficient adult recipients.

11. Prior to imaging tissue from weaned pups, it is possible to screen for chimerism levels in the peripheral blood (PB) if the donor cells can be discriminated from the recipient by flow cytometry. PB reconstitution is generally a good indicator of the chimerism in the tissue of interest (Fig. 2).
12. The subcapsular region/cortex of an intact thymic lobe can be imaged at this stage with the thymic lobe, concave side down, directly glued to a coverslip as per Subheading 3.5, **step 13**. In order to perform imaging of the deeper, central medulla region, however, it is necessary to slice the tissue prior to imaging.
13. Several imaging parameters will need to be empirically determined based on the sample to be imaged. We suggest setting all PMTs to the highest sensitivity while adjusting the laser to identify the optimal wavelength for the experiment and then subsequently adjusting the PMT sensitivity with the laser power at minimum. Lower exposure time and laser power as much as possible to levels that allow acquisition of images of suitable quality while limiting tissue damage/photobleaching. Notably, some acquisition software allows for adjustment of laser power at different *z* planes to compensate for fluorescence attenuation with tissue depth.

Acknowledgments

This work was supported by a Natural Sciences and Engineering Research Council grant (RGPIN-2019-05053) and Canadian Institutes of Health Research new investigator award (MSH-141967) to H.J.M., as well as a Fonds de recherche du Québec-Santé doctoral fellowship to M.D.

References

1. Germain RN, Robey EA, Cahalan MD (2012) A decade of imaging cellular motility and interaction dynamics in the immune system. Science 336(6089):1676–1681. https://doi.org/10.1126/science.1221063
2. Labrecque N, Dong M, Sood A, Melichar HJ (2018) In situ analysis of T cell receptor signals during positive selection. In: Soboloff J, Kappes DJ (eds) Signaling mechanisms regulating T cell diversity and function. CRC press/Taylor & Francis, Boca Raton, FL, pp 17–40 https://doi.org/10.1201/9781315371689-2
3. Au-Yeung BB, Shah NH, Shen L, Weiss A (2018) ZAP-70 in signaling, biology, and disease. Annu Rev Immunol 36:127–156. https://doi.org/10.1146/annurev-immunol-042617-053335
4. Neier SC, Ferrer A, Wilton KM, Smith SEP, Kelcher AMH, Pavelko KD, Canfield JM, Davis TR, Stiles RJ, Chen Z, McCluskey J, Burrows SR, Rossjohn J, Hebrink DM, Carmona EM, Limper AH, Kappes DJ, Wettstein PJ, Johnson AJ, Pease LR, Daniels MA, Neuhauser C, Gil D, Schrum AG (2019) The early proximal alphabeta TCR signalosome specifies thymic selection outcome through a quantitative protein interaction network. Sci Immunol 4(32):eaal2201. https://doi.org/10.1126/sciimmunol.aal2201

5. Daniels MA, Teixeiro E, Gill J, Hausmann B, Roubaty D, Holmberg K, Werlen G, Hollander GA, Gascoigne NR, Palmer E (2006) Thymic selection threshold defined by compartmentalization of Ras/MAPK signalling. Nature 444 (7120):724–729. https://doi.org/10.1038/nature05269
6. Dower NA, Stang SL, Bottorff DA, Ebinu JO, Dickie P, Ostergaard HL, Stone JC (2000) RasGRP is essential for mouse thymocyte differentiation and TCR signaling. Nat Immunol 1(4):317–321. https://doi.org/10.1038/79766
7. Kortum RL, Sommers CL, Pinski JM, Alexander CP, Merrill RK, Li W, Love PE, Samelson LE (2012) Deconstructing Ras signaling in the thymus. Mol Cell Biol 32(14):2748–2759. https://doi.org/10.1128/MCB.00317-12
8. Priatel JJ, Teh SJ, Dower NA, Stone JC, Teh HS (2002) RasGRP1 transduces low-grade TCR signals which are critical for T cell development, homeostasis, and differentiation. Immunity 17(5):617–627
9. Mariathasan S, Zakarian A, Bouchard D, Michie AM, Zuniga-Pflucker JC, Ohashi PS (2001) Duration and strength of extracellular signal-regulated kinase signals are altered during positive versus negative thymocyte selection. J Immunol 167(9):4966–4973
10. Alberola-Ila J, Hogquist KA, Swan KA, Bevan MJ, Perlmutter RM (1996) Positive and negative selection invoke distinct signaling pathways. J Exp Med 184(1):9–18
11. Zou Q, Jin J, Xiao Y, Hu H, Zhou X, Jie Z, Xie X, Li JY, Cheng X, Sun SC (2015) T cell development involves TRAF3IP3-mediated ERK signaling in the Golgi. J Exp Med 212 (8):1323–1336. https://doi.org/10.1084/jem.20150110
12. Kane LP, Hedrick SM (1996) A role for calcium influx in setting the threshold for CD4+CD8+ thymocyte negative selection. J Immunol 156(12):4594–4601
13. Lo WL, Donermeyer DL, Allen PM (2012) A voltage-gated sodium channel is essential for the positive selection of CD4(+) T cells. Nat Immunol 13(9):880–887. https://doi.org/10.1038/ni.2379
14. Macian F (2005) NFAT proteins: key regulators of T-cell development and function. Nat Rev Immunol 5(6):472–484. https://doi.org/10.1038/nri1632
15. Melichar HJ, Ross JO, Herzmark P, Hogquist KA, Robey EA (2013) Distinct temporal patterns of T cell receptor signaling during positive versus negative selection in situ. Sci Signal 6(297):ra92. https://doi.org/10.1126/scisignal.2004400
16. Au-Yeung BB, Melichar HJ, Ross JO, Cheng DA, Zikherman J, Shokat KM, Robey EA, Weiss A (2014) Quantitative and temporal requirements revealed for Zap70 catalytic activity during T cell development. Nat Immunol 15(7):687–694. https://doi.org/10.1038/ni.2918
17. Ross JO, Melichar HJ, Au-Yeung BB, Herzmark P, Weiss A, Robey EA (2014) Distinct phases in the positive selection of CD8+ T cells distinguished by intrathymic migration and T-cell receptor signaling patterns. Proc Natl Acad Sci U S A 111(25):E2550–E2558. https://doi.org/10.1073/pnas.1408482111
18. Ebert PJ, Ehrlich LI, Davis MM (2008) Low ligand requirement for deletion and lack of synapses in positive selection enforce the gauntlet of thymic T cell maturation. Immunity 29(5):734–745. https://doi.org/10.1016/j.immuni.2008.09.014
19. Bhakta NR, Oh DY, Lewis RS (2005) Calcium oscillations regulate thymocyte motility during positive selection in the three-dimensional thymic environment. Nat Immunol 6 (2):143–151. https://doi.org/10.1038/ni1161
20. Wei SH, Safrina O, Yu Y, Garrod KR, Cahalan MD, Parker I (2007) Ca2+ signals in CD4+ T cells during early contacts with antigen-bearing dendritic cells in lymph node. J Immunol 179 (3):1586–1594
21. Skokos D, Shakhar G, Varma R, Waite JC, Cameron TO, Lindquist RL, Schwickert T, Nussenzweig MC, Dustin ML (2007) Peptide-MHC potency governs dynamic interactions between T cells and dendritic cells in lymph nodes. Nat Immunol 8(8):835–844. https://doi.org/10.1038/ni1490
22. Le Borgne M, Raju S, Zinselmeyer BH, Le VT, Li J, Wang Y, Miller MJ, Shaw AS (2016) Real-time analysis of calcium signals during the early phase of T cell activation using a genetically encoded calcium biosensor. J Immunol 196 (4):1471–1479. https://doi.org/10.4049/jimmunol.1502414
23. Mues M, Bartholomaus I, Thestrup T, Griesbeck O, Wekerle H, Kawakami N, Krishnamoorthy G (2013) Real-time in vivo analysis of T cell activation in the central nervous system using a genetically encoded calcium indicator. Nat Med 19(6):778–783. https://doi.org/10.1038/nm.3180
24. Perez Koldenkova V, Nagai T (2013) Genetically encoded Ca(2+) indicators: properties and

evaluation. Biochim Biophys Acta 1833 (7):1787–1797. https://doi.org/10.1016/j.bbamcr.2013.01.011

25. Shulman Z, Gitlin AD, Weinstein JS, Lainez B, Esplugues E, Flavell RA, Craft JE, Nussenzweig MC (2014) Dynamic signaling by T follicular helper cells during germinal center B cell selection. Science 345(6200):1058–1062. https://doi.org/10.1126/science.1257861
26. Thestrup T, Litzlbauer J, Bartholomaus I, Mues M, Russo L, Dana H, Kovalchuk Y, Liang Y, Kalamakis G, Laukat Y, Becker S, Witte G, Geiger A, Allen T, Rome LC, Chen TW, Kim DS, Garaschuk O, Griesinger C, Griesbeck O (2014) Optimized ratiometric calcium sensors for functional in vivo imaging of neurons and T lymphocytes. Nat Methods 11 (2):175–182. https://doi.org/10.1038/nmeth.2773
27. Paredes RM, Etzler JC, Watts LT, Zheng W, Lechleiter JD (2008) Chemical calcium indicators. Methods 46(3):143–151. https://doi.org/10.1016/j.ymeth.2008.09.025
28. Melichar HJ, Li O, Herzmark P, Padmanabhan RK, Oliaro J, Ludford-Menting MJ, Bousso P, Russell SM, Roysam B, Robey EA (2011) Quantifying subcellular distribution of fluorescent fusion proteins in cells migrating within tissues. Immunol Cell Biol 89(4):549–557. https://doi.org/10.1038/icb.2010.122
29. Marangoni F, Murooka TT, Manzo T, Kim EY, Carrizosa E, Elpek NM, Mempel TR (2013) The transcription factor NFAT exhibits signal memory during serial T cell interactions with antigen-presenting cells. Immunity 38 (2):237–249. https://doi.org/10.1016/j.immuni.2012.09.012
30. Azar GA, Lemaitre F, Robey EA, Bousso P (2010) Subcellular dynamics of T cell immunological synapses and kinapses in lymph nodes. Proc Natl Acad Sci U S A 107(8):3675–3680. https://doi.org/10.1073/pnas.0905901107
31. Friedman RS, Beemiller P, Sorensen CM, Jacobelli J, Krummel MF (2010) Real-time analysis of T cell receptors in naive cells in vitro and in vivo reveals flexibility in synapse and signaling dynamics. J Exp Med 207 (12):2733–2749. https://doi.org/10.1084/jem.20091201
32. Ross JO, Melichar HJ, Halkias J, Robey EA (2016) Studying T cell development in thymic slices. Methods Mol Biol 1323:131–140. https://doi.org/10.1007/978-1-4939-2809-5_11
33. Sood A, Dong M, Melichar HJ (2016) Preparation and applications of organotypic thymic slice cultures. J Vis Exp 114. https://doi.org/10.3791/54355
34. Lancaster JN, Ehrlich LI (2017) Analysis of thymocyte migration, cellular interactions, and activation by multiphoton fluorescence microscopy of live thymic slices. Methods Mol Biol 1591:9–25. https://doi.org/10.1007/978-1-4939-6931-9_2
35. Calvo-Asensio I, Barthlott T, von Muenchow L, Lowndes NF, Ceredig R (2017) Differential response of mouse thymic epithelial cell types to ionizing radiation-induced DNA damage. Front Immunol 8:418. https://doi.org/10.3389/fimmu.2017.00418
36. Chung B, Barbara-Burnham L, Barsky L, Weinberg K (2001) Radiosensitivity of thymic interleukin-7 production and thymopoiesis after bone marrow transplantation. Blood 98 (5):1601–1606
37. Zhang SL, Wang X, Manna S, Zlotoff DA, Bryson JL, Blazar BR, Bhandoola A (2014) Chemokine treatment rescues profound T-lineage progenitor homing defect after bone marrow transplant conditioning in mice. Blood 124(2):296–304. https://doi.org/10.1182/blood-2014-01-552794
38. Dzhagalov IL, Melichar HJ, Ross JO, Herzmark P, Robey EA (2012) Two-photon imaging of the immune system. Curr Protoc Cytom Chapter 12: Unit12 26. https://doi.org/10.1002/0471142956.cy1226s60
39. Ladi E, Herzmark P, Robey E (2008) In situ imaging of the mouse thymus using 2-photon microscopy. J Vis Exp 11. https://doi.org/10.3791/652
40. Witt CM, Raychaudhuri S, Schaefer B, Chakraborty AK, Robey EA (2005) Directed migration of positively selected thymocytes visualized in real time. PLoS Biol 3(6):e160. https://doi.org/10.1371/journal.pbio.0030160
41. Le Borgne M, Ladi E, Dzhagalov I, Herzmark P, Liao YF, Chakraborty AK, Robey EA (2009) The impact of negative selection on thymocyte migration in the medulla. Nat Immunol 10(8):823–830. https://doi.org/10.1038/ni.1761
42. Hadjantonakis AK, Macmaster S, Nagy A (2002) Embryonic stem cells and mice expressing different GFP variants for multiple non-invasive reporter usage within a single animal. BMC Biotechnol 2:11
43. Lindquist RL, Shakhar G, Dudziak D, Wardemann H, Eisenreich T, Dustin ML, Nussenzweig MC (2004) Visualizing dendritic cell networks in vivo. Nat Immunol 5 (12):1243–1250. https://doi.org/10.1038/ni1139

44. Wright DE, Cheshier SH, Wagers AJ, Randall TD, Christensen JL, Weissman IL (2001) Cyclophosphamide/granulocyte colony-stimulating factor causes selective mobilization of bone marrow hematopoietic stem cells into the blood after M phase of the cell cycle. Blood 97(8):2278–2285
45. Ehrlich LI, Oh DY, Weissman IL, Lewis RS (2009) Differential contribution of chemotaxis and substrate restriction to segregation of immature and mature thymocytes. Immunity 31(6):986–998. https://doi.org/10.1016/j.immuni.2009.09.020
46. Kurts C, Heath WR, Carbone FR, Allison J, Miller JF, Kosaka H (1996) Constitutive class I-restricted exogenous presentation of self antigens in vivo. J Exp Med 184(3):923–930
47. Randall TD, Weissman IL (1997) Phenotypic and functional changes induced at the clonal level in hematopoietic stem cells after 5-fluorouracil treatment. Blood 89 (10):3596–3606
48. Swift S, Lorens J, Achacoso P, Nolan GP (2001) Rapid production of retroviruses for efficient gene delivery to mammalian cells using 293T cell-based systems. Curr Protoc Immunol Chapter 10: Unit 10 17C. https://doi.org/10.1002/0471142735.im1017cs31

Chapter 19

An Integrated Strategy for Identifying Targets of Ubiquitin-Mediated Degradation in CD4+ T Cells

Natania S. Field, Claire E. O'Leary, Joseph M. Dybas, Hua Ding, and Paula M. Oliver

Abstract

Ubiquitination is a crucial component of many immune processes. While ubiquitin-mediated degradation is essential to T cell activation via T cell receptor signaling, the specific E3 ligases and substrates involved are not well-understood. Here, we describe a strategy integrating RNA, protein, and posttranslational modification datasets to identify targets of ubiquitin-mediated degradation. When integrated, these assays can provide broad insight into how this posttranslational modification regulates protein function and influences T cell biology.

Key words Ubiquitin, T cell, Lymphocyte, Diglycine remnant profiling, Signaling

1 Introduction

Ubiquitin-mediated degradation is required for many aspects of T cell function. Mutations in E3 ubiquitin ligases or their activators lead to dramatic changes in T cell differentiation [1–3]. Furthermore, T cell activation is blocked when the proteasome is inhibited by MG132 [4]. However, our understanding of proteins that undergo ubiquitin-mediated degradation has been limited by lack of technologies that broadly query these events.

Our laboratory recently developed a strategy for identifying and validating targets of ubiquitin-mediated degradation using combined transcriptomic and proteomic analyses. By comparing whole-cell proteomic changes to whole-cell transcriptomic changes, we can generate a list of proteins that are downregulated at the protein but not mRNA level, indicating they are likely to undergo protein degradation. Additionally, Steven Carr's group has developed a technique to identify ubiquitinated proteins following trypsin digest, which reveals a diglycine (K-ε-GG) remnant; K-ε-GG-containing proteins can then be enriched using anti-

Chaohong Liu (ed.), *T-Cell Receptor Signaling: Methods and Protocols*, Methods in Molecular Biology, vol. 2111, https://doi.org/10.1007/978-1-0716-0266-9_19,

diglycine antibody, and identified using mass spectrometry [5, 6]. Cross-referencing these ubiquitination candidate proteins with degradation candidates generates a list of potential targets of ubiquitin-mediated degradation. These targets can be validated experimentally, for example by measuring protein half-life using cycloheximide treatment followed by Western blotting. Furthermore, performing a pulldown with a tandem ubiquitin-binding entity (TUBE) and blotting for the protein of interest can reveal whether the protein is ubiquitinated.

While we have used this strategy to screen for substrates degraded in stimulated CD4 T cells as compared to cells in a resting state [4], it can be adapted to screen for targets of specific E3 ligases by comparing wild-type cells to cells deficient for the ligase of interest. These studies enable us to understand the biological function of E3 ubiquitin ligases by revealing their substrates.

2 Materials

2.1 T Cell Preparation

1. Dissection tools (forceps and scissors).
2. 60 mm Dishes.
3. 70 μm Filters.
4. 1 ml Syringes.
5. 50 ml Conical tubes.
6. 10 cm Tissue culture plates.
7. L3T4 CD4 T cell enrichment kit (Miltenyi, 130-117-043).
8. Phosphate-buffered saline (PBS).
9. Fetal calf serum (our lab uses premium FCS from Atlanta Biologicals; *see* **Note 1**).
10. MACS buffer: phosphate-buffered saline, 0.5% FCS, 2 mM ETDA.
11. LS columns (Miltenyi 130-042-401).
12. 15 ml Conical tubes.
13. MACS magnet.
14. Trypan blue.
15. Hemocytometer.
16. Light microscope.
17. Red blood cell (RBC) lysis buffer (0.15 M NH_4Cl, 10 mM $KHCO_3$, 1 mM EDTA, pH 7.3).
18. T cell culture media: RPMI 1640 (HyClone, SH30027.01), 10% FCS, 100 U/ml penicillin-streptomycin (Gibco, 15140122), 1× Glutamax (ThermoFisher, 35050061), 1× nonessential amino acids (Gibco, 11140-050), 2 mM HEPES (Gibco, 15630-080), 1 mM sodium pyruvate (Corning,

20115013), 0.008 μl/ml 2-mercaptoethanol (Sigma, M6250).

19. Recombinant human IL-2 (PeproTech 200-02).
20. Anti-CD28 (Clone 37.51; BioLegend).
21. Anti-CD3 (Clone 145-2C11; BioLegend).
22. Dynabeads (ThermoFisher, 11456D).

2.2 K-ε-GG Pulldown

Note that the materials required for the K-ε-GG pulldown are described in Udeshi et al. [5]. Our protocol requires the following additional reagents:

1. K-ε-GG lysis buffer: 8 M urea, 75 mM NaCl, 50 mM Tris–HCl pH 8.0, 1 mM EDTA, 2 μg/ml aprotinin (Sigma, A6103), 10 μg/ml leupeptin (Roche, 11017101001), 1 mM PMSF (Sigma, 78830), 10 mM NaF, 25 μM PR619, 5 mM iodoacetamide (Sigma, A3221).
2. Pierce micro BCA assay kit (ThermoFisher, 23225).
3. Diglycine antibody (Cell Signaling, 5562).
4. RapiGest (Waters, 186001860).
5. Vacuum manifold (Waters, 186001831).
6. 96 well Oasis HLB μElution plate (Waters, 186001828BA).

2.3 Whole-Cell Proteome Preparation

Details on performing an in-gel digest can be found in Shevchenko et al. [7]. Preparing samples requires the following reagents:

1. Laemmli sample buffer (4×).
2. DTT.
3. 10% NuPage™ acrylamide gel (ThermoFisher, NP0301PK2).
4. NuPage™ running buffer (ThermoFisher, NP0001).
5. Western blotting apparatus.
6. Power supply.
7. Excision template.
8. Unstained protein ladder.
9. Coomassie stain.

2.4 RNA-Seq Preparation

1. Qiagen RNeasy Plus Mini kit (Cat: 74134).
2. RNaseZap™ wipes (ThermoFisher, AM9786).
3. Filter pipette tips.
4. Agilent bioanalyzer.

2.5 Data Analysis

1. R software, including the following packages:
   (a) DESeq2 [8].
   (b) ggplot2 [9].
2. STAR aligner [10] (https://github.com/alexdobin/STAR).

2.6 Validation

(*See* Subheading 2.1 for materials needed for T cell extraction and culture.)

1. Lysis buffer for TUBE pulldown: RIPA (ThermoFisher, 89900) supplemented with 50 mM Tris–HCl, 0.15 M NaCl, 1 mM EDTA, 1% NP40, 10% glycerol, 1× Complete EDTA-free protease inhibitor dissolved in 300 μl Milli-Q water for a 20× stock (Roche, 11873580001), 0.4 mM, PR619, 200 mM 1,10-phenyanthroline (*o*-PA).
2. Lysis buffer for cycloheximide chase: RIPA (ThermoFisher, 89900) supplemented with 1× HALT phosphatase inhibitor (ThermoFisher, 78420) and 1× Complete EDTA-free protease inhibitor (*See* **item 1**).
3. Cycloheximide (Sigma, C4859).
4. MG132 (10 mM for a 1:1000 stock solution).
5. Chloroquine (Sigma, C6628—dissolve 0.41 g in 10 ml PBS for a 1:1600 stock).
6. UbiTest Agarose-TUBE Elution Kit (LifeSensors, UM411).
7. Uncoupled agarose (LifeSensors, UM400).
8. Deubiquitinating enzymes (DUBs): USP2 core pan-DUB (LifeSensors, DB501) and/or Otub1 K48-specific DUB (LifeSensors, DB201).
9. Tris-buffered saline with 0.1% tween (TBS-T).
10. SDS sample buffer.
11. 4–12% NuPage™ gels.
12. NuPage™ running buffer (ThermoFisher, NP0001).
13. NuPage™ Transfer buffer (ThermoFisher NP0006).
14. Western blotting apparatus.
15. Power supply.
16. Odyssey blocking buffer (Li-Cor, 927-50000).
17. Antibodies for protein(s) of interest.

3 Methods

3.1 Preparing Samples for Mass Spectrometry and RNA Sequencing

1. Dishes should be prepared prior to starting dissection, using sterile technique in a biosafety cabinet, and kept on ice. *See* Fig. 1 for a workflow of sample preparation.
2. Harvest spleen and peripheral lymph nodes (LNs) from mice on C57BL6J background euthanized by CO_2 asphyxiation following IACUC-approved standards. Mice should be older than 8 weeks of age, depending on experimental design. For ubiquitin remnant profiling, multiple mice are required. Place

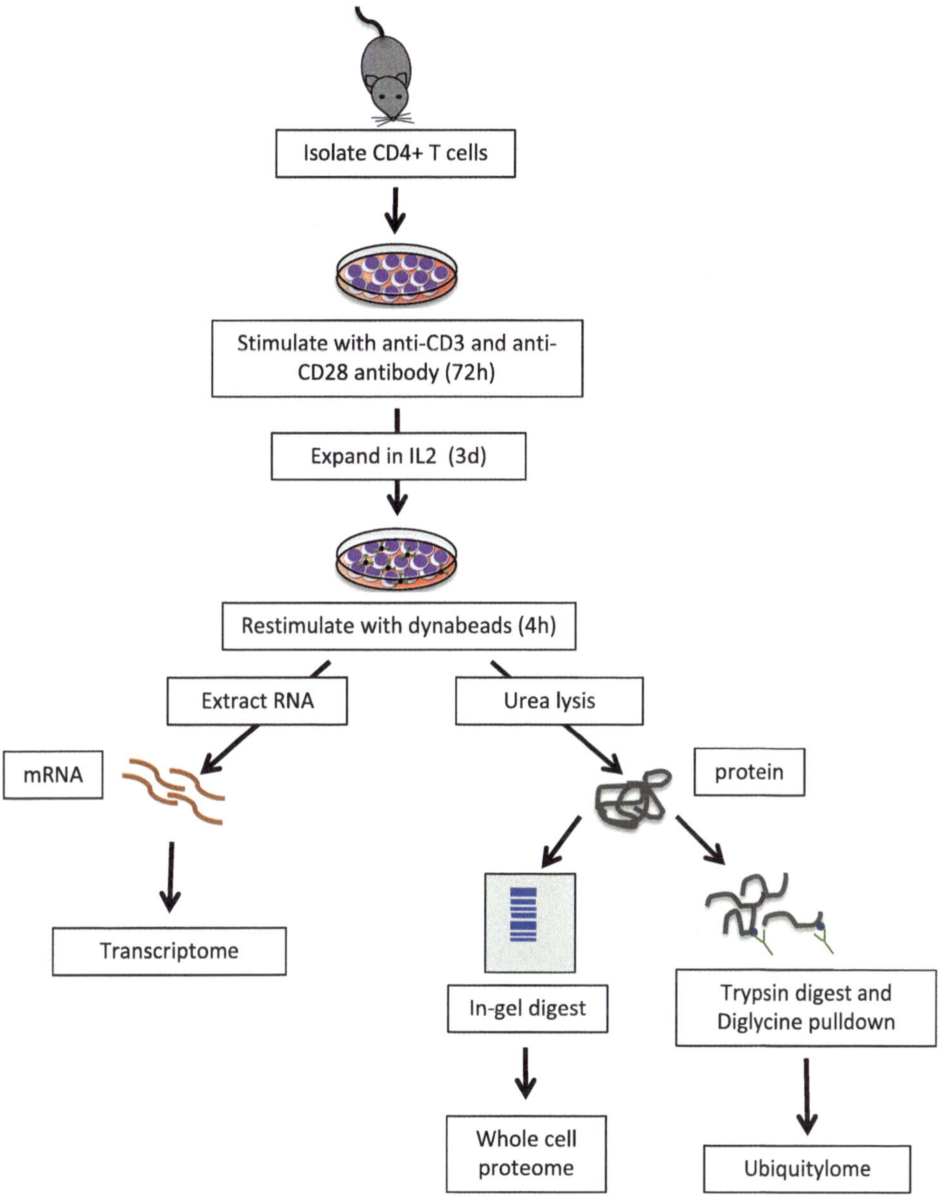

Fig. 1 Workflow for sample preparation

harvested spleen and LNs into 70 μM cell strainer which has been placed in a plastic 60 mm petri dish with 5 ml of RPMI supplemented with 1% fetal calf serum (FCS) (*see* **Notes 1–6**).

3. Coat tissue-culture-coated plates with anti-CD3 and anti-CD28 at 5 ng/ml in sterile PBS. Use enough solution to cover the bottom of the plates. Coat plates at least 2 h at 37 °C in a tissue culture incubator. This incubation can take place during cell preparation.

4. In sterile biosafety hood, mash the tissue through the strainer with the flat end of a plunger from a 1 ml syringe (opened sterilely in hood). All following steps should be performed under sterile conditions, with pre-sterilized serological pipets, tips, etc.
5. Pipet the cell solution through the filter into a 50 ml conical tube. Wash the filter twice with 5 ml of RPMI supplemented with 1% FCS.
6. Pellet cells at 1300 rpm for 5 min and 4 °C in a tabletop centrifuge.
7. Decant supernatant. LNs may be resuspended directly into CD4+ T cell isolation media (i.e., MACS buffer). Lyse red blood cells in spleen samples by resuspending the cells in 1–2 ml of red blood cell (RBC) lysis buffer. Quench lysis with MACS buffer after 30 s, and pellet cells immediately. Resuspend the splenocytes using LN cell solution for the same set of animals, filter through a 70 μm, filter, and pellet cells again.
8. Proceed to CD4+ T cell enrichment using CD4+ T cell microbeads (L3T4) following Miltenyi manufacturer instructions for enrichment with LS columns or automated selection (max capacity of total cells: 2×10^9; max capacity of labeled cells: 1×10^8). The spleen and LNs from about three uninflamed wild-type mice can be used in one column. MACS buffer should be prepared as indicated in the enrichment protocol, including degassing (*see* **Note 7**).
9. (*Recommended*) Stain an aliquot of cells with a lineage cocktail (CD11b, CD11c, FCERG1, Ter119, CD8b, GR1, Ly6G, Ly6C), CD3 (17A2), CD4, CD8a, and viability dye to assess purity by flow cytometry.
10. Resuspend cells in complete T cell media containing 50 U/ml of IL-2. Remove a small aliquot (about 10 μl) to count cells. Adjust volume of the cell resuspension to $1–1.5 \times 10^6$ cells/ml.
11. Aspirate the PBS from the coated plate, rinse plate once with sterile PBS, and add the cell mixture. Cell mixture should be added immediately to avoid drying the plate. Incubate at 37 °C 5% or 10% CO_2 (according to the laboratory's standard culturing conditions) for 72 h, or as called for in experimental workflow.
12. After 72 h, examine cells visually for signs of blasting and proliferation using a light microscope. Gently harvest cells from plates by gentle pipetting through a serological pipette. Wash plate with cold HBSS or PBS to collect residual cells. Examine the plate by microscope to ensure removal of cells.

13. Pellet cells, and resuspend in fresh culture media with IL-2 (50 U/ml). Count cells and dilute to 1–1.5 × 10^6/ml before plating on 10 or 15 cm tissue-culture-treated plates. If polarizing conditions were used, verify T cell differentiation by performing a brief restimulation of a cell aliquot with PMA/ionomycin in the presence of Brefeldin A for intracellular cytokine staining.
14. Split cells 1:1 with media containing IL-2 on the two consecutive days (*see* **Note 8**).
15. On the third day, harvest cells, count, and resuspend at 4 × 10^6/ml for short restimulation or other experimental modifications. Restimulation should be performed using Dynabeads, at a minimum of 1:3 beads/cell ratio in complete T cell media (*see* **Notes 9** and **10**).
16. On an aliquot of rested and restimulated samples, stain for CD3, CD4, and CD69 to obtain purity and activation status.
17. Pellet cells and beads in microcentrifuge tubes (3–5 min ~8000 rpm spin in microcentrifuge). Aspirate the supernatant, and freeze pellets at −80 °C for lysis. Be sure to freeze in appropriate aliquots for experimental purposes. 1 × 10^6 is sufficient for RNA analysis, and K-ε-GG analysis requires ~250–300 million cells for sufficient protein.

3.2 K-ε-GG Pulldown

This K-ε-GG analysis method closely follows the method described by Steven Carr's group [6] with modifications described here. Prior to beginning enrichment, crosslink antibody to beads and verify by gel electrophoresis.

1. Perform cell lysis using a slightly adapted lysis buffer (*see* Subheading 2.2, **item 1**).
2. Perform a BCA analysis on lysate to determine protein concentration. Use the manufacturer's protocol.
3. Save about 30 μg of protein from the lysate for proteomic analysis (*see* **Note 11**).
4. Proceed to the immunoprecipitation, following the published protocol with the modifications described below. Note that 31ug of crosslinked antibody was used for our experiments, but this ratio may need to be determined empirically for different amounts of input protein.
5. After the immunoprecipitation, wash the bead/antibody complex as described; however for the third wash, supplement the IAP with 0.05% RapiGest. Continue with three PBS washes.
6. Perform elution as described.
7. Condition a 96 well Oasis HLB μElution plate using 100% acetonitrile (200 μl per well) under a Waters vacuum manifold.

8. Equilibrate the plate using 2× 200 µl washed of water with 0.1% trifluoroacetic acid (TFA).
9. Apply the eluted peptides to the plate.
10. Wash 3× 200 µl of water with 0.1% TFA.
11. Elute the peptides with 3× 80 µl 80% acetonitrile with 0.1% TFA.

3.3 Whole-Cell Proteome Preparation

Details on performing an in-gel digest can be found in Shevchenko et al. [7]. We use the following method to prepare the samples:

1. Mix proteomic sample (about 30 µg lysate) 1:1 with 4× Laemmli buffer containing DTT.
2. Fractionate using a 10% acrylamide gel (without a stacking gel). Mark gel cassette at the tenth fraction using a discarded excision template (~2 cm past well).
3. Run sample with unstained protein ladder at 120 V until dye front has passed the tenth fraction mark to get approximately 10–12 fractions. Additional fractions will yield a more complex proteome.
4. Stain gel in high-sensitivity Coomassie [11].

3.4 RNA-Seq Preparation

1. Protect RNA samples from degradation. Wipe down all surfaces with the RNaseZap wipes to remove any RNases, and do not allow samples to leave the clean area. Use filter pipette tips for all RNA isolation steps, and always use gloves when handling any of the reagents.
2. Perform RNA isolation using the Qiagen RNeasy Plus Mini kit, following the kit protocol exactly.
3. RNA quality can be assessed using an Agilent bioanalyzer. RIN greater than eight is optimal. Quantity should be at least 200 ng per sample at a concentration of at least 10 ng/µl in RNase-free water.
4. RNA-seq can be carried out with Illumina technology using paired-end reads,100 base pair length, at a minimum of 30,000,000 read depth.

3.5 RNA-Seq Data Processing

There are a variety of alignment algorithms that can be implemented to align the RNA-seq reads to the reference genome [12]. This protocol describes the implementation of the STAR alignment program [10] and DESeq2 [8] to calculate differential expressions:

1. Obtain the reference genome to which the RNA-seq data will be aligned. The genome can be downloaded to a local directory from a genome browser such as Ensembl (https://www.ensembl.org/). The reference genome should match the species from which the RNA was harvested. The genomic

sequence data is stored in fasta (*.fa) files and comprises one file for each chromosome in the genome. The corresponding *.gtf annotation file must also be downloaded with the sequence information (*see* **Note 12**).

2. Generate genome indexes of the reference genome using the STAR program. The STAR program is run using the "genomeGenerate" switch for the "--runMode" option. The users must also specify the path to the read files and annotation file using the options "--genomeFastaFiles" and "--sjdbGTFfile," respectively.
3. Map the RNA-seq reads to the reference genome using the STAR program. The STAR program is run with the paths to the genome index directory and the read files specified using the options "--genomeDir" and "--readFilesIn," respectively. Additionally, it is advantageous to use the option "--quantMode GeneCounts," which will provide an output with the counts of aligned reads per gene. This data will be used to compute differential expression and associated statistics.
4. Input the read count data into the R statistical software package, and format the data to generate a count matrix where the columns are samples, the rows are the reference genes, and the data comprises the un-normalized read counts generated by STAR. For example, samples identified by the identifications "ctrl_rep_1," "ctrl_rep_2," "knockout_rep_1," and "knockout_rep_2" would be in the format:

#		ctrl_ rep_1	ctrl_ rep_2	knockout_ rep_1	knockout_ rep_2
#	ENSgene0001	count	count	count	count
#	ENSgene0002	count	count	count	count
#	ENSgene0003	count	count	count	count
#	ENSgene0004	count	count	count	count

5. Filter the count matrix to remove genes that contain counts below a specified threshold.
6. Generate a table relating the sample identification to the sample description ("condition"). For example, samples identified by identifications "control replicate 1," "control replicate 2," "knockout replicate 1," and "knockout replicate 2" for conditions "control" and "knockout" would be in the format:

#		Condition
#	control replicate 1	control
#	control replicate 2	control
#	knockout replicate 1	knockout
#	knockout replicate 2	knockout

7. Input the data into DESeq2 using the command "DESeqDataSetFromMatrix."
8. Calculate differential expression and associated statistics using the command "DESeq."
9. Generate the results of DESeq2 differential expression using the command "results." The results output the base mean, log2 fold change, log2 standard error, the Wald statistic, the Wald *p*-value, and Benjamini-Hochberg-adjusted *p*-value. The log2 fold changes and adjusted *p*-values are used to identify differentially expressed transcripts.
10. A volcano plot is a common and understandable way of representing differential expression data. A volcano plot graphs each identified transcript log2 fold change (*x*-axis) and corresponding log transformed adjusted *p*-value (*y*-axis) to visualize the magnitude of change and associated significance of expressed genes. Volcano plots can be generated using the software package ggplot2.

3.6 Whole-Cell Proteome Analysis

This protocol describes a mass spectrometry experiment consisting of two conditions with three biological replicates for each condition. Protein abundance is measured by label-free quantification. The raw data must be searched against the appropriate proteome with a program such as MaxQuant [13] or Open Mass Spectrometry Search Algorithm (OMSAA) [14]:

1. The output of the mass spectrometry search algorithm (e.g., the proteinGroups.txt file generated by MaxQuant) should be read into the preferred data manipulation/data analysis software (R, Excel, etc.). The output is composed of mass spectrometry data for which each identified protein is one row and the data for each protein is given in the respective columns. The relevant data for the differential protein expression analysis is the quantification parameter (iBAQ, LFQ, Intensity, etc.).
2. Remove the predicted contaminant proteins and reverse proteins from the dataset.
3. Convert all cases for which a protein was not quantified for the abundance measurement (a "blank" or a zero entry) to "NA" values.

4. Transform the protein abundance data into log base 2 form. NA values should remain NA.
5. Calculate the median or mean of the log2 transformed abundances for each sample in the experiment. NA values should not be included in this calculation as to not affect the average of the abundances of the identified proteins within the sample (*see* **Note 13**).
6. Normalize by adjusting the median (or mean) of the sample to "0." Subtract the sample median (or mean) from each log2 transformed abundance value within the respective sample. This normalization accounts for technical variation, such as amount of protein input, in the mass spectrometry data across each replicate. The normalization by sample median (mean) allows for protein abundance comparison across samples.
7. Remove proteins with less than two out of three replicates identified for both conditions (*see* **Note 14**).
8. Calculate a *z*-score of the abundance for each protein by subtracting the protein abundance from the sample mean abundance and dividing by the sample standard deviation. The *z*-score characterizes the relative abundance of each protein within a sample. Proteins with *z*-scores of approximately zero exhibit average abundance levels. Proteins with high or low (negative) *z*-score exhibit relatively high or low abundance, respectively, within the sample.
9. Average protein abundance across replicates within each condition. Protein abundance fold changes measure the differential abundance of a respective protein in each condition. Log2 fold changes are calculated by subtracting protein log2 transformed abundance values of a "control" sample from the protein log2 transformed abundance of a "test" sample. Fold changes calculated in the log2 can be interpreted as increased abundance in the "test" condition for positive log2 fold changes and increased abundance in the control condition for negative log2 fold changes. Log2 fold changes near zero indicate approximately equal abundance in each condition.
10. Perform a *t*-test (two-tailed) using log2 transformed abundance values for each identified replicate comparing the "control" and "test" conditions. The null hypothesis assumes no change in abundance between the conditions, and a *p*-value <0.05 is generally considered statistically significant (*see* **Note 15**).
11. Protein differential expression is assessed by considering fold change and corresponding statistical significance.
12. A volcano plot is a common and understandable way of representing differential expression data. A volcano plot graphs each

identified protein log2 fold change (x-axis) and corresponding log transformed p-value (y-axis) to visualize the magnitude of change and associated significance of expressed protein.

3.7 K-ε-GG Enrichment Analysis

The K-ε-GG analysis is very similar to the whole-cell proteome analysis. The major difference is that the K-ε-GG data abundance is generated on the peptide level rather than the protein level for whole-cell proteomics. However, the normalization, filtering, fold change, and statistics are computed the same way as described for the whole-cell proteome, except these calculations are done on the peptide level (i.e., using peptide abundance):

1. Peptide-based K-ε-GG modification abundance fold changes are converted to a protein-based fold change by calculating a weighted average of the peptide fold changes for each peptide identified within the respective protein. The peptide-based fold changes are weighted by modified peptide abundance, as measured by intensity level, such that the most abundant peptides constitute the highest impact to the protein fold change.
2. K-ε-GG modification fold changes are weighted by total protein changes, as determined from the corresponding whole-cell proteome, if a corresponding whole-cell proteome is available. The weighting is performed to determine whether the K-ε-GG fold change is driven by increases or decreases in total protein abundance. If there are increases or decreases in total protein abundance, the possibility of enriching for a K-ε-GG peptide from that protein is similarly increased or decreased. Differential changes in K-ε-GG abundance, after normalization for protein abundance, are used to classify increases or decreases in modification for purposes of identifying ubiquitination changes between the tested conditions.
3. Hypothesis testing is implemented to assess the significance of the changes observed in protein-based fold changes across conditions. The one sample t-test (two-tailed) is performed using the protein-based K-ε-GG fold change. The null hypothesis assumes no change in abundance between the conditions, and a p-value <0.05 is generally considered statistically significant (*see* **Note 15**).
4. A volcano plot is a common and understandable way of representing differential expression data. A volcano plot graphs each identified protein log2 fold change (x-axis) and corresponding log transformed p-value (y-axis) to visualize the magnitude of change and associated significance of expressed protein.

3.8 Validating Substrate Ubiquitination

We have used the UbiTest Agarose-TUBE elution kit from LifeSensors to detect ubiquitination. The enriched ubiquitinated proteins can be treated with deubiquitinating enzymes (DUBs) that cleave all ubiquitin chains, allowing for a single band on a gel to identify a protein. The enriched proteins can also be treated with cleavage-specific DUBs to identify the type of ubiquitin chains. This protocol includes an option for a K48-specific DUB. For the enrichment, we recommend the manufacturer's protocol with the following modifications:

1. Isolate and stimulate cells as described in Subheading 3.1, corresponding to the methods used for RNA-seq and proteomics data collection. Harvest cells in cold PBS, and wash three times. Pellet cells, and lyse immediately, or store at −80 °C.
2. Lyse cells in a Pierce RIPA (ThermoFisher)-based lysis buffer (*see* Subheading 2.6, **item 1**). Lyse cells in approximately 500 μl lysis buffer per 50×10^6 cells.
3. Lyse for 30 min on ice; vortex every 10 min.
4. Following incubation, centrifuge lysate for 20 min at maximum speed. Transfer supernatant to a clean tube (this is your lysate).
5. Perform BCA assay to test protein concentration.
6. Use 3–4 mg total protein as input for the pan-TUBE enrichment (approximately $100–150 \times 10^6$ cells)—keep in mind this is for three enrichments.
7. Equilibrate 150 μl slurry for the uncoupled agarose according to the manual. It is crucial to keep spin speed to 1000 × *g*.
8. Dilute input lysate 1:10 in PBS (or MilliQ). Add 0.1 mM PR619 and 50 μM *o*-PA and protease inhibitors at 1× concentrations.
9. Equilibrate 150 μl slurry for the TUBE-agarose according to the manual. It is crucial to keep spin speed to 1000 × *g*.
10. Add TUBE-agarose to the diluted lysate, and incubate at 4 °C, with rotation, for approximately 12 h.
11. Following incubation, spin agarose 1000 × *g* for 5 min. Collect supernatant for the unbound fraction if desired.
12. Wash agarose 3× with 1–2 ml TBST (0.1% tween). Following each wash collect beads by centrifuging at 1000 × *g* for 5 min, discard supernatant.
13. Resuspend agarose in 200–400 μl of supplied wash buffer.
14. Use desktop vortexer set to lowest speed to shake for 15 min at room temperature.
15. Centrifuge at 1000 × *g* for 5 min; discard supernatant.
16. Resuspend agarose in 150 μl of supplied elution buffer, and add 0.1 mM PR619 and 50 μM *o*-PA.
17. Use desktop vortexer set to lowest speed to shake for 25 min at room temperature.

18. Centrifuge at 4500 × *g* for 5 min and collect supernatant—this is your enrichment.
19. Neutralize supernatant by adding supplied 10× neutralization buffer at appropriate volume.
20. If using DUBs, split sample into three equal parts:
    (a) Leave one sample untreated.
    (b) Add USP2core pan-DUB to second sample.
    (c) Add Otub1 K48-specific DUB to third sample.
21. Incubate samples at 30 °C for 1.5 h.
22. Stop reaction by adding SDS sample buffer and boiling at 100 °C for 3–5 min. Your samples are now ready to load (see next section for Western blotting recommendations).

3.9 Validating Substrate Degradation

Since the proteomics studies require cells that have been expanded in vitro, using expanded, restimulated cells for initial validation studies is recommended. However, if you are interested in measuring degradation at earlier time points, it is also possible to treat with protein synthesis and degradation inhibitors within the first 24 h after activation:

1. Coat a plate with anti-CD3 anti-CD28 antibody as described in Subheading 3.1, **step 3**. We use 500 μl of this mixture per well of a 24 well plate or 200 μl per well of a 48 well plate. Incubate the plate at 37 °C for at least 2 h (this can be during the cell isolation).
2. Harvest lymph nodes and spleens from mice as described in Subheading 3.1 (**steps 1–6**), and proceed to magnetic separation in **step 7**:
    (a) When focusing on time points after the first 48 h of initial activation, we recommend using the Miltenyi CD4+ microbeads (*see* **Note 7**).
    (b) If testing for degradation within the first 48 h of activation, we recommend using the Miltenyi naïve CD4+ negative selection kit, in order to synchronize activation of the isolated cells.
3. Resuspend the cells in complete media with 50 U/ml IL-2 at a concentration of 1–1.5 million cells per ml.
4. Remove the coated plate, and aspirate off PBS.
5. Add the cell mixture to the plate (1–2 ml for a 24 well plate; 500 μl to 1 ml for a 48 well plate).
6. Spin the plate at 300 × *g* for 1 min to bring the cells to the bottom of the plate.
7. Place in the incubator for 72 h (if expanding/restimulating the cells) or for the desired time point.

8. Inhibiting protein synthesis or degradation:
   (a) For each experiment, at least four conditions are needed:
      i. An untreated sample collected prior to any drug treatments.
      ii. An untreated sample allowed to remain in culture for the duration of the experiment (any protein changes between samples i and ii are reflective of the "normal" behavior of this protein).
      iii. A cycloheximide-treated sample (to measure protein changes when new protein synthesis is blocked. Use cycloheximide stock at 1:10,000).
      iv. An MG132- and chloroquine-treated sample (to measure protein changes when protein degradation is inhibited). Use 10 mM MG132 at 1:1000 (for 10 μM final concentration) and chloroquine at 1:1600 (dissolved as described in Subheading 2.6, **item 5**).
      v. Since chloroquine inhibits lysosomal degradation and MG132 inhibits proteasomal degradation, you can opt to use these drugs in two separate conditions to differentiate between these mechanisms.
   (b) It is predicted that targets of ubiquitin-mediated degradation will decrease in abundance (relative to untreated) when treated with inhibitors of protein synthesis (e.g., cycloheximide) but will remain stable or increase when treated with inhibitors of degradation (e.g., MG132 and chloroquine) (*see* **Note 16**).
   (c) Do not treat samples with drugs for longer than 6 h as cells will start to lose viability at this time.
   (d) After desired treatment time (between 2 and 6 h), collect samples, and pellet for Western blot. They can be stored at −80 °C for several months.
9. Lyse samples in RIPA buffer supplemented with HALT phosphatase inhibitors (1:100) and 1× Roche Complete ETDA-free tablet (Described in Subheading 2.6, **item 2**).
10. Lyse on ice for 20 min; vortex occasionally.
11. Spin samples down at maximum speed for 10 min, and transfer supernatant to a new tube (this is your lysate).
12. Perform BCA assay to measure protein content.
13. Add 4× or 6× Laemmli sample buffer with 1:5 DTT added to each sample. Boil at 95 °C for 5 min. Your samples are now ready to load.
14. We typically use NuPage 4–12% precast gels, but other gel types may be appropriate depending on the size of your protein

of interest. Run gel for 2 h at 120 V using NuPage running buffer diluted to 1×.

15. Transfer: We typically use PVDF membranes activated in methanol and transfer in 1× NuPage transfer buffer with 10% methanol on ice. Run transfer for 75 min at 45 V (may need to adjust for your protein).
16. Block in Odyssey blocking buffer diluted 1:1 with PBS for at least 30 min.
17. Blot with primary antibody overnight in Odyssey blocking buffer diluted 1:1 with PBS at 4 °C.
18. Wash 3× for 10 min each with PBS+ 0.1% tween.
19. Incubate with secondary antibody for 1 h at room temperature.
20. Repeat wash step.
21. Expose membrane using a Li-Cor imager or traditional film methods.

4 Notes

1. It is recommended to test reagents before expanding a large number of cells. Over the course of the stimulation and expansion, CD3+CD4+ T cells should retain their phenotype with minimal presence of contaminating populations (less than 5% desired), and cells should expand approximately tenfold in number.
2. Quantification of proteomic data can be through labeling or label-free methods. Common labeling strategies include stable isotope labeling with amino acids (SILAC) or alternative labeling such as TMT [15]. Note that no ratiometric quantification is possible if desired protein/peptide is absent from one sample. Additionally, for SILAC labeling, it is recommended to perform a label switch experiment as modified carbon isotopes can impact cellular phenotypes.
3. Consider the reproducibility of experimental manipulations to determine which stage is appropriate for sample pooling.
4. For harvesting lymph nodes, we typically use inguinal, cervical, axial, brachial, popliteal, and lumbar. Using lymph nodes from obese mice can result in significant cell death during preparation due to excess adipose. Younger mice are recommended. Depending on experimental conditions, co-housing of different genotypes (greater than 2 weeks) or use of littermates is highly preferred. Comparing mice from different colonies/facilities should be avoided.

5. Lymph nodes and spleens from 2 to 3 mice can be processed together as one sample, but spleens should be kept separate until after RBC lysis.
6. Fetal calf serum (FCS) contains cytokines that may interfere with T cell differentiation. If the experiment involves polarizing conditions, cells should be harvested in serum-free media or plain Hank's balanced salt solution (HBSS).
7. While positive selection using directly conjugated microbeads has been reliable, cost effective, fast and provides high level of purity after culture, there are several alternatives for enriching for CD4+ T cells. We have used negative selection using Miltenyi CD4+ T cell isolation kit, positive selection using CD4-PE and anti-PE microbeads, and positive selection using anti-CD4 hybridoma and anti-FC magnetic beads. Sorting is not recommended for large-scale cultures due to cell loss during preparation and the large starting numbers of cells required for the expanded cultures.
8. When culturing T cells, keep the cell concentration between one and two million cells per ml of media. Lower cell concentrations will result in loss of proliferation, and higher cell concentrations will lead to overgrowth.
9. Restimulation should be done with bead-bound anti-CD3/CD28 as opposed to plate-bound antibody to increase ease of harvesting cells after stimulation and reduce cell death; agitation in the presence of cross-linked soluble anti-CD3 and anti-CD28 may also be sufficient. A suboptimal bead/cell ratio of 1:3 in restimulation reduces costs and achieves approximately 70% stimulation, but note the manufacturer recommended restimulation condition is 1:1 bead/cell.
10. Protein modifications other than ubiquitination (such as neddylation) can leave behind diglycine remnants. Consider adding a Nedd8 inhibitor for the final 1–2 h of the restimulation if neddylation is expected to be a complicating factor in the experiment.
11. If using label-based quantification for proteomics, mix the comparator sample and experimental sample together after lysing based on protein quantification. Start by mixing a small fraction of each sample, and validate the mixture using LC/MS before scaling up.
12. The alignment of RNA-seq data to the reference genome must be performed on a high-performance computing cluster implementing a scheduling protocol, such as Sun Grid Engine, that allows for submission of jobs to compute nodes.

13. For proteomic data, lack of detection of a particular protein should not be interpreted as an abundance of zero. Typically, these values are assigned "NA" to indicated lack of data.
14. Reproducibility can be assessed in a variety of ways including by identification of unique peptides, number of peptides identified, or repeated abundance quantification across biological replicates. This protocol assumes reproducibility is assessed by quantification across biological replicates.
15. Depending on the power of the dataset and the number of proteins identified, it may be appropriate to perform a multiple testing correction [16].
16. MG132, chloroquine, and cycloheximide are nonspecific and may interfere with protein expression indirectly.

References

1. Fang D et al (2002) Dysregulation of T lymphocyte function in itchy mice: a role for Itch in TH2 differentiation. Nat Immunol 3:281–287
2. Layman AAK et al (2017) Ndfip1 restricts mTORC1 signalling and glycolysis in regulatory T cells to prevent autoinflammatory disease. Nat Commun. https://doi.org/10.1038/ncomms15677
3. Layman AAK, Oliver PM (2016) Ubiquitin ligases and deubiquitinating enzymes in CD4 + T cell effector fate choice and function. J Immunol 196:3975–3982
4. Dybas JM et al (2019) Integrative proteomics reveals that CD4+ T cell activation promotes predominantly non-degradative ubiquitylation. Nat Immunol 20(6):747–755
5. Udeshi ND et al (2013) Refined preparation and use of anti-diglycine remnant (K-e-GG) antibody enables routine quantification of 10,000s of ubiquitination sites in single proteomics experiments. Mol Cell Proteomics 132:825–831
6. Mertins P et al (2013) Integrated proteomic analysis of post-translational modifications by serial enrichment. Nat Methods 10:634–637
7. Shevchenko A, Tomas H, Havlis J, Olsen JV, Mann M (2007) In-gel digestion for mass spectrometric characterization of proteins and proteomes. Nat Protoc. https://doi.org/10.1038/nprot.2006.468
8. Love MI, Huber W, Anders S (2011) Targeted analysis of nucleotide and copy number variation by exon capture in allotetraploid wheat genome. Genome Biol 15:550
9. Wickham H (2016). ggplot2: Elegant Graphics for Data Analysis. Springer-Verlag New York. ISBN 978-3-319-24277-4
10. Dobin A et al (2013) Sequence analysis STAR: ultrafast universal RNA-seq aligner. Bioinformatics 29:15–21
11. Dyballa N, Metzger S (2009) Fast and sensitive colloidal Coomassie G-250 staining for proteins in polyacrylamide gels. Part 1: Two-dimensional (2-D) gel electrophoresis using cup-loading Part 2: Colloidal Coomassie staining with CBB G-250. J Vis Exp. https://doi.org/10.3791/1431
12. Baruzzo G et al (2017) Simulation-based comprehensive benchmarking of RNA-seq aligners HHS Public Access. Nat Methods 14:135–139
13. Tyanova S, Temu T, Cox J (2016) The MaxQuant computational platform for mass spectrometry-based shotgun proteomics. Nat Protoc 11:2301–2319
14. Geer LY et al (2004) Open mass spectrometry search algorithm, J Proteome Res. https://doi.org/10.1021/PR0499491
15. Tan H et al (2017) Integrative proteomics and phosphoproteomics profiling reveals dynamic signaling networks and bioenergetics pathways underlying T cell activation. Immunity 46:488–503
16. Pascovici D, Handler DCL, Wu JX, Haynes PA (2016) Multiple testing corrections in quantitative proteomics: a useful but blunt tool. Proteomics 16:2448–2453. https://doi.org/10.1002/pmic.201600044

Chapter 20

Radioisotope-Based Protocol for Determination of Central Carbon Metabolism in T Cells

Xuyong Chen, John William Sherman, and Ruoning Wang

Abstract

T lymphocytes are the major components of the adaptive immune system. It's been known that T cells are able to engage a diverse range of metabolic programs to meet the metabolic demands during their life cycle from early development, activation to functional differentiation. Central carbon metabolic pathways provide energy, reducing power, and biosynthetic precursors to support T cell homeostasis, proliferation, and immune functions. As such, quantitative or semiquantitative analysis of central carbon metabolic flux activities offers mechanistic details, as well as insights into the regulation of metabolic pathways and the impact of changing metabolic programs on T cell life cycle. Global profiling of cellular metabolites by mass spectrometry-based metabolomics and metabolic flux analysis (MFA) using radioactive and nonradioactive/stable isotope approaches are powerful tools for determination of central carbon metabolic pathway activity. Here, we describe in detail the procedure for the radioisotope-based approach of analyzing central carbon metabolic fluxes in T cells.

Key words T lymphocytes, Radioactive isotope, Central carbon metabolism

1 Introduction

Upon antigen stimulation, T cells rapidly transit from a quiescent to an active state that is concomitant with cell growth and proliferation [1]. Activated and proliferating T cells can differentiate into various functional populations, which are largely determined by antigen stimulation and the surrounding cytokine milieu [1]. T cells are known to engage a variety of metabolic pathways, including glycolysis, glutaminolysis, and fatty acid metabolism, to meet their bioenergetics and biosynthetic requirements [2–5]. In addition, T cells that are in different states of their life cycle often display distinct metabolic characteristics. Naïve T cells rely on oxidative phosphorylation (OXPHOS), to generate energy and meet the basic requirements of cellular function and survival [1]. In activated T cells, aerobic glycolysis and glutaminolysis are the major catabolic pathways. These central carbon metabolic pathways support ATP

Chaohong Liu (ed.), *T-Cell Receptor Signaling: Methods and Protocols*, Methods in Molecular Biology, vol. 2111, https://doi.org/10.1007/978-1-0716-0266-9_20,

generation, maintain redox balance, and provide biosynthetic intermediates for the synthesis of amino acids, nucleotides, and lipids [1, 4–7].

The deployment and establishment of a robust and systemic T cell-mediated adaptive immune response takes days or even weeks and are bound to increase catabolic activities at both the cellular and whole-organism level. In addition, T cells are able to traffic to various inflamed sites, as well as expand and acquire effector functions in these sites with a wide range of nutrimental states. Understanding the T cell metabolic landscape and how T cells respond and adapt to changing metabolic conditions is fundamental to immunology and medicine. Changes in the metabolic pathway fluxes in living T cells that result from genetic or environmental perturbations can be quantified through metabolomics and metabolic flux analysis (MFA) [8, 9]. Global profiling of cellular metabolites by mass spectrometry-based metabolomics provides the steady-state composition of metabolites [10]. MFA based on the measurement of stable isotopomers using either nuclear magnetic resonance (NMR) or mass spectrometry is the most powerful tool that permits the high-resolution mapping of stable isotopomer incorporation in a given pathway and allows the determination of quantitative relationship between precursor and products [11].

However, stable isotope tracer based assays require extensive professional and technical expertise in instrument operation and computational modeling, which are critical in collecting, analyzing and refining the results. In addition, this approach requires a relatively large number of cells and is expensive and time consuming. Thus, a radioactive tracer based assay provides a complementary semi-quantitative tool that permits a rapid determination of central carbon metabolic flux activities in T cells [12]. Generally, T cell culture is pulsed with a radioactive tracer (radiolabeled metabolite), which can be transported into the cells and subsequently distributed by the metabolic reactions that define the fate of radioisotopomer transition. The rate of metabolic flux activity can be indirectly calculated by tracking the depletion of the radiolabeled substrates and/or the accompanying accumulation of the radiolabeled products. The detection of the radioactivity by scintillation counting is sensitive enough to detect small quantities of radioisotope, permitting a pulse-chase with a small amount of radiolabeled substrate in a short-term experiment.

Here, we describe the procedure for using radioisotopes to determine the rate of central carbon metabolic flux including glucose, glutamine, pyruvate, and long-chain fatty acid in resting and active T cells. Metabolic activity from glycolysis is determined by the detritiation of [5-^{3}H]-glucose during the reactions catalyzed by triose phosphate isomerase and enolase. Metabolic activity from fatty acid oxidation (FAO) is determined by the detritiation of [9,10-^{3}H]-palmitic acid, which is conjugated to bovine serum albumin (BSA). The metabolic flux rate is calculated from the radioactivity of ^3H_2O, which is separated from [^{3}H]-labeled substrates and intermediate metabolites by evaporation diffusion [13–17]. Metabolic activity from glutamine and pyruvate oxidation through the tricarboxylic acid (TCA) cycle is determined by the rate of $^{14}CO_2$ released from [U-$^{14}C_6$]-glutamine and [2-^{14}C]-pyruvate, respectively. Metabolic activity from glucose oxidation flux through the pentose phosphate pathway (PPP) is determined by the rate of $^{14}CO_2$ released from [1-^{14}C]-glucose. While the difference in the rate of $^{14}CO_2$ released from [1-^{14}C]-glucose and [6-^{14}C]-glucose was used to determine the PPP activity, we consistently found that the $^{14}CO_2$ production from [6-^{14}C]-glucose was close to the background in both resting and active T cells. The metabolic flux rate is calculated from the radioactivity of $^{14}CO_2$, which is generated through oxidation, largely trapped in the medium as carbonate, and then is collected by using alkaline solution (such as NaOH) after acidifying the medium with a strong acid (such as HCl) in a sealed system [18–21].

2 Materials

Prepare all solutions using distilled, deionized water (ddH_2O) or 1×PBS. Prepare all reagents at room temperature. Diligently follow all waste disposal regulations when disposing of waste materials.

2.1 Radiolabeled Tracers (Store at −20 °C or 4 °C for Short-Term Storage)

1. *[5-^{3}H]-Glucose*: stock concentration is 5 mCi/mL.
2. *[9,10-^{3}H]-Palmitic acid*: stock concentration is 5 mCi/mL, conjugated with 5% W/V BSA (fraction V/no lipids) (*see* **Note 1**).
3. *[U-$^{14}C_6$]-Glutamine*: stock concentration is 250 μCi/mL.
4. *[2-^{14}C]-Pyruvate*: stock concentration is 250 μCi/mL.
5. *[1-^{14}C]-Glucose*: stock concentration is 1 mCi/mL.

2.2 Other Reagents and Equipment

1. *T Cell Culture Medium*: RPMI 1640 media supplemented with 10% (v/v) heat-inactivated fetal bovine serum (FBS), 2 mM L-glutamine, 0.05 mM *2-mercaptoethanol*, 100 units/mL penicillin, and 100 μg/mL streptomycin. Store at 4 °C.
2. 1 μg/mL IL-7 cytokine. Store at −80 °C.
3. 1 μg/mL IL-2 cytokine. Store at −80 °C.
4. 1 mg/mL anti-CD3 antibody. Store at −80 °C.
5. 1 mg/mL anti-CD28 antibody. Store at −80 °C.
6. 1×PBS.
7. 0.2 N NaOH.

8. 5 N HCl.
9. MojoSort™ Mouse CD3 cell isolation kit. Store at 4 °C.
10. Forty-eight well cell culture plate.
11. 0.2 mL PCR tube.
12. 1.5 mL Centrifuge tube.
13. 6.5 mL Scintillation vials.
14. 20 mL Scintillation vials.
15. ScintiSafe 30% Scintillation Cocktail.
16. Radioactive Decontaminant Surface Cleaner.
17. 1 mL Insulin syringe.
18. 7 mL Clear vial.
19. Polypropylene hole cap 15 mm with PTFE/Silicone Septa.
20. *$^{14}CO_2$ Collection*: 0.2 mL PCR tube (with the lid cut off and 50 μL of 0.2 N NaOH is added before the experiment) is placed via hot glue to the inner wall (around 1 cm above the bottom) of a 7 mL clear vial with polypropylene hole cap, 15 mm with PTFE/Silicone Septum (Fig. 1).
21. *$^{3}H_2O$ Collection*: 1.5 mL centrifuge tube (with the lid cut off and 50 μL of 5 N HCl is added before the experiment) is placed in the bottom of a 20 mL scintillation vial with 0.5 mL water added before the experiment (Fig. 2).

3 Methods

Carry out all procedures at room temperature unless otherwise specified.

3.1 Pre-coat 48-Well Plates at Day −1

Dilute anti-CD3 and anti-CD28 antibodies in 1×PBS with a final concentration of 2 μg/mL, add 200 μL antibody solution per well in a 48 well plates (pre-coated plate), and incubate at 4 °C overnight. For other multiwell culture plate, the required volume of antibody solution and cell number are proportional to the surface area of well.

3.2 T Cell Isolation at Day 0

1. Enrich total mouse T cells from the spleens and lymph nodes by MojoSort™ Mouse CD3 cell isolation kit following the manufacturer's instructions.
2. Maintain freshly isolated total CD3 T cells in T Cell Culture Medium. For naïve T cells, 0.5 mL (1×10^6/mL) of T cells are supplemented with 5 ng/mL IL-7 and cultured in a non-pre-coated plate. For active T cells, 0.5 mL (1×10^6/mL) of T cells are supplemented with 5 ng/mL IL-2 and cultured in a pre-coated plate (discard the antibody solution before seeding

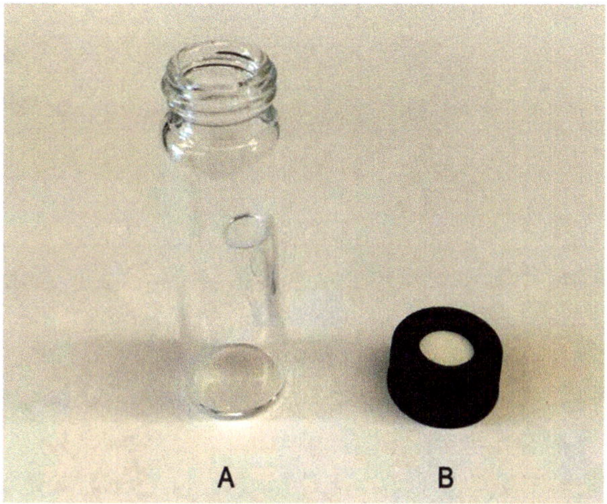

Fig. 1 (**a**) A PCR tube (with cap cut off) adhered to the inner side of a screw-cap vial. (**b**) Septum cap

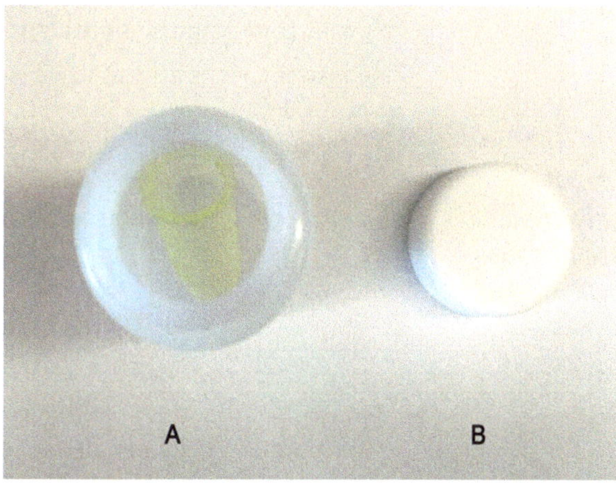

Fig. 2 (**a**) 1.5 mL centrifuge tube with lid cut off and placed in the bottom of the vial. (**b**) Vial's lid

the cells). For differentiated T cell subsets, please follow the individual instructions to generate a differentiated cell population and maintain the population utilizing the proper cytokines.

3. Culture cells at incubator with 37 °C in 5% CO_2.

3.3 Incubate T Cells in Flux Reactions at Day 1 or Other Selected Time Points

1. Collect cells and wash the cells with 1×PBS twice.
2. Count cell number. For ^3H_2O-based assay, prepare cell suspension with 1×10^6 per sample in 0.5 mL T Cell Culture Medium; to run samples in triplicate, prepare a total of 3×10^6 cells in 1.5 mL medium per treatment group. For

$^{14}CO_2$-based assay, prepare cell suspension with 5 × 10^6 per sample in 0.5 mL T Cell Culture Medium; to run samples in triplicate, prepare a total of 15 × 10^6 cells per treatment group in 1.5 mL medium.

Perform following steps in the designated radioactive area:

3.4 ^{3}H-Based Glycolysis and FAO (in 48 Well Plate) (See Note 2)

1. Prepare a 48 well plate.
2. For each treatment group, add 0.5 mL of ^{3}H cell suspension to a well for each sample; use medium alone as a background control.
3. Add 1 μCi radiolabeled compound per a well.
4. Incubate at 37 °C for 2 h.
5. Add 0.5 mL H_2O to the bottom of the 20 mL scintillation vials, as described in Subheading 2.2. Add a 1.5 mL centrifuge tube (with the lid cut off) to the bottom of each vial, and add 50 μL of 5 N HCl into it (Fig. 2) (*see* **Note 3**).
6. Add the cell suspension into the 1.5 mL centrifuge tube, but avoid touching the scintillation vials.
7. Close the vial's lid and leave at room temperature overnight (*see* **Note 4**).

3.5 ^{14}C-Based Glutamine, Pyruvate, and Glucose Oxidation (in Septum Glass Vials)

1. Prepare three septum glass vials (Fig. 1) for each treatment group, as described in Subheading 2.2.
2. Add 50 μL 0.2 N NaOH to the PCR tube in each vial (*see* **Note 5**).
3. Add 0.5 mL cell suspension with 0.3 μCi radiolabeled substrates in it to the bottom of each vial, and avoid touching the PCR tube; then seal the vial with a septum cap. Use cell-free medium as a background control.
4. Incubate at 37 °C for 2 h.
5. Stop the reaction and release CO_2 by using a 1 mL insulin syringe to inject 50 μL 5 N HCl into the cell suspension of each vial (*see* **Note 6**).
6. Incubate vials at room temperature overnight.

3.6 Prepare Scintillation Solution of ^{3}H-Based Samples at Day 2

1. Remove the 1.5 mL centrifuge tube from scintillation vials and dispose them as radioactive trash.
2. Add 10 mL scintillation fluid into the scintillation vial, close lid, and shake well (*see* **Note 7**).
3. Clean the forceps (*see* **Note 8**).

3.7 Prepare Scintillation Solution of ^{14}C-Based Samples at Day 2

1. Prepare a 6.5 mL scintillation vials by adding 5 mL scintillation fluid to each vial.
2. Transfer the NaOH solution from the PCR tube to the scintillation vials (*see* **Note 9**). Recap the septum vial and dispose them as radioactive trash.
3. Close lid of scintillation vials and shake well.
4. Clean the pipettor (*see* **Note 10**).

3.8 Read the Results

1. Read the samples by scintillation counter.
2. Subtract background CPMA values from sample values, and then normalize to the control sample for each treatment group.

4 Notes

1. Long-chain free fatty acids (FFAs) are important metabolic substrates for energy production. Two of the most common fatty acids in vertebrates are the long-chain saturated FFA palmitate (C16:0) and the monounsaturated oleate (C18:1). To overcome the low solubility of long-chain FFAs in aqueous solutions, FFAs are conjugated to BSA, allowing the preparation of solutions in the physiological concentration range. Lipids stored in chloroform or ethanol are first dried down under a very low pressure stream of inert gas (nitrogen) in a glass vial (Fisher 03-339-23A or Wheaton 223693 2 mL serum vials) until all of the solvent is removed and only a film of lipid remains on the bottom of the tube. The lipids are then hydrated in 5% W/V BSA (fraction V/no lipids) and are sonicated for 3 × 5 min in ultrasonic bath (Branson Ultrasonic Cleaner). The solution will be completely clear.
2. The rate of mitochondrial-dependent beta-oxidation can be calculated as the difference between oxidation counts in the presence or absence of 100 μM etomoxir. Etomoxir is an irreversible inhibitor of carnitine palmitoyltransferase I, the rate-limiting enzyme for mitochondrial import of long-chain fatty acids. The same dose of etomoxir is also able to inhibit the activity of the electron transportation chain (ETC) [22, 23].
3. The HCl was used for lysing cells and stopping the metabolic reactions.
4. ^3H_2O is separated from unmetabolized radiolabeled substrate or intermediate metabolites by evaporation diffusion for 12–24 h at room temperature (the step of vapor-phase equilibration between unlabeled H_2O in the bottom of the capped scintillation vial and ^3H_2O in 1.5 mL tube, Fig. 2). A cell-free sample containing 1 μCi radiolabeled substrate needs to be included as a background control.

5. NaOH solution is used to capture the CO_2 released from the cell. While adding NaOH into PCR tube, please avoid splashing the NaOH solution onto the outer vial.
6. The HCl should be injected directly into the cell suspension in the bottom of glass vial to lyse the cells and release CO_2. Please avoid injecting into the PCR tube.
7. The well-mixed water and scintillation fluid solution should appear clear.
8. To decontaminate the forceps, spray with Radioactive Decontaminant Surface Cleaner and rinse with water, and then spray with 70% ethanol and rinse with water.
9. To minimize radiation exposure, very carefully use one hand to open the lid of septum vial, and use the other hand to pipette NaOH from the PCR tube into the scintillation fluid in the scintillation vials.
10. To decontaminate the pipettor, spray with Radioactive Decontaminant Surface Cleaner and rinse with water, and then spray with 70% ethanol and rinse with water.

Acknowledgments

This work was supported by 1R01AI114581 from the National Institute of Health and 128436-RSG-15-180-01-LIB from the American Cancer Society (R.W.).

References

1. Slack M, Wang T, Wang R (2015) T cell metabolic reprogramming and plasticity. Mol Immunol 68(2 Pt C):507–512. https://doi.org/10.1016/j.molimm.2015.07.036
2. Pearce EL, Poffenberger MC, Chang C-H, Jones RG (2013) Fueling immunity: insights into metabolism and lymphocyte function. Science (New York, NY) 342 (6155):1242454–1242454. https://doi.org/10.1126/science.1242454
3. Verbist KC, Guy CS, Milasta S, Liedmann S, Kamiński MM, Wang R, Green DR (2016) Metabolic maintenance of cell asymmetry following division in activated T lymphocytes. Nature 532(7599):389–393. https://doi.org/10.1038/nature17442
4. Wang R, Green DR (2012) Metabolic reprogramming and metabolic dependency in T cells. Immunol Rev 249(1):14–26. https://doi.org/10.1111/j.1600-065X.2012.01155.x
5. Wang R, Green DR (2012) Metabolic checkpoints in activated T cells. Nat Immunol 13:907. https://doi.org/10.1038/ni.2386
6. Wang R, Dillon CP, Shi LZ, Milasta S, Carter R, Finkelstein D, McCormick LL, Fitzgerald P, Chi H, Munger J, Green DR (2011) The transcription factor Myc controls metabolic reprogramming upon T lymphocyte activation. Immunity 35(6):871–882. https://doi.org/10.1016/j.immuni.2011.09.021
7. Wang R, Green DR (2012) The immune diet: meeting the metabolic demands of lymphocyte activation. F1000 Biol Rep 4:9. https://doi.org/10.3410/B4-9
8. Guijas C, Montenegro-Burke JR, Warth B, Spilker ME, Siuzdak G (2018) Metabolomics activity screening for identifying metabolites that modulate phenotype. Nat Biotechnol 36 (4):316–320. https://doi.org/10.1038/nbt.4101
9. Buescher JM, Antoniewicz MR, Boros LG, Burgess SC, Brunengraber H, Clish CB,

DeBerardinis RJ, Feron O, Frezza C, Ghesquiere B, Gottlieb E, Hiller K, Jones RG, Kamphorst JJ, Kibbey RG, Kimmelman AC, Locasale JW, Lunt SY, Maddocks ODK, Malloy C, Metallo CM, Meuillet EJ, Munger J, Nöh K, Rabinowitz JD, Ralser M, Sauer U, Stephanopoulos G, St-Pierre J, Tennant DA, Wittmann C, Vander Heiden MG, Vazquez A, Vousden K, Young JD, Zamboni N, Fendt S-M (2015) A roadmap for interpreting (13)C metabolite labeling patterns from cells. Curr Opin Biotechnol 34:189–201. https://doi.org/10.1016/j.copbio.2015.02.003

10. Lu W, Su X, Klein MS, Lewis IA, Fiehn O, Rabinowitz JD (2017) Metabolite measurement: pitfalls to avoid and practices to follow. Annu Rev Biochem 86:277–304. https://doi.org/10.1146/annurev-biochem-061516-044952
11. Fan TWM, Lane AN (2016) Applications of NMR spectroscopy to systems biochemistry. Prog Nucl Magn Reson Spectrosc 92–93:18–53. https://doi.org/10.1016/j.pnmrs.2016.01.005
12. Wolfe RR, Chinkes DL (2005) Isotope tracers in metabolic research: principles and practice of kinetic analysis. Wiley, Hoboken, NJ
13. Goodwin GW, Cohen DM, Taegtmeyer H (2001) [5-3H]glucose overestimates glycolytic flux in isolated working rat heart: role of the pentose phosphate pathway. Am J Physiol Endocrinol Metab 280(3):E502–E508. https://doi.org/10.1152/ajpendo.2001.280.3.E502
14. Ashcroft SJ, Weerasinghe LC, Bassett JM, Randle PJ (1972) The pentose cycle and insulin release in mouse pancreatic islets. Biochem J 126(3):525–532
15. Brivet M, Slama A, Saudubray J-M, Legrand A, Lemonnier A (1995) Rapid diagnosis of long chain and medium chain fatty acid oxidation disorders using lymphocytes. Ann Clin Biochem 32(2):154–159. https://doi.org/10.1177/000456329503200204
16. Moon A, Rhead WJ (1987) Complementation analysis of fatty acid oxidation disorders. J Clin Invest 79(1):59–64. https://doi.org/10.1172/JCI112808
17. Ando S, Tomita-Yamaguchi M, Santoro TJ (1993) Long chain fatty acid utilization of T-cells from autoimmune MRL-lpr/lpr mice. Biochim Biophys Acta (BBA) - Mol Basis Dis 1181(2):141–147. https://doi.org/10.1016/0925-4439(93)90103-8
18. Brand K, Williams JF, Weidemann MJ (1984) Glucose and glutamine metabolism in rat thymocytes. Biochem J 221(2):471–475
19. Willems HL, de Kort TF, Trijbels FJ, Monnens LA, Veerkamp JH (1978) Determination of pyruvate oxidation rate and citric acid cycle activity in intact human leukocytes and fibroblasts. Clin Chem 24(2):200
20. Katz J, Wood HG (1960) The use of glucose-C14 for the evaluation of the pathways of glucose metabolism. J Biol Chem 235 (8):2165–2177
21. Wood HG, Katz J, Landau BR (1963) Estimation of pathways of carbohydrate metabolism. Biochem Z 338:809–847
22. Yao C-H, Liu G-Y, Wang R, Moon SH, Gross RW, Patti GJ (2018) Identifying off-target effects of etomoxir reveals that carnitine palmitoyltransferase I is essential for cancer cell proliferation independent of β-oxidation. PLoS Biol 16(3):e2003782. https://doi.org/10.1371/journal.pbio.2003782
23. Raud B, Roy DG, Divakaruni AS, Tarasenko TN, Franke R, Ma EH, Samborska B, Hsieh WY, Wong AH, Stüve P, Arnold-Schrauf C, Guderian M, Lochner M, Rampertaap S, Romito K, Monsale J, Brönstrup M, Bensinger SJ, Murphy AN, McGuire PJ, Jones RG, Sparwasser T, Berod L (2018) Etomoxir actions on regulatory and memory T cells are independent of Cpt1a-mediated fatty acid oxidation. Cell Metab 28(3):504–515.e507. https://doi.org/10.1016/j.cmet.2018.06.002

Chapter 21

Studying Peripheral T Cell Homeostasis in Mice: A Concise Technical Review

Moutuaata M. Moutuou, Simon-David Gauthier, Nicolas Chen, Dominique Leboeuf, and Martin Guimond

Abstract

For several years, it was believed that the thymus was entirely responsible for maintaining T cell homeostasis. Today, it is well-known that homeostatic peripheral mechanisms are essential in order to maintain T cell numbers and diversity constant in the periphery. Naïve and memory T cells require continual access to self-peptide MHC class I and II molecules and/or cytokines to survive in the periphery. Under normal conditions, homeostatic resources are low, and lymphocytes undergo very slow proliferation and survive. Following T cell depletion, the bioavailability of homeostatic resources is significantly increased, and T cell proliferation is dramatically augmented. The development of lymphopenic mouse models has helped our current understanding of factors involved in the regulation of peripheral T cell homeostasis. In this minireview, we will give a brief overview about basic techniques used to study peripheral T cell homeostasis in mice.

Key words Lymphopenia, Homeostasis, Lymphocytes, CD4, CD8

1 Introduction

Homeostasis of T lymphocytes is regulated not only by thymopoiesis but also by homeostatic peripheral mechanisms that control mature T cell persistence. Early in life, the thymus is responsible for the production of T cells, providing immunocompetence. Later, as thymic involution occurs, homeostatic peripheral mechanisms become more important in the maintenance of T cell counts and diversity [1]. Several studies have now demonstrated that lymphocyte counts are tightly regulated in the periphery and that T cell maintenance depends largely on the niche size and homeostatic resources available in the periphery. For instance, increasing thymic output through the addition of prenatal thymus lobes in adults results only in a small increase in peripheral T cell numbers [2]. Similarly, thymectomy performed during adulthood does not lead to a substantial decrease in T cell counts [3]. Thus, despite the essential

Chaohong Liu (ed.), *T-Cell Receptor Signaling: Methods and Protocols*, Methods in Molecular Biology, vol. 2111,
https://doi.org/10.1007/978-1-0716-0266-9_21,

role played by the thymus early in life to generate a highly diversified T cell repertoire, later in life, T cell maintenance relies largely on homeostatic peripheral mechanisms.

1.1 Homeostasis of T Cells in Nonlymphopenic Hosts

In normal hosts, mature lymphocytes circulate and undergo very slow proliferation called homeostatic cycling (HC). HC of naïve T cells occurs in response to interleukin 7 (IL-7) and T cell receptor (TCR) stimulation by self-peptide MHC class I or II for $CD8^+$ and $CD4^+$ T cells, respectively. While antigen-presenting cells provide MHC class II and are essential for the peripheral maintenance of naïve $CD4^+$ T cells [4], naïve $CD8^+$ T cells can also receive TCR stimulation from non-hematopoietic stromal cells through MHC class I present on all stromal cells [5]. Factors required for $CD4^+$ and $CD8^+$ T cell HC vary considerably between memory and naïve cells. For instance, memory $CD8^+$ T cells do not require continuous access to MHC class I as they can survive and undergo HC in hosts lacking MHC class I [6]. In addition, interleukin 15 (IL-15) is important for HC and accumulation of $CD8^+$ T cells during lymphopenia [7] and is essential, along with IL-7, for memory $CD8^+$ T cell HC and persistence. While memory $CD4^+$ T cells may require TCR stimulation by self-peptide MHC class II molecules, recent studies have provided functional data to support a greater dependency to IL-7 and IL-15 [8, 9]. Based on these data, cytokine stimulation appears to be the main driving force for memory T cell HC, whereas TCR stimulation is absolutely required for the persistence of naïve subsets.

1.2 Homeostasis of T Lymphocytes in Lymphopenic Hosts

In most clinical settings of human lymphopenia, T cell regeneration occurs through homeostatic proliferation (HP). While the requirements for HP are similar to HC, the level of proliferation depends on the degree of lymphopenia and the bioavailability of MHC class I and II and/or IL-7 and IL-15. Following postnatal T cell depletion, the bioavailability of IL-7 increases significantly due to the fact that there are little to no T cells present to use it [10, 11]. In addition, competition between lymphocytes for accessibility to MHC class I and II on antigen-presenting cells is diminished. These important changes in systemic IL-7 levels and antigen-presenting cells availability that occur during lymphopenia contribute to increase T cell HP [4, 5, 12]. Lymphocytes undergoing HP experience significant changes and adopt an activated phenotype [13] although whether these phenotypic changes are reversible remains unknown. While HP of $CD8^+$ T cells is typically robust, HP of naïve $CD4^+$ T cells is weak, and normalization of the $CD4^+$ T cell counts requires thymopoiesis [1]. Thus, in all but the youngest, T cell regeneration after lymphopenic insult is a long process that can take several months or years to occur.

1.3 Studying T Cell Homeostasis in Mice

Although in vitro studies have been used to study T cell homeostasis [14], these models do not recapitulate the complexity of the in vivo microenvironment that normally provides support to lymphocytes. The development of genetically modified mouse models has provided important insights about factors involved in T cell homeostasis. However, many of these lymphopenic models present a dysfunction of the thymus and of the peripheral niche. The development of congenic strains and monoclonal antibodies specific for congenic markers has greatly facilitated the study of the peripheral T cell niche in mice. For example, while it is well-documented that IL-$7^{-/-}$ mice have a dysfunction of the thymus, the adoptive transfer of naïve T cells into IL-$7^{-/-}$ recipients has provided functional data to demonstrate the essential role of IL-7 on naïve $CD4^+$ and $CD8^+$ T cell survival and proliferation [15]. In order to quantify T cell proliferation in mice, many groups have used bromodeoxyuridine (BrDU), an analog of thymidine that labels replicating DNA [16–19]. Today however, researchers are preferring carboxyfluorescein diacetate succinimidyl ester (CFSE) or Cell Trace Violet (CTV) to quantify T cell proliferation because it is more accurate than BrDU [20, 21]. For these analyses, lymphocytes must be isolated and labelled with either CFSE or CTV prior to their adoptive transfer in recipient mice. Hours to days following T cell transfer, recipient mice are sacrificed, and lymphocytes are analyzed by flow cytometry to evaluate T cell proliferation based on CFSE or CTV dilution. Unlike BrDU that requires the use of anti-BrDU monoclonal antibody for its detection, CFSE and CTV are both fluorescent and can be readily detected by flow cytometry. CFSE and CTV analyses are normally informative when cells undergo few cell divisions (between one and seven divisions). When the number of divisions is higher, the dye is often completely diluted, and quantification of cell division is typically less accurate. If cells are dividing slowly, CFSE or CTV can be detected even after several weeks or months post-adoptive transfer.

1.4 Source of T Cells to Study T Cell Homeostasis

T cells can be isolated from lymph nodes (LNs) or from the spleen. While the proportion of naïve T cells is higher in LNs of young mice, the number of activated and/or memory T cells is typically higher in the spleen (Fig. 1). This difference may affect the degree of cell proliferation as activated, memory, and naïve T cells possess their own niche which, in some cases, could minimally overlap with each other. Naïve T cells can be isolated based on the expression of CD45rb [22]. In contrast, memory cells express much lower levels of CD45rb and express higher levels of CD44 [23, 24]. The number of T cells infused in lymphopenic hosts is also important when studying T cell homeostasis. During lymphopenia, homeostatic resources are higher, but infusing too many lymphocytes can increase competition between T cells and diminish HP. Similarly, the transfer of monoclonal TCR transgenic T cells can be subjected

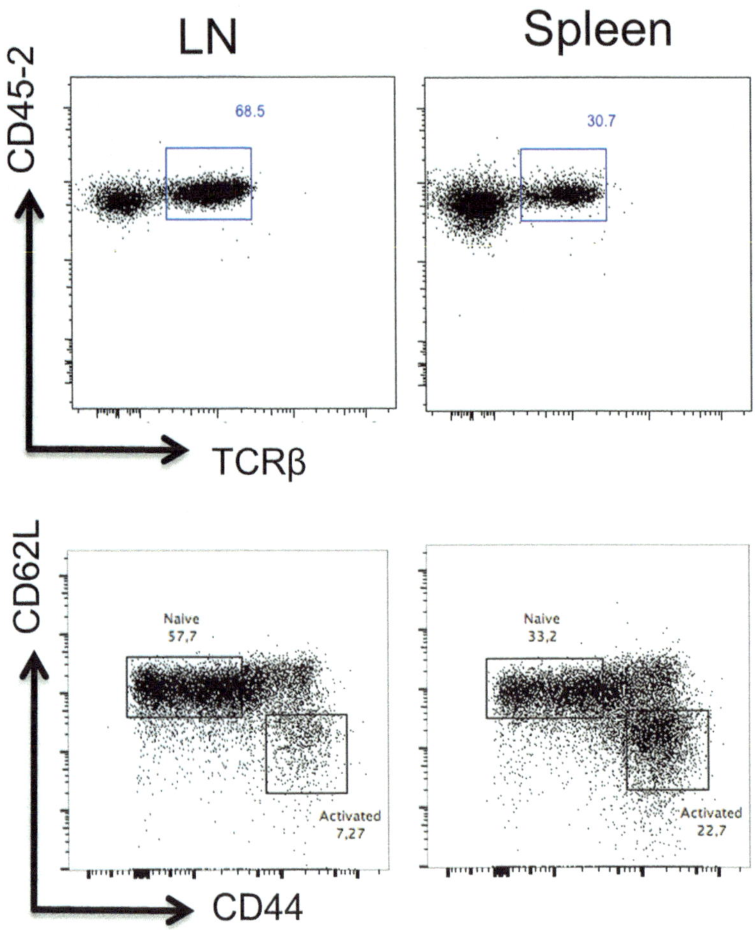

Fig. 1 Lymphocytes from LNs and the spleen. (**a**) Proportion of TCRβ$^+$ cells in LNs and the spleen. (**b**) Proportion of naïve and memory-activated lymphocytes in LNs and the spleen based on CD62L and CD44 expression. Naïve cells are CD62L^{hi}CD44^{low}, whereas memory-activated cells are CD44$^{mid/hi}$

to intra-clonal competition which in turn can negatively impact HP of these cells [25–28]. Thus, by transferring few lymphocytes into recipient mice, we ensure that competition between lymphocytes is minimal and the degree of proliferation occurs in response to the bioavailability of homeostatic resources found in lymphopenic recipients.

1.5 Mouse Models to Study T Cell Homeostasis

Several lymphopenic mouse models are currently available to study in vivo T cell homeostasis. C57BL6/RAG$^{-/-}$ mice are probably the most well-versed model to study HP of T cells during lymphopenia as these mice do not have T and B lymphocytes due to the deletion of the Rag-1 or Rag-2 genes [29, 30]. C57BL6/RAG$^{-/-}$ mice have been widely used to study HP and thymic-independent T cell regeneration in lymphopenic hosts. Other lymphopenic mouse models include IL-$7^{-/-}$, IL-7R$\alpha^{-/-}$, and $\gamma C^{-/-}$. These mice are profoundly lymphopenic although few mature T cells can be detected in the spleen and LNs. While IL-$7^{-/-}$ mice do not support HP of naïve T cells [12, 31], IL-7R$\alpha^{-/-}$ and $\gamma C^{-/-}$ hosts supported efficient HP, especially CD4$^+$ T cell HP [11, 32, 33]. To study lymphopenia-driven CD8$^+$ proliferation and accumulation [7], IL-$15^{-/-}$ mice have been used, although they are not lymphopenic [34]. MHC class $I^{-/-}$ and $II^{-/-}$ mice have been used to study CD8$^+$ and CD4$^+$ T cell homeostasis [35, 36]. MHC class $I^{-/-}$ mice do not have CD8$^+$ T cells, and MHC class $II^{-/-}$ mice do not have CD4$^+$ T cells due to lack of positive selection inside the thymus. Both MHC class $I^{-/-}$ and $II^{-/-}$ mice are partially lymphopenic and must be bred to $Rag^{-/-}$ or CD3$\varepsilon^{-/-}$ background to obtain lymphopenic hosts in order to study the requirement of these molecules in T cell HP. Finally, several congenic strains have been developed (B6S.SJL CD45.1$^+$, B6PLCD90.1$^+$) for adoptive T cell transfer. Using these models, studies have delineated critical factors controlling CD4$^+$ and CD8$^+$ T cell homeostasis under normal and lymphopenic conditions. In this mini technical review, we will explain the methodology of adoptive transfer to study lymphocyte survival and proliferation in lymphopenic and nonlymphopenic hosts.

2 Materials

2.1 Buffers and Equipment

See Table 1.

3 Methods

Adoptive transfer of T lymphocytes is performed between congenic donors (C57BL6/SJL CD45.1$^+$) and recipient mice (C57BL6 CD45.2$^+$). Lymph node T cells are preferred over spleen T cells because the proportion of naïve T cells is higher in LNs (Fig. 1). To avoid contamination by microbes, these experimental procedures are performed using sterile equipment and reagents under a laminar flow hood.

3.1 Tissue Collection and Preparation of Cell Suspension

Lymphocytes are obtained from lymph nodes of congenic C57BL6/SJL CD45.1$^+$ mice. Before the organ collection, mice must be sacrificed according to the guidelines established by the animal care ethic committee of your institute. In our studies, we use sex-matched mice aged between 6 and 8 weeks to avoid potential immune activation by the HY minor antigen [37]:

Table 1
Materials and reagents. List of reagents, buffers, medium, and instruments required for in vivo and in vitro procedures

Materials	Company	Catalogue number
RPMI 1640	Wisent	350-000-CL
Fetal bovine serum	Wisent	085-150
Penicillin-streptomycin	Wisent	450-201-EL
Phosphate-buffered saline 1×	Gibco	
Red blood cell lysis buffer	Gibco	
Ethanol 70%	HMR	
Dissecting instruments (scissors, scalpels, forceps)	Roboz Instrument	
Conical tubes 15 ml, 50 ml	Sarstedt	62.554.002 62.547.205
Sterile polystyrene round-bottom tube 5 ml	Fisher Scientific	352054
Sterile petri dishes 100 × 15 mm	Fisher Scientific	351029
Cell stainer 70 μm	Fisher Scientific	542070
Tips 1000 μl, 10 μl, 200 μl, 1–2 μl	Ultident	87-A1000B 87-A200Y 75-37650
Serological pipettes 5 ml, 10 ml, 25 ml	Sarstedt	86.1253.001 86.1254.001 86.1685.001
Syringes 1 ml with needle $27G^{1/2}$	Terumo	SS-01T2713
Syringes 10 ml	BD Syringe	302995
Needles $18G^{1/2}$	Terumo	NN-2516R
Needles $25G^{1/2}$	BD PrecisionGlide	305196
Glass pestle	VWR	
Pipettes P1000, P2, P20, P200	Gilson	
Pipette gun	Drummond	D113262
Mouse T cell Enrichment Kit	STEMCELL Technologies	19751A
Magnet	STEMCELL Technologies	18000
RoboSep Buffer	STEMCELL Technologies	20104
CTV (A) + DMSO (B)	Life Technologies	C34557
Micro tube 1.5 ml	Sarstedt	72.690.301
FACS polystyrene round-bottom tube 5 ml	Falcon A Corning Brand	352008
Anti-mouse CD4 (clone RM4-5)	BioLegend	100516

(continued)

Table 1
(continued)

Materials	Company	Catalogue number
Anti-mouse CD8α (clone 53-6.7)	BioLegend	100708
Anti-mouse TCRβ (clone 5H10-1)	BioLegend	100804
Anti-mouse CD45-1 (clone A20)	BioLegend	110728
Anti-mouse CD45-2 (clone 104)	BioLegend	109830
Compensation beads	Invitrogen	01-2222-42
Microscope	Zeiss	Leica CME model
Trypan blue dilution 1/25	Invitrogen	15250
Hemocytometer	Hausser Scientific Co.	3200
Centrifuge	Eppendorf	Centrifuge 5810R
Flow cytometer	BD Biosciences	LSRFortessa
Flow cytometer analysis software	BD Biosciences	FlowJo V10

1. *Harvesting lymph nodes*: LNs are collected in 15 ml conical tube containing RPMI media with and 1% (v/v) Pen/Strep. After the animal is euthanized, death is confirmed by pinching the skin between the toes to verify the absence of the pedal withdrawal reflex. In the laminar flow hood, place the mouse on a gauze, and soak the hair with 70% ethanol. With scissors, perform an incision in the medial part of abdominal skin, and gently remove the skin. Inguinal and axillary LNs are superficial and located in front of the hindlimbs and forelimbs, respectively (Fig. 2). With tweezers, remove and transfer the four inguinal and axillary LNs to a 15 ml conical tube containing 2–5 ml RPMI media. In older mice, LNs can be difficult to find since they can be confounded with body fat. However, when LNs are transferred into a liquid, they normally sink to the bottom of the tube, whereas body fat floats on the surface of the liquid.
2. *Cell suspension from lymph nodes*: Transfer the LNs and the media into a sterile petri dish. Manually disrupt the LNs using a glass pestle by pressing the LNs against the petri dish. Using a 1 ml pipette, dissolve cell aggregates by pipetting up and down before passing the cell suspension through a 70 μm mesh nylon strainer on top of a 50 ml conical tube to remove debris and undissolved aggregates. Add media up to 15–20 ml. Centrifuge the cells at 288 × *g* for 10 min, 4 °C. Remove the supernatant, and resuspend in 5 ml RPMI. Keep cells on ice.

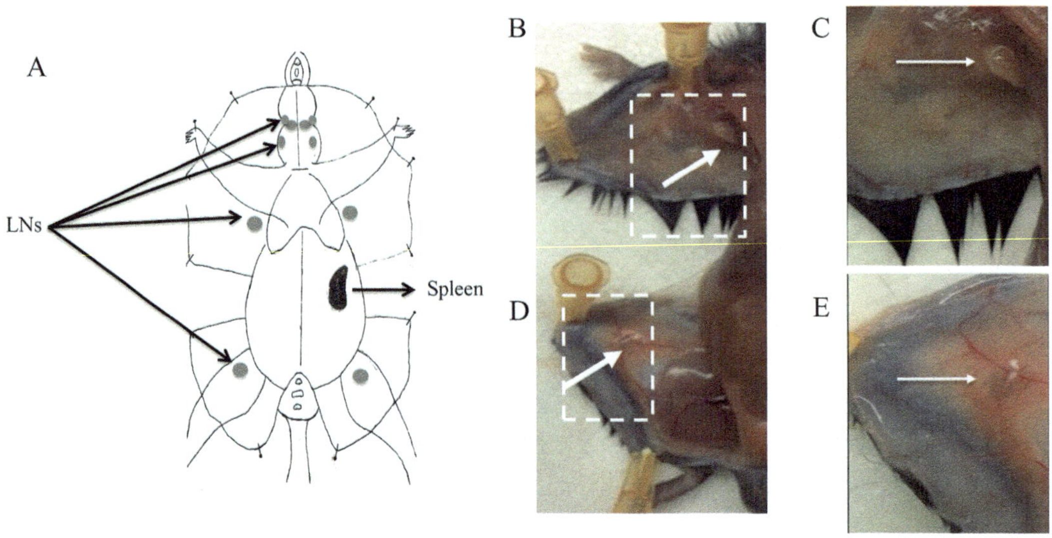

Fig. 2 Spleen and LNs isolation. (**a**) Schematic representation of the location of LNs and the spleen in mice. (**b** and **c**) The white arrows identify axillary LNs (**d** and **e**). The white arrows identify inguinal LNs

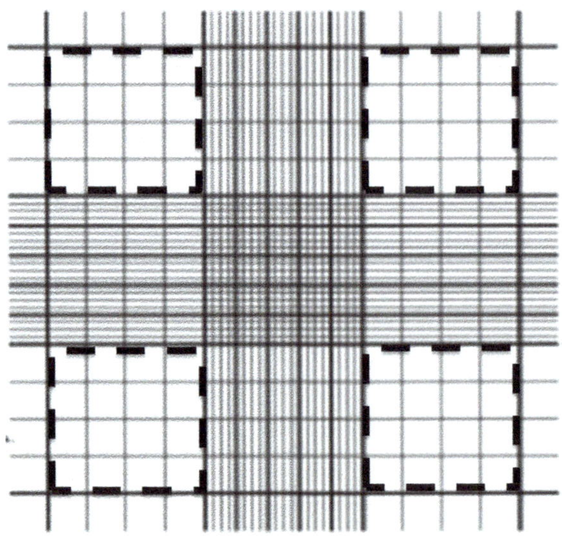

Fig. 3 Hemocytometer grid. The four large corner squares containing 16 small squares are used for cell counting

3.2 Cell Counting and Viability Analysis

In a microcentrifuge tube, add either 90 or 190 μl of diluted Trypan blue in phosphate-buffered saline (PBS), and add 10 μl of cell suspension for 1/10 or 1/20 dilution, and mix well. Take 10 μl of diluted cells, and load on a hemocytometer. With a light microscope (10× or 40×), count the cells in each of the four corner squares. Do not count the blue cells since they are dead (Fig. 3).

3.3 Enumeration of LN Cells

The total cell count is obtained with the following equation:

$$\text{Total viable cells count} = \left[\frac{\text{Total number of live cells counted}}{\text{Number of squares counted}}\right] \times 10^3 \times \text{dilution factor} \times \text{resuspension volume of cells}$$

After counting the cells, centrifuge cells at $288 \times g$ for 10 min. Discard the supernatant, and adjust the cell concentration by adding the appropriate volume of RoboSep Buffer in order to obtain a final concentration of 10^8 cells/ml. Gently agitate the tube to resuspend the cells, and remove 100–200 μl of cells that will be used later to evaluate T cell purity before T cell enrichment.

3.4 T Cell Enrichment with the Stem Cell Kit

3.4.1 Kit Reagents Stored at 4 °C

Normal Rat Serum.

EasySep Mouse T Cell Enrichment Kit.

EasySep Mouse T Cell Enrichment Cocktail.

EasySep Biotin.

EasySep magnetic particles.

RoboSep Buffer.

EasySep Magnet.

3.4.2 Procedure

1. Prepare a single cell suspension at 10^8 cells/ml in RoboSep media in a polystyrene round-bottom tube of 5 or 10 ml if up to 2×10^8 cells. To prevent unspecific binding of antibodies, add rat serum at 50 μl/ml before adding the EasySep Mouse T Cell Enrichment Cocktail containing biotinylated antibodies at 50 μl/ml of cells. Mix well, and incubate at 4 °C for 15 min.
2. Add the EasySep Biotin Selection Cocktail at 100 μl/ml of cells. The EasySep Biotin Selection Cocktail will bind cells labeled with biotinylated antibodies. Mix well, and incubate at 4 °C for 15 min.
3. Before pipetting magnetic particles, vortex the aliquot for 30 s, and add magnetic particles to the sample at 75 μl/ml. Mix well, and incubate at room temperature (RT) for 5 min.
4. Add RoboSep buffer to a final volume of 2.5 ml. Mix gently by pipetting up and down three times. Place the tube (without lid) into the magnet for 5 min at RT. Keep the tube in the magnet, and gently invert the cell suspension in a new 5 ml polystyrene tube. If lymphocytes are isolated by a negative selection, lymphocytes are not labelled with the magnetic particles. As a result, the cells of interest have been transferred into the new 5 ml polystyrene tube. Magnetic cell sorting to deplete nontarget cells involves magnet attracting the cells that have bound

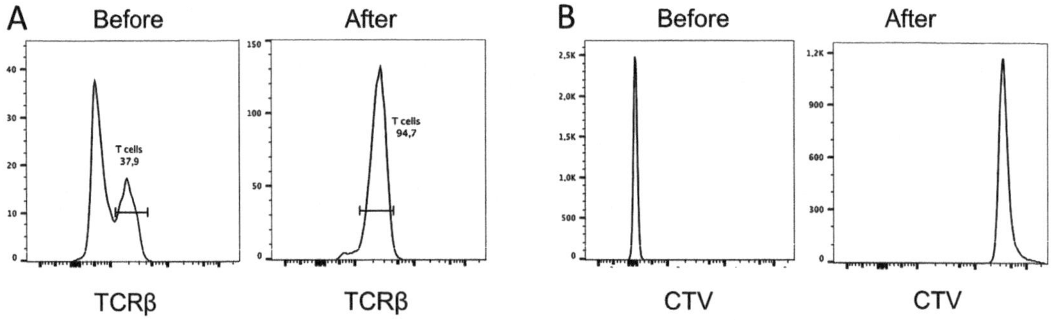

Fig. 4 Evaluation of the purity of enriched cells. Lymphocytes were enriched from donor CD45.1$^+$ mice. (**a**) Histogram showing T cells before and after T cell enrichment. (**b**) Histogram showing CTV staining (left unstained and right stained)

the antibodies coupled to the biotin microbeads and keeping them on the wall of the tube.

5. Place the new tube (without lid) in the magnet, and incubate at RT for 5 min. Keep the tube in the magnetic, and gently invert the cell suspension in a new 5 ml polystyrene tube to obtain the enriched T cell suspension.

3.5 Evaluation of T Cell Purity

The purity of the enriched fraction relates to the percentage of target cells in the isolate compared to the initial heterogeneous cell preparation and is evaluated by flow cytometry (Fig. 4a). Stain an aliquot of the initial cell preparation as well as an aliquot of the enriched fraction with an appropriately diluted anti-TCR antibody conjugated to a fluorophore. Evaluate the percentage of TCR$^+$ cells before and after enrichment: the purity of the TCR$^+$ population should reach approximately 95%.

3.6 Evaluation of T Cell Recovery

The recovery is a measure of the efficiency of cell separation. It is calculated as the absolute number of cells obtained after cell separation compared to the number of cells enumerated in the original suspension. This parameter is informative about the number of cells lost during cell separation.

3.6.1 Labelling of Lymphocytes with CFSE or CTV

CFSE or CTV is used to evaluate lymphocyte proliferation. Based on the dilution of the dye, it is possible to estimate the number of divisions achieved by lymphocytes. CFSE is excited by a 488 nm laser and has an emission of 520 nm. CTV is chemically related to CFSE but is excited by a 405 nm violet laser and is detected using a 450/40 nm band-pass filter:

1. CellTrace™ Violet should be prepared and used according to the manufacturer protocol. Briefly, resuspend cells at a concentration of 10×10^6 cells/ml in PBS. Add 1 μl of 5 mM CellTrace™ Violet stock solution to each ml of cells (final

concentration of 5 μM). Incubate the cells for 20 min at 37 °C, and protect cells from light.

2. Wash the cells by adding 10–30 ml of RPMI media. Centrifuge and resuspend cells in fresh media.
3. To ensure that cells have incorporated CTV, the staining must be verified by flow cytometry. It is important to note that CTV staining is expected to be very bright. In order to detect the fluorescence, the voltage of the laser may need to be decreased (Fig. 4b).

3.7 Adoptive Transfer in Lymphopenic and Genetically Modified Recipients

1. Based on the purity and proportion of T cells obtained after T cell enrichment, resuspend cells in PBS at a final concentration of 2×10^6/ml.
2. Inject 1×10^6 (500 μl) CTV-labelled lymphocytes through the lateral tail vein using a 1 ml syringe and a 27 1/2 gauge needle (Fig. 5).

3.8 Evaluation of Transferred T Cells by Flow Cytometry

3.8.1 Sacrifice of Mice

In our experiments, we normally evaluate T cells at day +6 post-transfer. Spleens and LNs are retrieved as previously described. Prepare cell suspension from LNs and the spleen, and resuspend cells at 10×10^6 cells/ml in RPMI media.

3.8.2 Antibody Staining

Stain cells with a cocktail of properly diluted primary antibodies consisting of CD4, CD8, TCR, CD45.1, and CD45.2, tagged to different fluorochromes. Do not use fluorochromes such as Pacific Blue, BV421, or eFluor450 if CTV was used to label T cells or

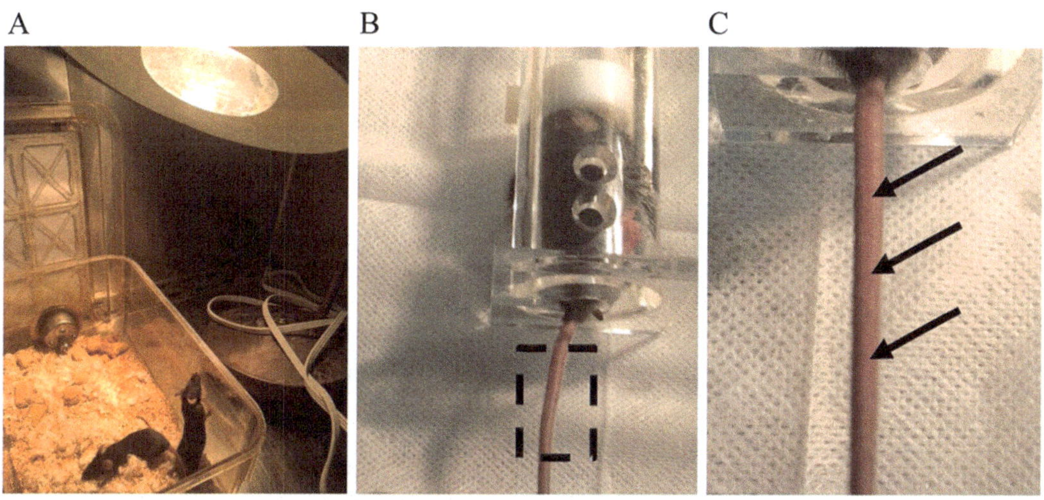

Fig. 5 Intravenous injection of T cells in recipient mice. (**a**) The cage of mice is placed under a heating lamp for a maximum of 5 min to dilate the lateral tail vein. (**b**) The mouse is then placed in a restrainer cage prior to IV injection. (**c**) Location of the lateral tail veins of the mouse

FITC and AF488 if CFSE was used. Stain 2–10 × 10^6 cells per organ, depending on the expected outcome: if less proliferation is expected, then more cells should be stained. Incubate for 30 min in the dark at 4 °C. Add PBS 1× to wash the excess of antibodies, centrifugation at 288 × *g* for 10 min, and resuspend in 100–200 μl of PBS. Keep the cells at 4 °C until analysis. Avoid exposure to light.

In order to obtain peaks of proliferation, it is preferable to label more cells. In these experiments, we normally acquire between 2 and 10 millions events per sample by flow cytometry.

3.8.3 Flow Cytometry

When several fluorochromes are used simultaneously, overlap or interference between fluorochromes can occur. Compensation is a mathematical process to reduce the overlap of emission between fluorochromes. Compensation is performed using compensation beads. Prepare individual 5 ml polystyrene round-bottom tube by adding 1 μl of antibody to 10 μl of universal compensation beads + 100 μl of PBS. In addition, reserve a tube of unstained cells for compensation. After compensations have been calculated manually or automatically, experimental tubes can be processed. For CTV or CFSE analysis, a minimum of 1 × 10^6 events must be acquired.

The acquisition of data by flow cytometry is complex and requires to draw a gate around the cell population of interest. The parameters forward and side scatter (FSC and SSC) are used to delineate the lymphocyte population based on size and granularity (Fig. 7a). Once lymphocytes are identified, their staining properties can be analyzed separately. Two-dimensional dot plots are often used to analyze the fluorescence of the cell population that has been stained with antibodies conjugated to fluorophores (Fig. 7b–f).

4 Results and Discussion

Adoptive transfer of T cells in different mouse models is routinely used to evaluate HP of T lymphocytes in vivo. For these experiments, transfer of enriched total T cells, individual $CD4^+$ or $CD8^+$ T cell populations, or even specific T cell populations like naïve or memory T cells is possible. Several companies have commercial kits available for positive and negative selection of these specific sub-types. The complexity of the process and the time to isolate cells may vary considerably. Although purities are normally high, the yield of recovery can also vary considerably between companies. In our laboratory, negative selection is preferred since positive selection includes binding of antibodies to the target cells, which could trigger activation and can affect the function of the cells. However, depending on the cell type isolated, negative selection kits are not always available. For these experiments, positive selection or cell sorting by flow cytometry can be used.

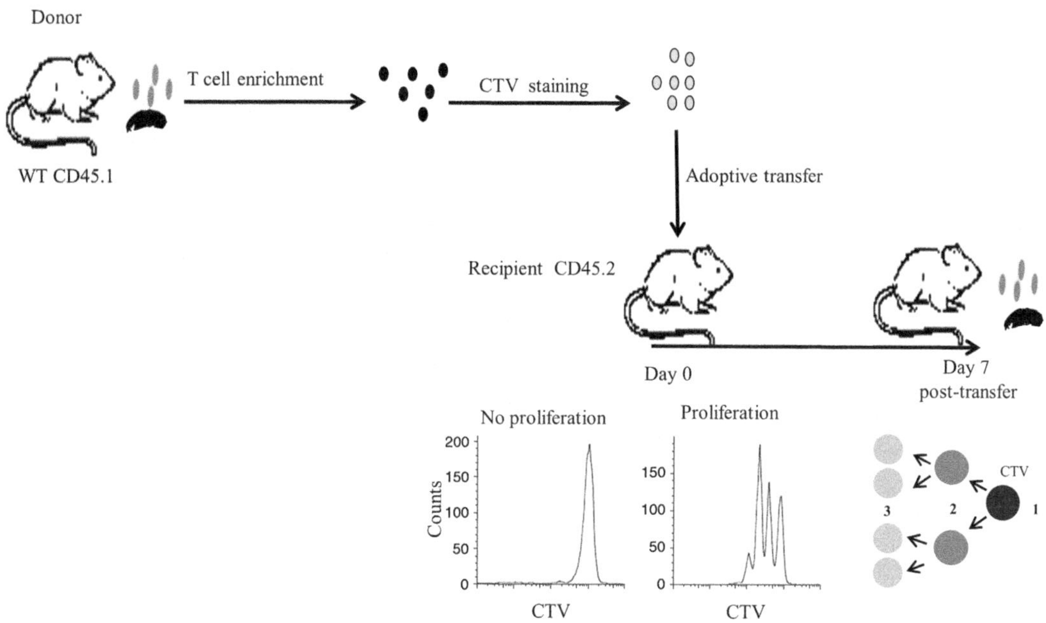

Fig. 6 Schematic representation of CTV staining. T cells from CD45.1$^+$ donor mouse were enriched from LNs using the EasySep Mouse T Cell Enrichment Kit from STEMCELL Technologies. Enriched T cells are stained with CTV and IV injected in CD45.2$^+$-recipient mice. Six days later, mice are sacrificed and LNs are retrieved and donor CD45.1$^+$ lymphocyte evaluated by flow cytometry for evidence of proliferation based on CTV dilution

In this minireview, we describe how to enrich T cells from LNs using the EasySep Mouse T Cell Enrichment Kit from STEMCELL Technologies. The procedure takes between 30 and 45 min, the purity is normally above 90–95%, and 15 × 10^6 T cells can be obtained from 12 LNs from a single 7–10-week-old wild-type (WT) mouse (Fig. 4a).

In the representative experiment illustrated in Fig. 6, we transferred enriched CTV-labeled polyclonal CD45.1$^+$ T cells into different CD45.2$^+$ recipient mice (Fig. 6). Six days later, LNs and spleens were retrieved from the recipients and T cells analyzed by flow cytometry. The use of congenic marker (CD45.1 vs. CD45.2) is essential in order to distinguish between donor and recipient T cells. Although donor T cells have been labelled with CTV, brisk proliferation can completely dilute CTV content, and the congenic CD45 marker becomes essential for the identification of donor cells.

For data analysis, multiple gates are necessary in order to evaluate proliferation of donor lymphocytes (Fig. 7a). First, the lymphocyte population is identified using forward and side scatter. After gating on lymphocytes, TCR$^+$ cells are identified (Fig. 7b). TCR$^+$ cells are composed of donor and recipient lymphocytes. After gating on TCR$^+$ cells, we can discriminate between donor and

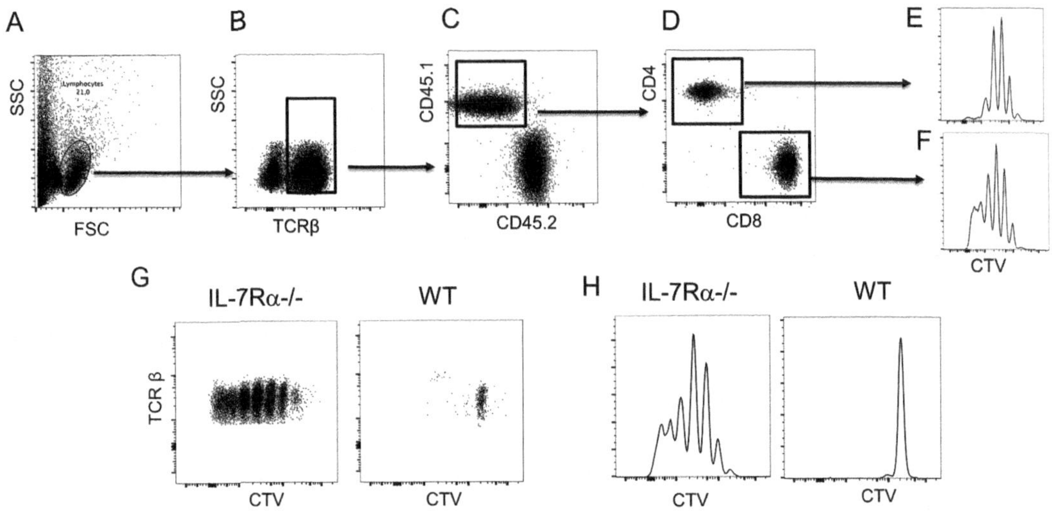

Fig. 7 Gating strategy to evaluate lymphocyte proliferation using CTV. CTV-labeled CD45.1$^+$-enriched T cells were transferred into CD45.2$^+$ IL-7Rα$^{-/-}$ recipient, and 6 days later, LN T cells were evaluated by flow cytometry. (**a**) Lymphocyte population was identified based on FSC and SSC gating. (**b**) TCRβ$^+$ T cells were identified based on SSC and TCRβ positive cells gated on lymphocytes in (**a**). (**c**) Recipient and donor T cells were identified based on CD45.1 and CD45.2 positive cells gated on TCRβ$^+$ cells in (**b**). (**d**) Donor CD4$^+$ and CD8$^+$ T cells were identified based on CD4 and CD8 receptors gated on CD45.1$^+$ cells in (**c**). (**e**, **f**) CTV profile of CD4$^+$ and CD8$^+$ T cells was visualized by gating on CD4$^+$ or CD8$^+$ receptors, respectively. (**g-h**) Dot plot and histograms were used to evaluate CTV profile of T cells in lymphopenic IL-7Rα$^{-/-}$ (left) and wild-type hosts (right)

recipient lymphocytes based on CD45.1 and CD45.2 (Fig. 7c). Donor lymphocytes are CD45.1$^+$, and they include both CD4$^+$ and CD8$^+$ lymphocytes (Fig. 7d). By gating on CD4$^+$ cells, we can visualize CD4$^+$ T cell proliferation based on CTV dilution (Fig. 7e). Similarly, by gating on CD8$^+$ lymphocytes, we can visualize CD8$^+$ T cell proliferation (Fig. 7f). Dot plots or histograms are commonly used to represent proliferation (Fig. 7g-h).

The proliferation of transferred cells is illustrated by the dilution of CTV. Each peak corresponds to one cell division. After several days in a lymphopenic host, the number of divisions is expected to be higher than in a normal mouse which already has a full niche (Fig. 7g). This is also illustrated in Fig. 8, which includes proliferation profiles of T cells transferred into WT, IL-7Rα$^{-/-}$, Rag$^{-/-}$, and IL-7$^{-/-}$ hosts. As demonstrated, CD4$^+$ and CD8$^+$ T cell proliferation is not supported in WT and IL-7$^{-/-}$ hosts (Fig. 8). In contrast, CD8$^+$ proliferation is robust in both lymphopenic IL-7Rα$^{-/-}$ and Rag$^{-/-}$ hosts. However, CD4$^+$ T cell proliferation is robust in only the IL-7Rα$^{-/-}$ host and is very modest in Rag$^{-/-}$. These differences may relate to the level of systemic IL-7 or MHC class II expression by antigen-presenting cells in Rag$^{-/-}$ versus IL-7Rα$^{-/-}$ mice [11].

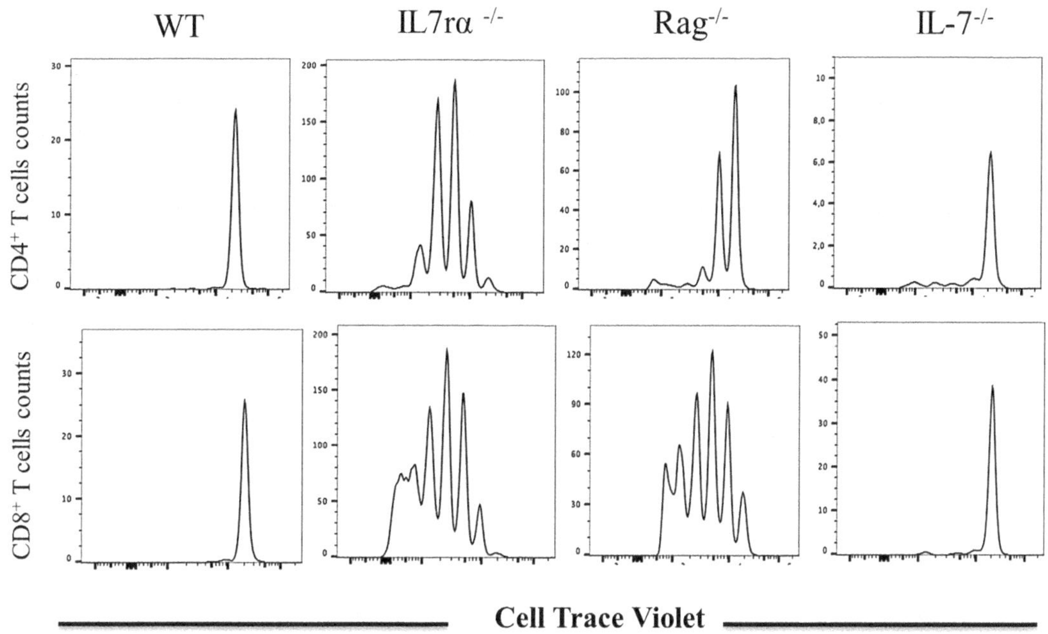

Fig. 8 CTV profile of $CD4^+$ and $CD8^+$ T cells in different mouse recipients. Congenic CTV-labelled enriched CD45.1$^+$ T cells (1×10^6) were intravenously injected into WT, IL-7R$\alpha^{-/-}$, Rag$^{-/-}$, and IL-7$^{-/-}$ mice, and proliferation was evaluated by flow cytometry 6 days post-transfer

In conclusion, the use of CFSE or CTV has considerably helped our understanding of critical factors involved in T cell homeostasis. With such technology, it is now possible to accurately measure the effect of diseases or drugs on T cell proliferation in vivo. Whether it is used in basic or preclinical studies, the use of CTV or CFSE to measure cell proliferation is not limited to lymphocytes but can be applied to many other cell types.

Acknowledgments

This work was supported by grants from the Cancer Research Society of Canada (grant no. 22669 and 24380 to M.G.) and in part by a grant from the Foundation de l'Hôpital Maisonneuve-Rosemont, Montreal, Quebec, Canada.

References

1. Mackall CL, Bare CV, Granger LA, Sharrow SO, Titus JA, Gress RE (1996) Thymic-independent T cell regeneration occurs via antigen-driven expansion of peripheral T cells resulting in a repertoire that is limited in diversity and prone to skewing. J Immunol 156 (12):4609–4616
2. Berzins SP, Godfrey DI, Miller JF, Boyd RL (1999) A central role for thymic emigrants in peripheral T cell homeostasis. Proc Natl Acad Sci U S A 96(17):9787–9791
3. Miller JFAP (1962) Immunological significance of the thymus of the adult mouse. Nature

195(4848):1318–1319. https://doi.org/10.1038/1951318a0

4. Gauthier SD, Leboeuf D, Manuguerra-Gagne-R, Gaboury L, Guimond M (2015) Stromal-derived factor-1alpha and interleukin-7 treatment improves homeostatic proliferation of naive CD4(+) T cells after allogeneic stem cell transplantation. Biol Blood Marrow Transplant 21(10):1721–1731. https://doi.org/10.1016/j.bbmt.2015.06.019
5. Gauthier SD, Moutuou MM, Daudelin F, Leboeuf D, Guimond M (2018) IL-7 is the limiting homeostatic factor that constrains homeostatic proliferation of CD8(+) T cells after allogeneic stem cell transplantation and graft-versus-host disease. Biol Blood Marrow Transplant. https://doi.org/10.1016/j.bbmt.2018.12.066
6. Murali-Krishna K, Lau LL, Sambhara S, Lemonnier F, Altman J, Ahmed R (1999) Persistence of memory CD8 T cells in MHC class I-deficient mice. Science (New York, NY) 286 (5443):1377. https://doi.org/10.1126/science.286.5443.1377
7. Sandau MM, Winstead CJ, Jameson SC (2007) IL-15 is required for sustained lymphopenia-driven proliferation and accumulation of CD8 T cells. J Immunol 179(1):120–125
8. Lenz DC, Kurz SK, Lemmens E, Schoenberger SP, Sprent J, Oldstone MB, Homann D (2004) IL-7 regulates basal homeostatic proliferation of antiviral CD4+T cell memory. Proc Natl Acad Sci U S A 101(25):9357–9362. https://doi.org/10.1073/pnas.0400640101
9. Purton JF, Tan JT, Rubinstein MP, Kim DM, Sprent J, Surh CD (2007) Antiviral CD4+ memory T cells are IL-15 dependent. J Exp Med 204(4):951–961. https://doi.org/10.1084/jem.20061805
10. Fry TJ, Connick E, Falloon J, Lederman MM, Liewehr DJ, Spritzler J, Steinberg SM, Wood LV, Yarchoan R, Zuckerman J, Landay A, Mackall CL (2001) A potential role for interleukin-7 in T-cell homeostasis. Blood 97 (10):2983–2990
11. Guimond M, Veenstra RG, Grindler DJ, Zhang H, Cui Y, Murphy RD, Kim SY, Na R, Hennighausen L, Kurtulus S, Erman B, Matzinger P, Merchant MS, Mackall CL (2009) Interleukin 7 signaling in dendritic cells regulates the homeostatic proliferation and niche size of CD4+ T cells. Nat Immunol 10(2):149–157. https://doi.org/10.1038/ni.1695
12. Tan JT, Dudl E, LeRoy E, Murray R, Sprent J, Weinberg KI, Surh CD (2001) IL-7 is critical for homeostatic proliferation and survival of naive T cells. Proc Natl Acad Sci U S A 98 (15):8732–8737. https://doi.org/10.1073/pnas.161126098
13. Goldrath AW, Bogatzki LY, Bevan MJ (2000) Naive T cells transiently acquire a memory-like phenotype during homeostasis-driven proliferation. J Exp Med 192(4):557–564
14. Abdelsamed HA, Zebley CC, Youngblood B (2017) In vitro homeostatic proliferation of human CD8 T cells. Bio Protoc 7(22). https://doi.org/10.21769/BioProtoc.2619
15. von Freeden-Jeffry U, Vieira P, Lucian LA, McNeil T, Burdach SE, Murray R (1995) Lymphopenia in interleukin (IL)-7 gene-deleted mice identifies IL-7 as a nonredundant cytokine. J Exp Med 181(4):1519–1526
16. Rocha B, Penit C, Baron C, Vasseur F, Dautigny N, Freitas AA (1990) Accumulation of bromodeoxyuridine-labeled cells in central and peripheral lymphoid organs: minimal estimates of production and turnover rates of mature lymphocytes. Eur J Immunol 20 (8):1697–1708. https://doi.org/10.1002/eji.1830200812
17. Tough DF, Sprent J (1994) Turnover of naive- and memory-phenotype T cells. J Exp Med 179 (4):1127–1135
18. Parretta E, Cassese G, Santoni A, Guardiola J, Vecchio A, Di Rosa F (2008) Kinetics of in vivo proliferation and death of memory and naive CD8 T cells: parameter estimation based on 5-bromo-2′-deoxyuridine incorporation in spleen, lymph nodes, and bone marrow. J Immunol 180(11):7230–7239
19. Dulude G, Roy DC, Perreault C (1999) The effect of graft-versus-host disease on T cell production and homeostasis. J Exp Med 189 (8):1329–1342
20. Lyons AB, Parish CR (1994) Determination of lymphocyte division by flow cytometry. J Immunol Methods 171(1):131–137
21. Quah BJ, Parish CR (2012) New and improved methods for measuring lymphocyte proliferation in vitro and in vivo using CFSE-like fluorescent dyes. J Immunol Methods 379 (1–2):1–14. https://doi.org/10.1016/j.jim.2012.02.012
22. Powrie F (1995) T cells in inflammatory bowel disease: protective and pathogenic roles. Immunity 3(2):171–174
23. Budd RC, Cerottini JC, Horvath C, Bron C, Pedrazzini T, Howe RC, MacDonald HR (1987) Distinction of virgin and memory T lymphocytes. Stable acquisition of the Pgp-1 glycoprotein concomitant with antigenic stimulation. J Immunol 138(10):3120
24. Desbarats J, Wade T, Wade WF, Newell MK (1999) Dichotomy between naïve and memory

CD4(+) T cell responses to Fas engagement. Proc Natl Acad Sci U S A 96(14):8104–8109
25. Moses CT, Thorstenson KM, Jameson SC, Khoruts A (2003) Competition for self ligands restrains homeostatic proliferation of naive CD4 T cells. Proc Natl Acad Sci U S A 100 (3):1185. https://doi.org/10.1073/pnas.0334572100
26. Ge Q, Bai A, Jones B, Eisen HN, Chen J (2004) Competition for self-peptide-MHC complexes and cytokines between naïve and memory CD8+ T cells expressing the same or different T cell receptors. Proc Natl Acad Sci U S A 101(9):3041. https://doi.org/10.1073/pnas.0307339101
27. Troy AE, Shen H (2003) Cutting edge: homeostatic proliferation of peripheral T lymphocytes is regulated by clonal competition. J Immunol 170(2):672–676
28. Min B, Yamane H, Hu-Li J, Paul WE (2005) Spontaneous and homeostatic proliferation of CD4 T cells are regulated by different mechanisms. J Immunol 174(10):6039–6044
29. Shinkai Y, Rathbun G, Lam KP, Oltz EM, Stewart V, Mendelsohn M, Charron J, Datta M, Young F, Stall AM et al (1992) RAG-2-deficient mice lack mature lymphocytes owing to inability to initiate V(D)J rearrangement. Cell 68(5):855–867
30. Mombaerts P, Mizoguchi E, Grusby MJ, Glimcher LH, Bhan AK, Tonegawa S (1993) Spontaneous development of inflammatory bowel disease in T cell receptor mutant mice. Cell 75 (2):274–282
31. Schluns KS, Kieper WC, Jameson SC, Lefrancois L (2000) Interleukin-7 mediates the homeostasis of naive and memory CD8 T cells in vivo. Nat Immunol 1(5):426–432. https://doi.org/10.1038/80868
32. Vivien L, Benoist C, Mathis D (2001) T lymphocytes need IL-7 but not IL-4 or IL-6 to survive in vivo. Int Immunol 13(6):763–768
33. Ramsey C, Rubinstein MP, Kim DM, Cho JH, Sprent J, Surh CD (2008) The lymphopenic environment of CD132 (common gamma-chain)-deficient hosts elicits rapid homeostatic proliferation of naive T cells via IL-15. J Immunol 180(8):5320–5326
34. Kennedy MK, Glaccum M, Brown SN, Butz EA, Viney JL, Embers M, Matsuki N, Charrier K, Sedger L, Willis CR, Brasel K, Morrissey PJ, Stocking K, Schuh JC, Joyce S, Peschon JJ (2000) Reversible defects in natural killer and memory CD8 T cell lineages in interleukin 15-deficient mice. J Exp Med 191 (5):771–780
35. Ge Q, Rao VP, Cho BK, Eisen HN, Chen J (2001) Dependence of lymphopenia-induced T cell proliferation on the abundance of peptide/MHC epitopes and strength of their interaction with T cell receptors. Proc Natl Acad Sci U S A 98(4):1728–1733. https://doi.org/10.1073/pnas.98.4.1728
36. Le Campion A, Bourgeois C, Lambolez F, Martin B, Leaument S, Dautigny N, Tanchot C, Penit C, Lucas B (2002) Naive T cells proliferate strongly in neonatal mice in response to self-peptide/self-MHC complexes. Proc Natl Acad Sci U S A 99(7):4538–4543. https://doi.org/10.1073/pnas.062621699
37. James E, Chai JG, Dewchand H, Macchiarulo E, Dazzi F, Simpson E (2003) Multiparity induces priming to male-specific minor histocompatibility antigen, HY, in mice and humans. Blood 102(1):388–393. https://doi.org/10.1182/blood-2002-10-3170

Chapter 22

Detection, Expansion, and Isolation of Human MAIT Cells

Yu Liu, Wei Wang, Xiongwen Wu, and Xiufang Weng

Abstract

Mucosal-associated invariant T (MAIT) cells are a novel subset of innate-like T cells that recognize vitamin B metabolites from a range of microbes presented by MHC class I-related molecules (MR1). The term mucosal-associated invariant T cells derives from the fact that MAIT cells are abundant in the liver and mucosal tissues, and human MAIT cells use a semi-invariant TCR Vα7.2 Jα33 paired with Vβ2 or Vβ13. Here, based on the interaction between MAIT cell and its ligand 5-OP-RU/MR1, we describe the protocols for identification, rapid expansion, and isolation of human MAIT cells.

Key words MAIT cells, MR1, 5-OP-RU, Semi-invariant TCR

1 Introduction

Similar to NKT cells, mucosal-associated invariant T (MAIT) cells have numerous features distinguishing them from conventional T cells, including semi-invariant TCR usage (Va7.2Ja33 paired with Vβ2 or Vβ13) and innate-like phenotypes [1–4]. They are readily detected in blood, mesenteric lymph nodes, and the gastrointestinal mucosa and enrich up to −40% in the liver and lungs in human [5–7]. MAIT cells are not reactive to peptides. They recognize vitamin B metabolites from microorganisms that are presented on major histocompatibility complex (MHC) class I-like molecules, called MR1 [2–4]. Kjer-Nielsen et al. [8] discovered that 6-formyl pterin (6-FP), a degradation product of folic acid that was present in tissue culture media, was able to bind to MR1 and allowed its refolding. While 6-FP was unable to activate MAIT cells, 5-OE-RU (5-(2-oxoethylideneamino)6-d-ribitylaminouracil) and 5-OP-RU (5-(2-oxopropylideneamino)-6-d-ribitylaminouracil), by-products of microbial riboflavin synthesis, were able to covalently bind to MR1 and activate MAIT cells. Upon stimulation, human MAIT cells display a preferential ability to produce Th1 cytokines and IL-17, which is in line with their predominantly expressed transcription factors Tbx21 and RORC (Th1 and Th17 specific,

Chaohong Liu (ed.), *T-Cell Receptor Signaling: Methods and Protocols*, Methods in Molecular Biology, vol. 2111,
https://doi.org/10.1007/978-1-0716-0266-9_22,

respectively) [9]. MAIT cells harbor multiple layers of heterogeneity in response to microbial antigens and innate cytokine stimuli. On this base, MAIT cells may play different roles in immune responses against different types of bacteria, fungi, and viruses [10–12]. Moreover, Won et al. [13] suggested that circulating MAIT cells may promote cytotoxicity or even directly kill tumor cells.

This unit details methods for detection, in vitro expansion, and isolation of these human MAIT cells. All human blood samples should be handled under the biological guidelines for human samples: investigators should work with the class 2 safety cabinet and wear gloves, and all materials should be disinfected with bleach before being discarded. For patient samples, the protocol should be approved by the Institutional Committee and the Committee for the Protection of Patients from Biological Risks.

2 Material

2.1 Detection of MAIT Cells

1. MR1 tetramer (1.2 mg/ml): Brilliant Violet™ (BV421) or allophycocyanin (APC)-labeled 5-OP-RU/MR1 tetramer was provided by the Tetramer Core Facility of the National Institutes of Health, USA (http://research.yerkes.emory.edu/tetra mer core/MR1-Tetramers.html).
2. Phosphate Buffered Saline (PBS) without calcium and magnesium (Sigma, D8537).
3. Red blood cell (RBC) lysing solution, 10× (Sigma, R7757).
4. Ficoll-Paque Plus (Tianjin Bayang Biological Products Technology, LTS1077).
5. Trypan blue staining solution, 0.4% (wt/vol) (GIBCO, 15250-061).
6. RPMI medium 1640 (1×) (GIBCO, 8119009).
7. Hemocytometer.
8. Antibodies for FACS analysis or sorting (*see* Table 1).
9. FACS staining buffer: PBS, 1% (vol/vol) flow cytometry (FCM), 0.09% (wt/vol) NaN3 (*see* **Note 1**).
10. Flow cytometer, FACS Lsr (BD Biosciences) for analysis.
11. Cell strainer, 100 mm (Falcon, 352360).

2.2 Expansion of MAIT Cells

1. Red blood cell (RBC) lysing solution, 10× (Sigma, R7757).
2. Ficoll-Paque Plus (Tianjin Bayang Biological Products Technology LTS1077).
3. Trypan blue stain, 0.4% (wt/vol) (GIBCO, 15250-061).

Table 1
Antibodies for FACS analysis or sorting

Labeled	Antibody clone	Supplier cat. no.
APC-CY7 anti-human CD3	HIT3a	BioLegend 300318
PE anti-human CD161	HP-3G10	BioLegend 339904
PE-CY7 anti-human TCRVα7.2	3C10	BioLegend 351712
BV510 anti-human CD8	SK1	BD horizon 583919
FITC anti-human CD3	UCHT1	BioLegend 300440
BV421 anti-human CD161	HP-3G10	BioLegend 339914

4. Phosphate Buffered Saline (PBS) without calcium and magnesium (Sigma, D8537).
5. Plastic tubes: 15 ml tube (Falcon, cat. no. 352196), 50 ml tube (Falcon, 352070), and 1.5 ml tube (Eppendorf, 3810).
6. Culture plates and dish: 96-well (Falcon, 353072).
7. Hemocytometer.
8. Heat-inactivated flow cytometry (FCM).
9. RPMI medium 1640 basic (1×) (GIBCO, 8119009).
10. Complete RPMI1640 culture medium: RPMI1640 medium plus 10%(vol/vol) FBS, penicillin–streptomycin (100 U/ml), 10 mM HEPES buffer solution, 0.1 mM MEM nonessential amino acids, 1 mM sodium pyruvate, and 5.5 mM 2-mercaptoethanol (2-ME).
11. Penicillin–streptomycin (GIBCO, 11360-070).
12. HEPES buffer solution (GIBCO, 15630-080).
13. MEM nonessential amino acids (GIBCO, 11140-050).
14. Sodium pyruvate (GIBCO, 11360-070).
15. 2-ME (GIBCO, 21985-023).
16. Sulfate Latex Beads: 8% (wt/vol), 3.5 μm (Thermo Fisher, S37224).
17. Brilliant Violet™ (BV421)-labeled 5-OP-RU MR1 tetramer (1.2 mg/ml).
18. Purified NA/LE mouse anti-human CD28 (BD Pharmingen™, 555725).
19. Bovine serum albumin (BSA) albumin fraction V (BioFroxx, 4240GR025).
20. Human recombinant interleukin, interleukin-2 (IL-2) (R&D Systems, 202-IL-050).
21. Water-jacketed CO_2 incubator.

2.3 Isolation of MAIT Cells

1. Anti-APC microbeads (Miltenyi Biotec, 130-090-855).
2. MS column (Miltenyi Biotec, 130-042-401).
3. Binding buffer: PBS, 2% (vol/vol) FBS, 2 mM EDTA.
4. Washing buffer: 0.5% BSA PBS.
5. Hemocytometer.
6. Magnetic bead sorting frame (Multi Stand).

3 Method

3.1 Detection of MAIT Cells

3.1.1 Preparation of Human Peripheral Blood Mononuclear Cells (PBMCs)

1. Transfer 5 ml of peripheral blood from volunteer donors into a heparinized tubes.
2. Dilute 5 ml of whole blood with equal volume of 1 × PBS.
3. Put 5 ml of Ficoll-Paque Plus into a 15 ml polypropylene centrifugal tube. Carefully add 5 ml of diluted blood along the wall of the tube onto Ficoll-Paque gently; do not destroy the liquid interface.
4. Centrifuge at 266 × *g* for 25 min at room temperature (*see* **Note 2**).
5. Human PBMCs are milky white middle layer visible at the lipid interface. Carefully remove the upper light yellow layer, and suck out PBMCs at the interface by a capillary pipette gently. Do not suck out the lower layer.
6. Wash the harvested PBMCs twice with 10 ml of PBS by centrifugation at 266 × *g* for 10 min.
7. Resuspend the PBMC pellet with 1 ml of RBC lysing solution (if need), and let it remain for 3–5 min at room temperature.
8. Wash PBMC pellet twice with PBS by centrifugation at 1000 rpm for 10 min.
9. Resuspend PBMCs with complete RPMI1640 culture medium.
10. Check cell viability by Trypan blue staining, and count live cell numbers on a hemocytometer under a light microscope (*see* **Note 3**).

3.1.2 Detection of Human MAIT Cells by Flow Cytometry

Human MAIT cells can be detected through flow cytometry by either MR1 tetramer loaded with 5-OP-RU or by co-staining with antibodies against CD161 and TCR Vα7.2 chain (Fig. 1). Generally, MAIT cells are 10–40% of the human liver, 0.1–10% in peripheral blood, 1–10% of the lung, 2–10% of the intestine, and <1% of lymphoid tissue among CD3 + T cells (healthy volunteer).

1. Resuspend PBMCs at a concentration of approximately 1×10^6 cells per 100 μl FACS staining buffer.

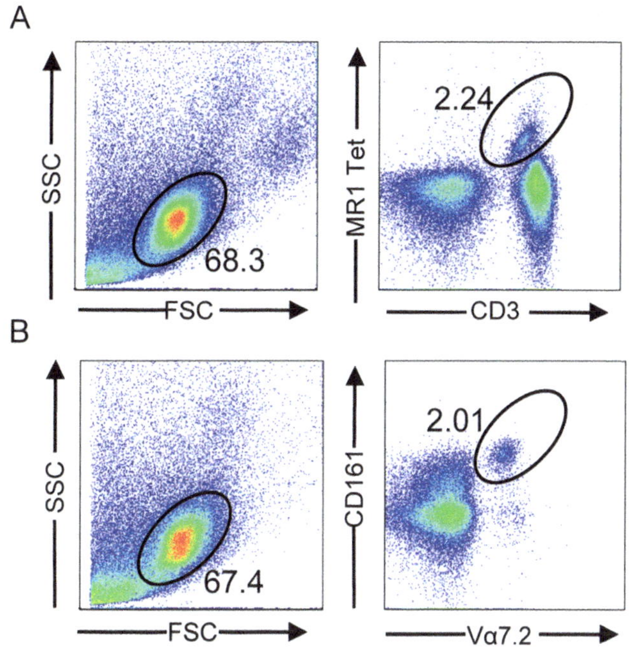

Fig. 1 Detection of human MAIT cells from peripheral blood mononuclear cells (PBMCs). Lymphocytes were gated, and MAIT cells were identified by their expression of CD3 and reactivity with the 5-OP-RU/MR1 tetramer (**a**) or expression of CD161 and TCRVα7.2 (**b**)

2. Add mAbs and/or 5-OP-RU/MR1 tetramer to the samples. For detection of human MAIT cells, two methods can be used:

   Tetramer staining: use APC-labeled human 5-OP-RU MR1 tetramer and APC-CY7-conjugated anti-human CD3.

   Substitution marking: PE-CY7-conjugated anti-human Va7.2, APC-CY7-conjugated anti-human CD3, and PE-conjugated anti-human CD161 (*see* **Note 4**).

3. Incubate for 45 min at 4 °C in the dark.
4. Wash the cells with FACS staining buffer by centrifuging at 266 × *g* for 5 min at 4 °C, and resuspend with 200–400 μl FACS staining buffer (final concentration, 5×10^6 cells/ml).
5. Filter the cells through cell strainers before transferring to an FACS tube. Keep the cells on ice until the flow cytometry analyzer is ready.
6. Analyze MAIT cells by flow cytometry (Fig. 1).

3.2 Expansion of Human MAIT Cells

3.2.1 Preparation of Artificial Antigen-Presenting Cells

1. Take one drop of Sulfate Latex Beads and put it into 15 ml tube. Wash with 1× PBS twice by centrifuging 10 min at a speed of 1660 × *g*. Add appropriate PBS to suspend the beads and adjust the concentration to 5 × 10^8/ml.
2. Suck out 100 μl (about 5 × 10^7) diluted beads above into another 15 ml tube. Add 1 μl anti-human CD28 (1 mg/ml) and 1 μl 5-OP-RU/ MR1 tetramer (1.5 mg/ml); incubate at 4 °C for 24 h with intermittent oscillation mixing.
3. Wash with 1 × PBS once by centrifuging 10 min at 1660 × *g*, discard supernatant completely, and add 500 μl 3% BSA PBS. Incubate at 4 °C for 12 h with intermittent oscillation mixing.
4. Wash with 1 × PBS once by centrifuging 10 min at 1660 × *g*, and discard supernatant completely. Add appropriate amount of PBS to suspend, adjust the concentration of beads to 5 × 10^5/μl, transfer it into a brown (light-shielded) EP tube, and store at 4 °C. Ready to use as 5-OP-RU/MR1 artificial antigen-presenting cells (aAPCs). The beads coated with only anti-human CD28 were used as control beads.

3.2.2 Expansion of Human MAIT Cells

1. Isolation of PBMCs from 10 ml blood (*see* Subheading 3.1.1). Adjust the cell concentration to 5 × 10^6/ml with complete RPMI1640 culture medium.
2. Seed the cells into 96-well plate, 100 μl/well, 5 × 10^5 cells/ well. Add 100 μl complete RPMI1640 culture medium or the medium containing 40 U/ml recombinant human IL-2 per well. The final volume of each well was 200 μl, and the final concentration of IL-2 was 20 U/ml (if added) (*see* **Note 5**).
3. Add beads (5-OP-RU/MR1 aAPCs) prepared above into each well, 0.5 μl/well (whirlpool mixing is required before adding). Mix the liquid in each well gently by pipetting. Put the cell culture plate into 37 °C, 5% CO_2 incubator.
4. On the third day of cell culture, half of the fluid is exchanged. The supernatant of 100 μl culture medium is extracted from each well and replenished with fresh complete RPMI1640 culture medium or the medium containing 20 U/ml IL-2, 100 μl/well.
5. Culture cells in the cell incubator for 7 days.
6. Suck out a small amount of the expanded cells to test the MAIT cell frequency by flow cytometry (Fig. 2).

3.3 Magnetic Bead Sorting of MAIT Cells

1. Collect the expanded cells into 15 ml tube. Wash the cells with washing buffer (e.g., 0.5% BSA PBS). Discard supernatant, and resuspend cell pellets with 200 μl complete RPMI1640 culture medium. Add 1 μl APC-labeled 5-OP-RU/MR1 tetramer; mix evenly. Incubate for 45 min on ice.

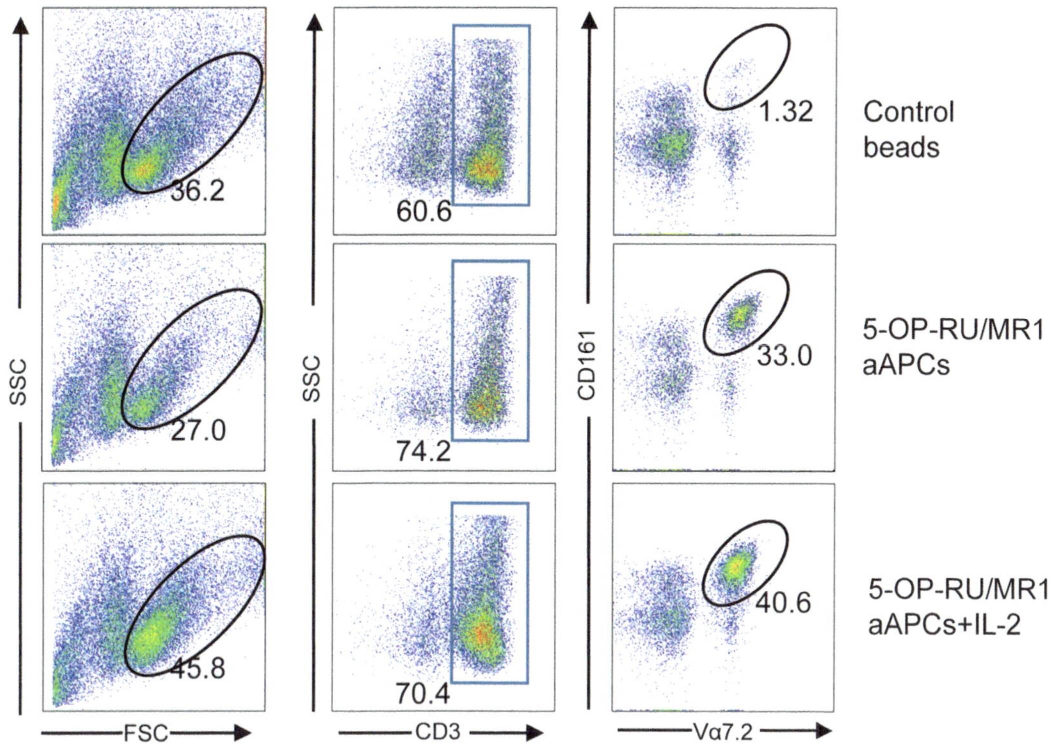

Fig. 2 Expansion of MAIT cells from PBMCs. PBMCs were obtained from healthy donor and cultured in vitro in the presence of control beads (upper panel), 5-OP-RU/MR1-coated beads (5-OP-RU/MR1 aAPCs, middle panel), or 5-OP-RU/MR1-coated beads with IL-2 supplement (5-OP-RU/MR1 aAPCs+IL-2, lower panel). The dot plots depict the gating strategy and the frequencies of MAIT cells in different coculture system

2. Wash the cells with washing buffer once by centrifuging 6 min at 266 × *g*, room temperature or chilled.
3. Resuspend the cells ($<10^7$) in 80 μl washing buffer, add 20 μl of anti-APC microbeads, incubate for 30 min on ice, and mix intermittently.
4. Wash the cells twice in 1–2 ml washing buffer, each time by centrifuging 10 min at 266 × *g*, room temperature or chilled, and resuspend in 1 ml binding buffer.
5. Choose an appropriate MACS column and corresponding MACS separator according to the number of total cells and MAIT cells. The size and separating capacity of different MACS columns are detailed in Table 2.
6. Prewash the column with 3 ml binding buffer and assemble on the magnet. Apply the cells to the column and wash the column three times, each time with 3 ml washing buffer.
7. Remove the column from the magnet and elute bound cells in 2 ml washing buffer.

Table 2
MACS column and MACS separator

Column	Max number of labeled cells	Max number of total cells	Separator
MS	10^7	2×10^8	MiniMACS, OctoMACS, VarioMACS, SuperMACS
LS	10^8	2×10^9	MidiMACS, QuadroMACS
XS	10^9	2×10^{10}	VarioMACS, SuperMACS SuperMACS

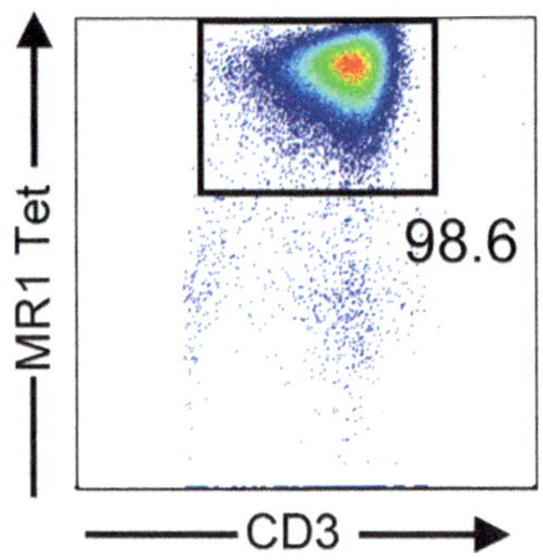

Fig. 3 Purity of MACS-enriched MAIT cells. MAIT cells were expanded from PBMCs by 5-OP-RU/MR1-coated beads supplemented with IL-2 stimulation in vitro. MACS enrichment was conducted using APC-labeled 5-OP-RU/MR1 tetramer and anti-APC magnetic beads. The dot plot depicts the frequency of enriched MAIT cells

8. To increase the purity of the magnetically labeled cells, the eluted fraction can be enriched over a second column. Repeat the magnetic separation procedure as described in **steps 5**–7 by using a new column (optional).
9. Wash the eluted cells by centrifuging 10 min at 266 × *g*, room temperature or chilled, and resuspend in 1 ml complete RPMI1640 culture medium.
10. Count the viable cells and check for the purity of MAIT cells by flow cytometry (Fig. 3).

4 Notes

1. NaN3 is known to be toxic. Avoid contact with the skin, eyes, and mucous membranes.
2. The accelerating and decelerating rates are all 4. Better separation effect can be obtained when centrifuging at room temperature.
3. To check viability, mix 10 μl of cell suspension with equal volume of Trypan blue solution, and count Trypan blue negatively stained live cells.
4. The amount of mAbs was determined according to the manufacturer's recommendations or after the titer test.
5. The supplement of IL-2 is optional, but adding IL-2 in coculture system tends to have higher yields of MAIT cells.

References

1. Tilloy F An invariant T cell receptor a chain defines a Novel TAP-independent major histocompatibility complex class Ib-restricted a/b T cell subpopulation in mammals. J Exp Med 189(12):1907–1921
2. Reantragoon R, Kjer-Nielsen L, Patel O, Chen Z, Illing PT, Bhati M et al (2012) Structural insight into MR1-mediated recognition of the mucosal associated invariant T cell receptor. J Exp Med 209(4):761–774
3. Martin E, Treiner E, Duban L et al (2009) Stepwise development of MAIT cells in mouse and human. PLoS Biol 7(3):e54
4. Treiner E, Duban L, Bahram S (2003) Selection of evolutionarily conserved mucosal-associated invariant T cells by MR1. Nature 422:164–168
5. Dusseaux M, Martin E, Serriari N et al (2011) Human MAIT cells are xenobiotic-resistant, tissue-targeted, CD161hi IL-17-secreting T cells. Blood 117:1250–1259
6. Huang S, Gilfillan S, Kim S, Thompson B, Wang X, Sant AJ et al (2008) MR1 uses an endocytic pathway to activate mucosal-associated invariant T cells. J Exp Med 205 (5):1201–1211
7. Le Bourhis L, Mburu YK, Lantz O (2013) MAIT cells, surveyors of a new class of antigen: development and functions. Curr Opin Immunol 25(2):174–180
8. Kjer-Nielsen L, Patel O, Corbett AJ, Le Nours J, Meehan B, Liu L et al (2012) MR1 presents microbial vitamin B metabolites to MAIT cells. Nature 491(7426):717–723
9. Maggi L, Santarlasci V, Capone M, Peired A, Frosali F, Crome SQ et al (2010) CD161 is a marker of all human IL-17-producing T-cell subsets and is induced by RORC. Eur J Immunol 40(8):2174–2181
10. Le Bourhis L, Martin E, Péguillet I, Guihot A, Froux N, Coré M et al (2010) Antimicrobial activity of mucosal-associated invariant T cells. Nat Immunol 11(8):701–708
11. Marrack P, Gold MC, Cerri S, Smyk-Pearson S, Cansler ME, Vogt TM et al (2010) Human mucosal associated invariant T cells detect bacterially infected cells. PLoS Biol 8(6): e1000407
12. van Wilgenburg B, Scherwitzl I, Hutchinson EC, Leng T, Kurioka A, Kulicke C et al (2016) MAIT cells are activated during human viral infections. Nat Commun 7:11653
13. Won EJ, Ju JK, Cho YN et al (2016) Clinical relevance of circulating mucosal-associated invariant T cell levels and their anti-cancer activity in patients with mucosal-associated cancer. Oncotarget 7:76274–76290

Index

Chaohong Liu (ed.), *T-Cell Receptor Signaling: Methods and Protocols*, Methods in Molecular Biology, vol. 2111, https://doi.org/10.1007/978-1-0716-0266-9,

Printed by Printforce, the Netherlands